Subsea Pipelines

Subsea Pipelines

Editors

Baiqiao Chen
Carlos Guedes Soares

Basel • Beijing • Wuhan • Barcelona • Belgrade • Novi Sad • Cluj • Manchester

Editors

Baiqiao Chen
Centre for Marine Technology
and Ocean Engineering
(CENTEC), Instituto Superior
Técnico
Universidade de Lisboa
Lisbon, Portugal

Carlos Guedes Soares
Centre for Marine Technology
and Ocean Engineering
(CENTEC), Instituto Superior
Técnico
Universidade de Lisboa
Lisbon, Portugal

Editorial Office
MDPI
St. Alban-Anlage 66
4052 Basel, Switzerland

This is a reprint of articles from the Special Issue published online in the open access journal *Journal of Marine Science and Engineering* (ISSN 2077-1312) (available at: https://www.mdpi.com/journal/jmse/special_issues/subsea_pipelines).

For citation purposes, cite each article independently as indicated on the article page online and as indicated below:

Lastname, A.A.; Lastname, B.B. Article Title. *Journal Name* **Year**, *Volume Number*, Page Range.

ISBN 978-3-0365-9921-2 (Hbk)
ISBN 978-3-0365-9922-9 (PDF)
doi.org/10.3390/books978-3-0365-9922-9

Contents

About the Editors

Baiqiao Chen

Dr. Baiqiao Chen is currently an Assistant Professor at the Centre for Marine Technology and Ocean Engineering (CENTEC) at Instituto Superior Tecnico (IST), University of Lisbon (UL). His research interests focus on the structural performance of ships and offshore structures, more specifically, on the topics of ultimate strength, photogrammetry, corrosion, collision and grounding, floating offshore wind, digital twin, etc. He is also a member of the International Ship and Offshore Structures Congress (ISSC).

Carlos Guedes Soares

Prof. Dr. Carlos Guedes Soares is a Distinguished Professor of Instituto Superior Tecnico and the Scientific Coordinator of the Centre for Marine Technology and Ocean Engineering (CENTEC). He has broad research interests in Ocean Engineering, ranging from modelling the marine environment to the hydrodynamics of floating structures, strength of marine structures, and safety and reliability.

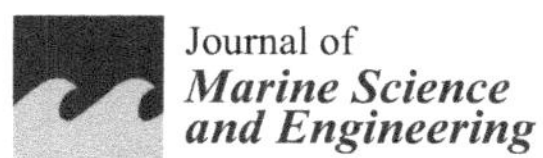

Journal of
**Marine Science
and Engineering**

Editorial

Subsea Pipelines

Bai-Qiao Chen and C. Guedes Soares *

Centre for Marine Technology and Ocean Engineering (CENTEC), Instituto Superior Técnico, Universidade de Lisboa, 1049-001 Lisbon, Portugal; baiqiao.chen@centec.tecnico.ulisboa.pt
* Correspondence: c.guedes.soares@centec.tecnico.ulisboa.pt; Tel.: +351-218-417-607

Citation: Chen, B.-Q.;
Guedes Soares, C. Subsea Pipelines. *J. Mar. Sci. Eng.* **2023**, *11*, 1249. https://doi.org/10.3390/jmse11061249

Received: 26 May 2023
Accepted: 29 May 2023
Published: 19 June 2023

A subsea pipeline (also known as an offshore pipeline or submarine pipeline) is a pipeline that is laid on the seabed or inside a specially constructed trench. The use of pipelines represents a reliable mode of transport of oil and gas. To prevent failure scenarios, such as improper pressurization, localized buckling, fatigue failure and instability, the limit strength, internal/external pressure, corrosion-resistant material selection, and stability management are the main concerns in the design, installation, and operation of submarine pipelines.

Fifteen recent research studies within the broad domain of structural design and analysis of subsea pipelines or risers are featured in this book. The applications of the digital twin in subsea pipelines are highlighted [1]. Classic structural topics including collapse strength [2–4], burst strength [5], tensile performance [6], buckling [7,8], fatigue [9–11], and bending [12] are investigated by numerical and experimental methods in this book. Computational fluid dynamics (CFD) simulations [13,14] and proportional integral derivative (PID) control [15] of the pipelines and risers are also included.

Digital twins are described as a concept that combines multi-physics modelling and data-driven analytics to replicate the behaviour of an entity in the real world. In [1], a review of the digital twins of the subsea pipelines is presented, which covers their present applications and the challenges associated with the design, construction, service, and assessments of life extension of the structures. The key opportunities in enhancing the integrity management of offshore assets using the digital twin include data contextualization, standardization, automated anomaly detection, and shared learning. However, the collection, interpretation, and sharing of data, and cybersecurity are among the main challenges identified.

The oil and gas industry is encountering difficulties in designing pipelines and risers that are well-insulated and capable of extreme operating conditions as they shift towards exploiting hydrocarbons in ultra-deep waters. Under these circumstances, the pipelines are exposed to extreme hydrostatic pressure and are primarily engineered to resist collapse. Li et al. [2] investigated the effect of ovality length on imperfect sandwich pipes (SP) using the three-dimensional finite element method (3D-FEM) in the scenario of local buckling of SPs subject to external pressure. Sandwich pipes are composite pipes consisting of three layers: an inner and outer layer made of steel or another material with high tensile strength, and a lightweight material such as foam or honeycomb between them. The sandwich structure provides better insulation and weight reduction compared to traditional pipes, making them suitable for use in harsh environments such as deep-sea oil and gas exploration. By analysing the results of 1200 cases, a predictive equation was developed in [2] to illustrate the correlation between the collapse strength and ovality length of imperfect SPs. The 3D-FEM was also used in [3] to analyse the collapse pressure of pipes with internal corrosion defects repaired with carbon-fibre-reinforced polymer (CFRP). The results showed that the collapse pressure of the composite-repaired pipe increased and the CFRP significantly reduced the strain in the defect region. In [4], it was assessed the structural safety of SPs for collapse failure and reliability-based design probabilistically. Results were distinguished in fully bonded, partially bonded, and unbonded core categories based on SP interlayer adhesion conditions. Through First Order Reliability Method (FORM) based sensitivity analysis, external pressure is found to be the most important parameter for safety, followed by Young's modulus of elasticity.

When pipelines corrode and are under internal pressure, they become vulnerable to burst failure. Therefore, accurate prediction of burst pressure is crucial for the safety of corroded pipelines. Aiming at providing vital guidance on the design and maintenance of different steel-grade pipelines, FORM was employed in [5] to estimate the partial safety factors and their uncertainty as a function of operational time. The probabilistic characteristics of the pipe burst strength were studied using Monte Carlo simulation (MCS) and the normal distribution is selected based on Chi-squared test results. The influence of the corrosion growth model on the evaluation of partial safety factors was assessed.

The tensile mechanical behaviour of a helically wound structure of marine flexible pipe/cables was investigated in [6] based on the curved beam theory. The deformation mechanisms of the flexible structures of umbilicals, flexible pipes, and cryogenic hoses with varying winding angles were discussed.

Subsea pipelines can experience lateral buckling when subjected to excessive axial compressive force resulting from high-temperature and high-pressure conditions. Taking the nonlinear pipe–soil interaction (NL-PSI) model into account, a mathematical model was proposed in [7] to investigate the lateral buckling of subsea pipelines triggered by a sleeper. A reduction in the deformation of the buckled pipeline and an increase in both the minimum critical temperature difference and the maximum stress along the buckled pipeline can be observed when the NL-PSI model is incorporated. The phenomenon of buckle propagation in a pipe-in-pipe (PIP) system under uniform external pressure was studied in [8]. It is revealed that the initial imperfections of PIP, i.e., the ovality and eccentricity of the inner pipe, have an insignificant influence on the buckle propagation pressure, and the relationship between the buckle propagation pressure and the ratio between the diameter and the thickness of pipes follows the format of power functions. A fitted formula was proposed for predicting the buckle propagation pressure of the PIP with good accuracy.

The experimental method plays a crucial role in the fatigue analysis of subsea pipelines by providing reliable data for validating numerical models and improving the understanding of the structural behaviour under cyclic loading. A full-scale fatigue test system for deepwater steel catenary risers, flexible pipes, and seabed pipelines was presented in [9]. The fatigue strength assessment of single-sided girth welds in offshore pipelines subjected to start-up and shut-down cycles was performed in [10]. The finite element analyses (FEAs) for the estimation of the effective notch stress were performed using ANSYS. For a specific study case, the plastic behaviour of the weld root was investigated to justify the use of the low cycle fatigue (LCF) approach, and the effect of weld root geometry on the notch stress factor was studied to identify the dominant geometrical parameters. A fracture mechanics-based fatigue reliability analysis of a submarine pipeline was investigated in [11] using the Bayesian approach based on limited experimental data. Bayesian updating method and Markov Chain MCS were used to estimate the posterior distribution of the parameters of the fatigue model regarding different sources of uncertainties. Failure load cycle distribution and the reliability-based performance assessment of API 5L X65 submarine pipelines as a case study were estimated for three different cases. The effects of the stress ratio, maximum load, uncertainties of stress range and initial crack size, and the corrosion-enhanced factor on the reliability of the investigated submarine pipeline were indicated through a sensitivity study.

FEM was also used in a parametric analysis model of an offshore screen pipe based on ABAQUS and Python software under pure bending load in [12]. The study identified and discussed the deformation patterns and mechanisms and developed an empirical formula for determining the ultimate moment of the screen pipe. Han et al. [13] presented a numerical investigation of turbulent flow in double-curved pipes with various spatial configurations, examining the impact of spatial angle and interval distance between bends on secondary flow through analysis of vector fields, velocity distributions, vortex developments, and dissipations of swirl intensity. Anwar et al. [14] performed a numerical analysis to investigate the impact of surface roughness on vortex-induced vibration (VIV) in the crossflow direction of a circular cylinder. It was concluded that roughness on a cylinder results in a reduction in amplitude response.

Under severe external marine stresses, the coupling effect between the offshore platform and the riser in the offshore platform-riser multi-body system can be significantly amplified. In [15], a new PID control approach based on the unscented Kalman filter (UKF) for the dynamic positioning (DP) system was created, and the DP control of a rigid–flexible fluid coupling system composed of an offshore platform and risers was realized by combining the OrcaFlex application program interface (API) with a PID-DP control statement in UKF mode based on Python programming. This method is feasible when considering the interaction between the riser and DP offshore platform in the overall analysis.

We hope that this set of papers can be seen as a contribution to the subject area by covering various relevant aspects related to submarine pipelines, and can be a useful reference to those who work in the field.

Author Contributions: Conceptualization, B.-Q.C. and C.G.S.; writing—original draft preparation, B.-Q.C.; writing—review and editing, C.G.S. All authors have read and agreed to the published version of the manuscript.

Funding: This work contributes to the Strategic Research Plan of the Centre for Marine Technology and Ocean Engineering financed by the Portuguese Foundation for Science and Technology (Fundação para a Ciência e Tecnologia—FCT) under contract UIDB/UIDP/00134/2020.

Acknowledgments: The editors wish to express sincere gratitude to all the authors and reviewers.

Conflicts of Interest: The authors declare no conflict of interest.

References

1. Chen, B.; Videiro, P.; Guedes Soares, C. Opportunities and Challenges to Develop Digital Twins for Subsea Pipelines. *J. Mar. Sci. Eng.* **2022**, *10*, 739. [CrossRef]
2. Li, R.; Chen, B.; Guedes Soares, C. Effect of Ovality Length on Collapse Strength of Imperfect Sandwich Pipes Due to Local Buckling. *J. Mar. Sci. Eng.* **2022**, *10*, 12. [CrossRef]
3. Yu, J.; Xu, W.; Yu, Y.; Fu, F.; Wang, H.; Xu, S.; Wu, S. CFRP Strengthening and Rehabilitation of Inner Corroded Steel Pipelines under External Pressure. *J. Mar. Sci. Eng.* **2022**, *10*, 589. [CrossRef]
4. Bhardwaj, U.; Teixeira, A.; Guedes Soares, C. Probabilistic Collapse Design and Safety Assessment of Sandwich Pipelines. *J. Mar. Sci. Eng.* **2022**, *10*, 1435. [CrossRef]
5. Bhardwaj, U.; Teixeira, A.; Guedes Soares, C. Uncertainty in the Estimation of Partial Safety Factors for Different Steel-Grade Corroded Pipelines. *J. Mar. Sci. Eng.* **2023**, *11*, 177. [CrossRef]
6. Wang, H.; Yang, Z.; Yan, J.; Wang, G.; Shi, D.; Zhou, B.; Li, Y. Prediction Method and Validation Study of Tensile Performance of Reinforced Armor Layer in Marine Flexible Pipe/Cables. *J. Mar. Sci. Eng.* **2022**, *10*, 642. [CrossRef]
7. Wang, Z.; Guedes Soares, C. Lateral Buckling of Subsea Pipelines Triggered by Sleeper with a Nonlinear Pipe-Soil Interaction Model. *J. Mar. Sci. Eng.* **2022**, *10*, 757. [CrossRef]
8. Li, R.; Chen, B.; Guedes Soares, C. Design Equation of Buckle Propagation Pressure for Pipe-in-Pipe Systems. *J. Mar. Sci. Eng.* **2023**, *11*, 622. [CrossRef]
9. Yu, J.; Wang, F.; Yu, Y.; Liu, X.; Liu, P.; Su, Y. Test System Development and Experimental Study on the Fatigue of a Full-Scale Steel Catenary Riser. *J. Mar. Sci. Eng.* **2022**, *10*, 1325. [CrossRef]
10. Dong, Y.; Ji, G.; Fang, L.; Liu, X. Fatigue Strength Assessment of Single-Sided Girth Welds in Offshore Pipelines Subjected to Start-Up and Shut-Down Cycles. *J. Mar. Sci. Eng.* **2022**, *10*, 1879. [CrossRef]
11. Kakaie, A.; Guedes Soares, C.; Ariffin, A.; Punurai, W. Fatigue Reliability Analysis of Submarine Pipelines Using the Bayesian Approach. *J. Mar. Sci. Eng.* **2023**, *11*, 580. [CrossRef]
12. Peng, Y.; Fu, G.; Sun, B.; Sun, X.; Chen, J.; Estefen, S. Bending Deformation and Ultimate Moment Calculation of Screen Pipes in Offshore Sand Control Completion. *J. Mar. Sci. Eng.* **2023**, *11*, 754. [CrossRef]
13. Han, F.; Liu, Y.; Lan, Q.; Li, W.; Wang, Z. CFD Investigation on Secondary Flow Characteristics in Double-Curved Subsea Pipelines with Different Spatial Structures. *J. Mar. Sci. Eng.* **2022**, *10*, 1264. [CrossRef]
14. Anwar, M.; Khan, N.; Arshad, M.; Munir, A.; Bashir, M.; Jameel, M.; Rehman, M.; Eldin, S. Variation in Vortex-Induced Vibration Phenomenon Due to Surface Roughness on Low- and High-Mass-Ratio Circular Cylinders: A Numerical Study. *J. Mar. Sci. Eng.* **2022**, *10*, 1465. [CrossRef]

15. Zhang, D.; Zhao, B.; Bai, Y.; Zhu, K. Dynamic Response of DP Offshore Platform-Riser Multi-Body System Based on UKF-PI Control. *J. Mar. Sci. Eng.* **2022**, *10*, 1596. [CrossRef]

Journal of
*Marine Science
and Engineering*

Review

Opportunities and Challenges to Develop Digital Twins for Subsea Pipelines

Bai-Qiao Chen [1], Paulo M. Videiro [1,2] and C. Guedes Soares [1,*]

[1] Centre for Marine Technology and Ocean Engineering (CENTEC), Instituto Superior Técnico, Universidade de Lisboa, 1049-001 Lisbon, Portugal; baiqiao.chen@centec.tecnico.ulisboa.pt (B.-Q.C.); paulovideiro@tecnico.ulisboa.pt (P.M.V.)
[2] Laboratory for Relaibility Analysis of Offshore Structures (LACEO), COPPE, Universidade Federal do Rio de Janeiro, Rio de Janeiro 21941-596, Brazil
* Correspondence: c.guedes.soares@centec.tecnico.ulisboa.pt

Abstract: A vision of the digital twins of the subsea pipelines is provided in this paper, with a coverage of the current applications and the challenges of the digital twins in the design, construction, service life, and assessments of life extension. Digital twins are described as a paradigm combining multi-physics modelling with data-driven analytics, which are used to mirror the life of its corresponding twin. Realistic virtual models of structural systems are shown to bridge the gap between design and construction and to mirror the real and virtual worlds. Challenges in properly using the new tools and how to create accurate digital twins considering data acquired during the construction phase are discussed. The key opportunities for improved integrity management using the digital twin are data contextualization, standardization, automated anomaly detection, and learning through sharing. The collection, interpretation and sharing of data, and cyber-security are identified as some of the main challenges.

Keywords: digital twin; subsea pipeline; IoT; maintenance

Citation: Chen, B.-Q.; Videiro, P.M.; Guedes Soares, C. Opportunities and Challenges to Develop Digital Twins for Subsea Pipelines. *J. Mar. Sci. Eng.* **2022**, *10*, 739. https://doi.org/10.3390/jmse10060739

Academic Editor: José-Santos López-Gutiérrez

Received: 17 April 2022
Accepted: 23 May 2022
Published: 27 May 2022

Publisher's Note: MDPI stays neutral with regard to jurisdictional claims in published maps and institutional affiliations.

1. Introduction

Subsea pipelines are the flowlines connecting a subsea wellhead to a manifold or a platform, or the export lines (trunk lines) used as a long-distance transportation system for the oil and gas [1,2]. Important considerations of the pipeline system are the efficient operation and the development of the system for future needs. Due to the low price of oil in recent years, oil and gas field operators are looking beyond the traditional operational maintenance strategy for the sake of reducing the downtime caused by the planned or sometimes unplanned preventive maintenance in the production field, thus reducing the operational cost (OPEX) [3,4].

In the digital solutions to subsea integrity management (SIM), the use and management of data bring benefits to the daily operations, e.g., the increased efficiency, optimization, reduction in cost, and safety [3]. New technologies and associated processes are required for increased safety and better monitoring of the subsea assets.

Recent advancements in information and communication technologies, including cloud computing, high-performance processors, high dimensional visualization capabilities, internet of things (IoT), wearable technologies, additive manufacturing (AM), big data analytics (BDA), artificial intelligence (AI), autonomous robotic systems, submarine drones, and blockchain technology, have catalysed digital adoption across industries. These new technologies have facilitated the cyber-physical integration by which data can be collected, analysed, and visualized to make informed decisions and to serve as a basis for simulations to optimize operations [5]. The concept of cyber-physical interaction and associated simulation is referred to as a Digital Twin.

The digital twin draws considerable interest as it can provide a cost-effective, reliable and intelligent maintenance strategy based on the machine learning (ML) algorithm [4] and assess extension life projects. The digital twin concept consists of three components: (1) the physical asset, (2) the virtual representation of the asset, and (3) the connection between the previous two components. The connection includes the information transferred from the physical asset to the digital twin and from the digital twin to the asset. A clear business purpose is a key principle in the development of digital twins to provide value [3].

The concept of the digital twin has also been used in offshore structures as platforms and subsea pipelines throughout the stages of construction and operation life. In this paper, the vision of the digital twin and its evolution, and the applications of the digital twin in subsea pipelines in design, construction and service life are reviewed. Further discussions and comments are also addressed.

2. Literature Review

2.1. Digital Twin

A digital twin is defined as one virtual representation of a system (or an asset) that can be used to calculate the states of the system and to make the information available. The integrated models and data of the digital twin can provide decision support over the life cycle of the system or asset. The idea to use a twin model can be dated back to the 1970s when two identical space vehicles were built in NASA's Apollo program to allow for mirroring of the conditions of the space vehicle [6].

Although the concept of digital twins was first put forward in 1991 [7], the model of the digital twin was first introduced in 2002 as a concept for product lifecycle management (PLM). After its initial names of mirrored spaces model (MSM) and information mirroring model (IMM), the model was finally referred to as the Digital Twin in 2011 [8].

The digital twin has been applied to a wide range of industries including aerospace [9,10], automotive [11,12], healthcare [13–16], manufacturing [17–19], and smart city [20,21]. Some early applications of the digital twin can be found in NASA's spacecraft [22–24] and in the US Air Force's jet fighters [25]. More recently, world-class vendors such as Dassault Systèmes, PTC, and Siemens use the digital twin concept in their PLM. The digital twin model was also proposed for the robust deployment of IoT [26]. Aiming at developing digital twins for all built cars, TESLA enabled synchronous data transmission between their cars and the factory [27]. From 2017 to 2019, the Gartner listed the digital twin as being among the top 10 strategic technology trends and indicated that billions of things would have their digital twins within 3–5 years [28–30].

Although the concept of the digital twin is not new, it was more a descriptive one and lacked auxiliary technologies in the first few years [31–33]. Figure 1 shows the number of results obtained by searching "digital twin" as a "topic" in the Web of Science database, indicating a tremendous increase of research interest in the digital twin both in industry and academia, especially since 2017/2018. In terms of the country of study, China is the leading country in publications on the digital twin, followed by Germany and the USA.

Even though the digital twin technologies are of interest to companies in the industry, a major part of the studies (about 79%) are conducted by researchers in academia, according to a survey [34]. Most academic research focuses on improving modelling techniques rather than optimizing data and implementing digital twins. Only 6% of the articles are initiated in the industry; the remaining are collaborative studies between academia and industry. There are few connections between industry and academia, particularly due to commercial secrecy.

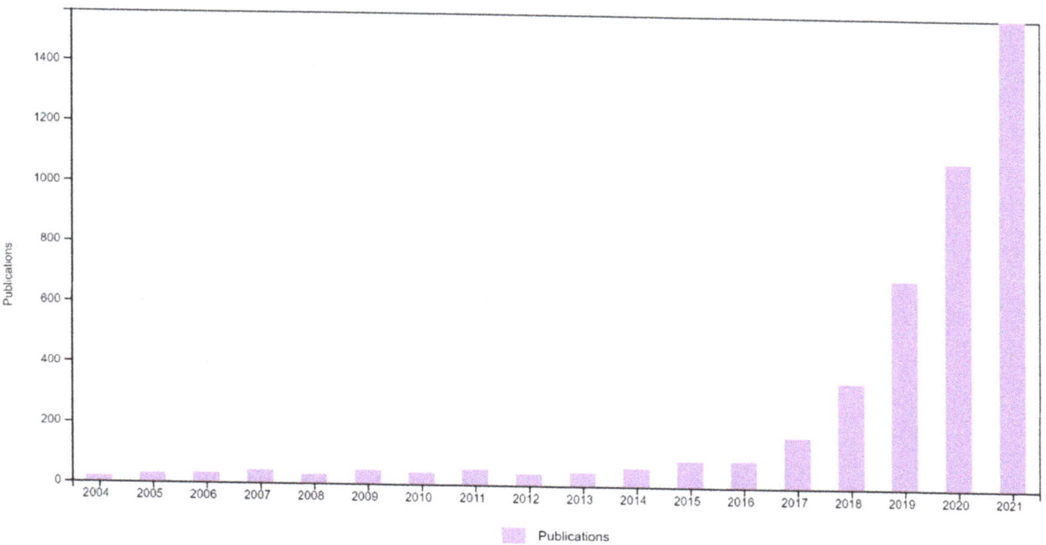

Figure 1. The number of published results about digital twins in the Web of Science.

To meet the new requirement from applications, Tao et al. [35] presented an extended five-dimension digital twin model, adding data and services to the initial three-dimensional digital twin concept. Figure 2 shows the key technologies for modelling each dimension of the digital twin. Qi et al. [36] classified the tools for the service applications of digital twins into platform service tools, simulation service tools, optimization service tools, and diagnostic and prognosis service tools. A list of the tools for each category is shown in Figure 3.

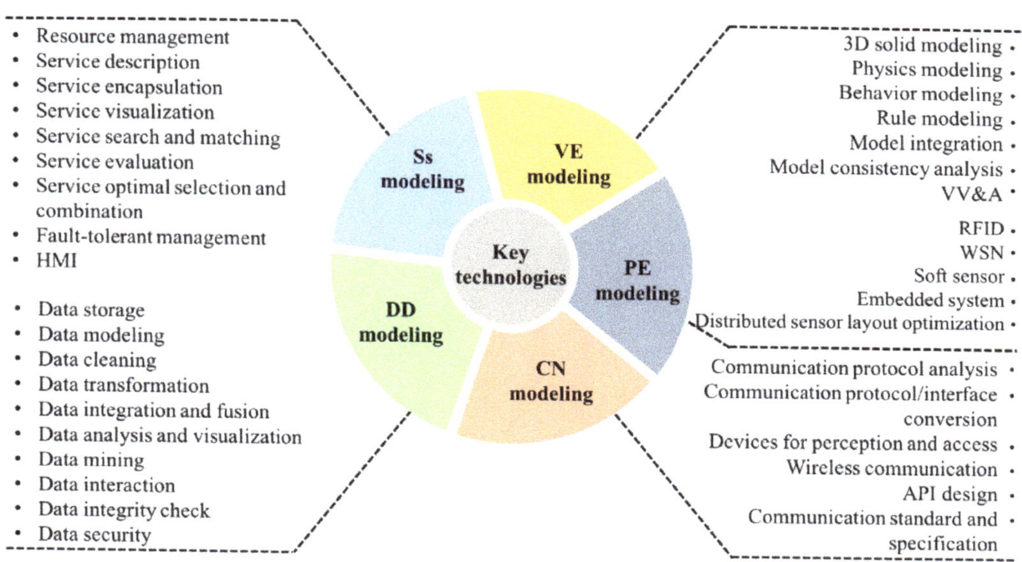

Figure 2. Five-dimensional digital twin model and the key technologies. PE—Physical Entity, VE—Virtual Entity, Ss—Services, DD—Digital twin data, CN—Connection. Reproduced from [35], with permission from Elsevier, 2022.

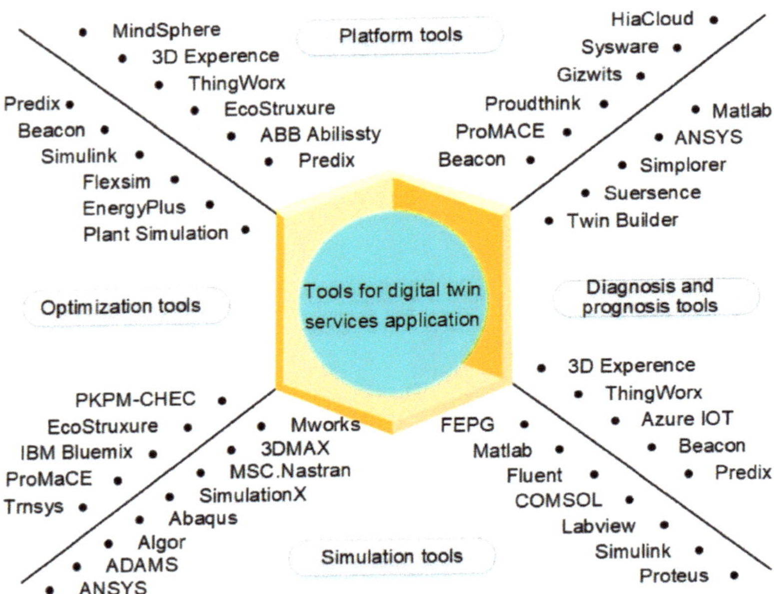

Figure 3. The tools for services applications of digital twins. Reproduced from [36], with permission from Elsevier, 2022.

ANSYS Twin Builder is one of the multi-physics simulation packages used to add physical realism to the digital twins [27]. It contains extensive application-specific libraries and features third-party tool integration and can enable engineers to quickly build, validate and deploy the digital models of physical assets at an appropriate level of detail.

As with all new concepts, there are also obstacles to the further application of the digital twin. For instance, it is sometimes difficult to collect the most important data from thousands of sensors that track vibration, temperature, environmental conditions, force, speed, or power. In addition, the data can be spread among different owners in various formats. Consequently, the digital twin may fail to echo what is going on in the real world, leading to poor decisions made by the managers accordingly [37]. In this regard, further research is needed for improved data collection and processing methods to implement the communication interface between the digital and the physical twins. In terms of standardization, the development of universal platforms and tools are also required for further applications of digital twins.

2.2. Subsea Pipelines and Application of Digital Twin

The first oil pipeline is widely believed to have been installed in the 1860s in the USA to transport crude oil [38]. Since then, subsea pipelines have become the most economical means of efficiently transporting crude oil, natural gas, and other products from offshore installations for the exploitation of subsea reservoirs. Figure 4 illustrates different uses of subsea pipelines associated with platforms and wells, including the infield flowlines and export pipelines.

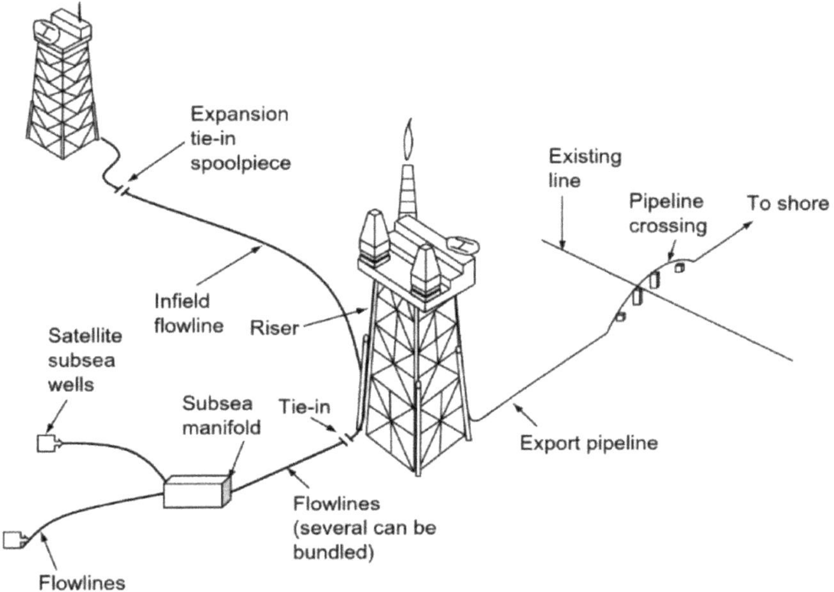

Figure 4. Uses of subsea pipelines. Reproduced from [38], with permission from Elsevier, 2022.

In a harsh sea environment, sufficient structural strength is necessary for designing, analysing, and operating subsea pipelines to guarantee the safety and integrity of subsea pipelines [38,39] throughout their life span.

Pipeline design is affected by multiple factors identified during the early stages (i.e., conceptualization, front end engineering and design stage, and detailed engineering phase). These factors include the site selection, route survey, local environmental conditions, material selection, wall-thickness design, pipeline protections, and the budget of the project [40,41]. During the pipelay process, the pipelines are deposited from installation vessels, with new pipe segments welded to form the pipelines. A proper pipe-lay is required to avoid excessive bending stresses that may cause fractures and buckling. The low temperature on the seafloor can sometimes cause a global contraction of the pipeline, while heat coming up from the reservoir fluid may induce local thermal expansion of the pipeline [42,43]. Temperature or pressure changes in the operation process can also cause "pipeline walking" [44,45] in the case of improper restraints of the pipelines. Corrosion and erosion, which occur due to chemical attacks and abrasion from the internal fluids, are one of the major limiting factors in the continued operation of subsea pipelines [46–50]. These factors necessitate the routine inspection of subsea pipelines and the use of monitoring technologies.

According to the B31G code [51] of the American Society of Mechanical Engineers (ASME), the geometry of the corrosion pits can be idealized as elliptical shapes, and a bulging factor can be applied when considering the defect geometry. The class society Det Norske Veritas (DNV) headquartered in Norway published the standard ST-F101 [52] and also one document of recommended practice RP-F101 [53] for pipeline applications. The American Petroleum Institute (API) Specification 5L [54] includes the requirements for manufacturing seamless and welded steel pipes in the transportation of oil and gas. One current challenge for subsea pipeline inspection is the often-deep waters.

More recently, sandwich pipes (SPs) were proposed as an effective alternative to the pipe-in-pipe (PIP) system for ultra-deepwater applications [55–57]. In the SPs, a polyester foam material with low density and heat-conducting ability is filled between two metal pipes, providing high structural resistance with thermal insulation capability [58–61].

As reported by DNV [3], the recent digitalization requires the reduced cost of sensor technology and computational power, as well as cloud storage and computing. Figure 5 illustrates the main technologies and enablers in digitalization for SIM: digital worker + support, inspection + data collection, and the analysis.

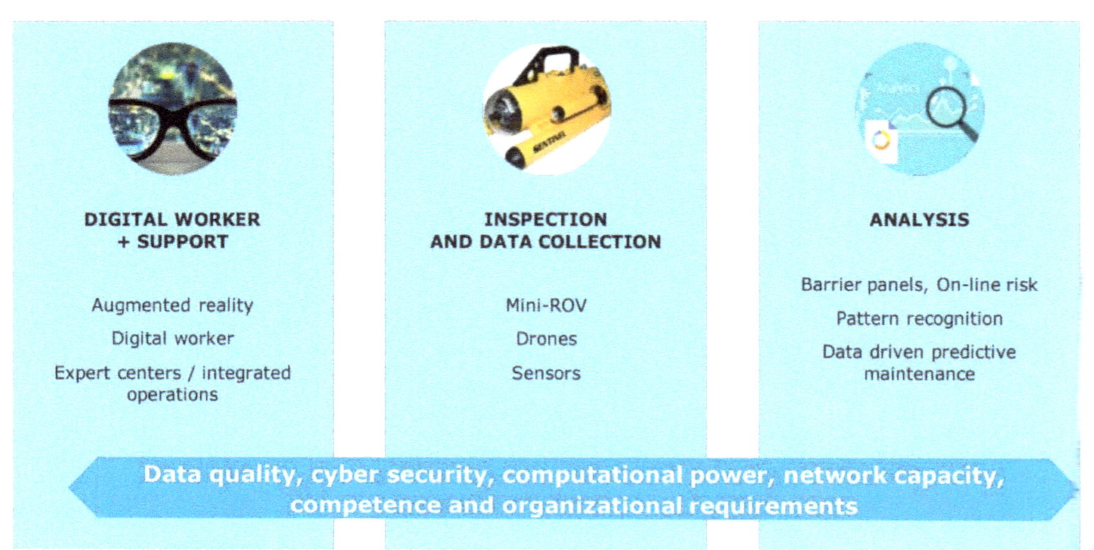

Figure 5. The key enablers for digitalization in subsea integrity management [3].

DNV define in the RP-A204 [62] six levels of capability (stages of the evolution) of digital twins, as illustrated in Figure 6. Real-time data streams are not required in Level 0 (standalone) but are in Level 1 (descriptive capability). Level 2 (diagnostic capability) provides support to monitor the conditions and detect the fault, together with troubleshooting. Health and condition indicators are further enriched in Level 3 (predictive) to support prognostic capabilities. Level 4 (prescriptive capability) can be used to provide recommended actions based on the predictions. In Level 5 (autonomous capability), the users can determine the functional element to perform actions or make decisions regarding the system.

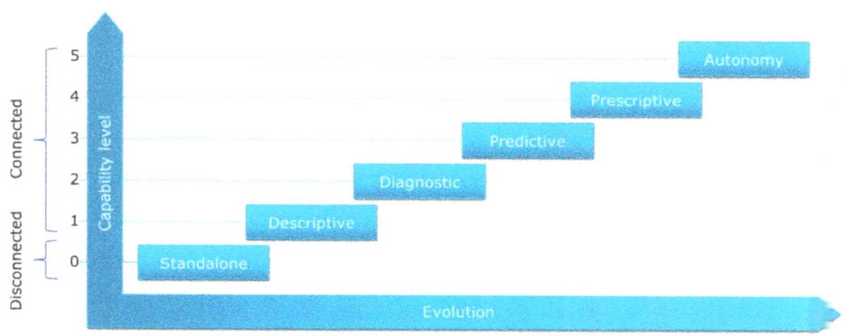

Figure 6. The capability levels of the digital twins at different stages of the evolution [62].

The digitalization of the data exchanges of the subsea pipelines plays an important role in avoiding the errors of the data during manual transcriptions. In order to simplify the data exchange between parties at the different stages of projects, the pipeline data exchange format (PDEF) collaborative initiative was proposed. It is an open-source joint industry project (JIP) involving many companies in the pipeline industry, aiming at developing a standard format for the exchange of the data and describing the data used in the design of subsea pipelines [3].

3. Automated Creation of the Digital Twin during Construction

The oil and gas industry has been adopting digital twins of asset life cycle management (ALCM) in recent years. Subsea pipelines are unique when compared to other major infrastructure assets due to the fact that they are buried immediately during construction and are used to transport hazardous and explosive contents at high pressures. These facts result in an enormous responsibility for all phases of a pipeline asset's life cycle, including construction, operation, and long-term integrity management.

To accurately trace the condition of a pipeline asset at any time in its lifecycle to previous environmental conditions, operational conditions or events during its life, a single source of truth representing the entire meter-by-meter condition of the asset (i.e., a digital twin), must be created. The creation of the pipeline digital twin during pipeline construction can lead to a full representation of the asset from creation to decommissioning, which construction management, quality control, engineering and integrity can all refer to [63]. The goal is to create the data required by inspectors during construction and capture it in such a manner that it can also be easily or instantly accessed and used by the operator during construction for quality management and operations and maintenance post-commissioning.

Field trials were completed to test and evaluate workflows and sensor platforms for the creation of digital twins for pipelines in 2017 [63]. After pipeline stringing, welding, and lowering, the open ditch and lowered pipe were scanned in the final resting position prior to backfilling. The laser imaging, detection, and ranging (LiDAR) system and downward-facing camera were suspended above the open ditch on the side-mounted boom. The system has a measurement rate of up to 700,000 measurements per second, high accuracy fibre optic gyro (FOG) inertial measurement units (IMU), and a high-end survey-grade global positioning system (GPS). The trials resulted in highly accurate pipeline centrelines, weld locations, depth to cover (DoC), and ditch geometry capture in digital formats. The resulting point clouds contain about 6.2 million highly-accurate points over a scan distance of 370 m. The weld location was added to the geographic information system (GIS) based on the imagery interpretation.

In addition to the remote sensing method, techniques of magnetic flux leakage and acoustic detection are commonly used to detect the cracks of subsea pipelines. One of the limitations of the LiDAR inspection is the assumption used in the algorithm of the software. Since visualization is one key feature of digital transformation, methods to obtain an accurate and useable result from the massive amount of measurement data need to be developed and improved in the further application of digital twins.

A pipeline design automation was introduced based on the cloud-based digital twin McDermott SubseaXD [64]. SubseaXD is collaborating with Dassault Systemes/Simulia to leverage their 3D Experience platform with a smart, collaborative PLM platform. The web-based graphical user interface (GUI) worked as an integrated system producing a 3D digital field diagram, together with all pipeline design calculations in one digital platform. Various calculations, including wall thickness calculations based on API/DNV/ASME code check, on-bottom stability analysis, pipeline span analysis, pipeline end expansion analysis, out of straightness analysis, and pipeline buckling analysis are performed sequentially and systematically in the cloud using the metadata information (i.e., pipe data, soil, environment) available through Python-based API from the digital field data. Abaqus and Orcaflex are integrated with the SubseaXD, which can be used for detailed finite element analysis (FEA). It was stated that the automated pipeline design can save hours with fewer errors, thus saving on cost.

Based on the study in [63], the following recommendations for further digital twin creations were provided:

- Refine the hardware used, and thus the field execution and subsequent data processing workflow.
- Compare the digital twin centreline to the in-line inspection and weld location surveys, and seek to explain variances.
- Make the results more readily available using cloud services and mobile communications to better expose the data to decision-makers.
- Investigate the schedule and cost savings in greater detail.
- Garner more operator feedback on the potential value from a construction quality control or quality assurance perspective.
- Examine long-term impacts on the integrity and general operations and maintenance, as well as on failure investigations.

4. Update of the Digital Twin with Information Acquired in Inspections

Notable offshore production regions include the Gulf of Mexico (GoM), the North Sea, Brazil, West Africa, the Persian Gulf, Atlantic Canada, the Gulf of Thailand, the East and the South China Sea, the Caspian Sea, and the Southern and Western Australia [65]. The GoM has the greatest number of offshore pipelines installed, followed by the North Sea. From 1952 to 2017, more than 72,000 km of pipelines have been installed in the GoM, and about 42,000 km of them are still active. The North Sea has the second-largest pipeline network with approximately 45,000 km of pipeline installed since 1966 [66].

However, the issues and obstacles make the inspection and monitoring of pipelines a challenging task. Problems may occur throughout the life of the pipe, and the environment of service is full of potential dangers [65]. In the use of digital twin technology in offshore structures, the twin model must be updated during the service life with the data of corrosion and other structural degradation gathered during service life inspections and could be updatable according to the accidental damages.

A pipe segment is often coated during the manufacturing process to protect it against corrosion or abrasion. Additional layers of coating can be added, depending on the requirements. Some subsea pipelines have outer concrete coatings for protection against corrosion and impacts and weight stabilization [67,68]. PIP designs may also be implemented with additional layers for protection and thermal insulation [69–71]. In addition to the inspection of the primary metallic body, the coating layers may also require inspections for damage and debonding from the pipelines. Depending on the task and the technology used in the inspection, part of the pipelines may need to be stripped of coatings to be fully inspected.

Based on an investigation by The US Department of Transportation, the major failures of subsea pipelines were categorized into five categories: mechanical, operational, corrosion, natural hazards, and third party. All of these possible failures necessitate the routine inspection of the pipeline and the use of permanent monitoring technologies. Nowadays, one of the challenges for subsea pipeline inspection is the extreme water depths. Many well proved inspection technologies cannot be delivered to the pipeline without costly equipment and procedures. In the cases of multi-layered pipelines, the cost of stripping away the coatings for routine inspections in deepwater is impractical. In-line inspection (ILI) may be used to inspect the pipeline and the inner from the inside. Pigging is one of the ILI techniques in which devices referred to as "pigs" or scrapers are inserted into pipelines to perform inspection activities. Pigging can be conducted on a variety of pipeline sizes without having to stop the flow of material through the line.

Given the increasing capacity of satellite links, the potential of cloud computing and deep learning (DL) algorithms, the digital twin models with millimetre precision move the subsea asset management into a new era, enabling engineers to incorporate less margin into their remediation advice which translates into more targeted maintenance. A solution to processing high-resolution data collected from pipeline inspections was presented by the geo-data specialist company Fugro using in-house Remote Observation, Automatic

Modelling, Economic Simulations (Roames) technology [71]. The resultant product is a web-based service that enables pipeline inspection data to be uploaded to the secure cloud environment in near real-time, processed using ML, verified by experts onshore and visualized in an intuitive 4D web viewer. This approach can greatly reduce the cost of infrastructure management practices, lower the risk exposure, and contribute to the extended life of the pipeline.

The project was executed which employed a Kongsberg Hugin autonomous underwater vehicle (AUV) operated at an average of 3 m altitude and 4.5 kt [72]. The point cloud was acquired with a Reson 7125 multi-beam echosounder (MBES). The pipeline position was automatically detected using a convolution neural network (CNN), which was trained to detect pipeline profiles from a cross-section of the point cloud. Ground truth data was generated from historical surveys, where the pipeline position was placed manually by a qualified data processor. The ground truth contains a pipe diameter marker, representing the position. Using the pipe and terrain model, the free-span and burial events can then be computed automatically. The web-based Roames Pipe Inspection tool can be used to inspect and adjust machine-learning-based point classification and pipeline positioning. The neural network-based pipe modelling positions the pipe within the correct location relative to the pipe points. The results of the neural network quantitative analysis showed a 97.2% accuracy in detecting pipe burials. The observed mean error in pipeline position was 0.46 cm.

While these new developments provide the operators with the desired deliverable, the large volume of information presents new challenges for inspection contractors and their onboard data processors working in a remote environment. Ongoing technological advancements in satellite communications present new opportunities for data to be transferred off the vessel in near real-time. This allows data processing to be performed securely in the cloud environment and validated by a global team of experts in offices around the world.

It is noted that the construction of the digital twins shall be started at the design phase of the offshore structure. The digital twin model must be updated with as-built and as-installed conditions and could be updatable in a fast way for accidental damages, permitting a quick evaluation of the asset safety and providing information.

5. Maintenance Planning Based on the Digital Twin

In recent years, the digital twins have been implemented in different industrial sectors, and in the design, production, manufacturing, and maintenance of the subsea assets. Among these application areas, maintenance is one of the most researched applications, as the execution of maintenance tasks may have a great influence on the business.

Different maintenance strategies might be used in the decision making, namely reactive maintenance, preventive maintenance, condition-based maintenance, predictive maintenance, and prescriptive maintenance [73–84]. Predictive maintenance combines conditioning monitoring integrating with an ML-based decision support system and can enhance economic efficiency and availability [4]. It has the capability to determine when to perform the maintenance based on the real conditions of the subsea pipelines. Once in place, these capabilities could reduce the unplanned maintenance downtime events and thus optimize the OPEX.

Figure 7 shows a digital twin system for predictive maintenance. One of the components is a computational model of the asset which is normally a finite element (FE) model. The computational model of the subsea pipeline is updated based on different field sensor data such as the motion sensor/accelerometer, the subsea strain gauge, the acoustic doppler current profiler (ADCP), the wave radar, the subsea pressure sensor, and the subsea temperature sensor (see Figure 8). Both the data-driven and FE-based models can be used to predict the remaining fatigue life (RFL) of the pipelines based on the stress range data, and are further used in the decision-making process. Knowledge of the RFL can enable efficient maintenance planning and avoid unpredicted shutdowns.

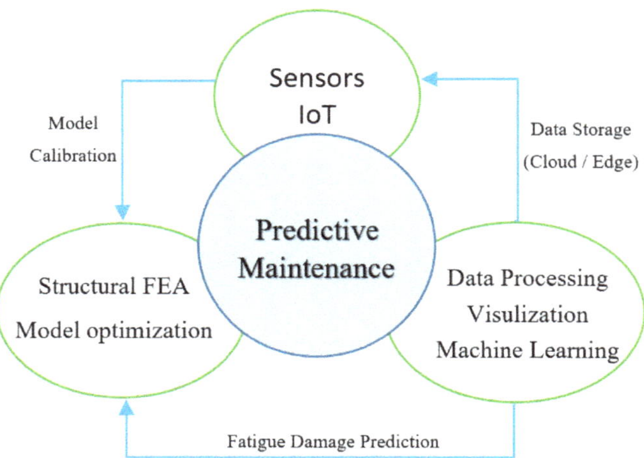

Figure 7. A digital twin system for maintenance [4].

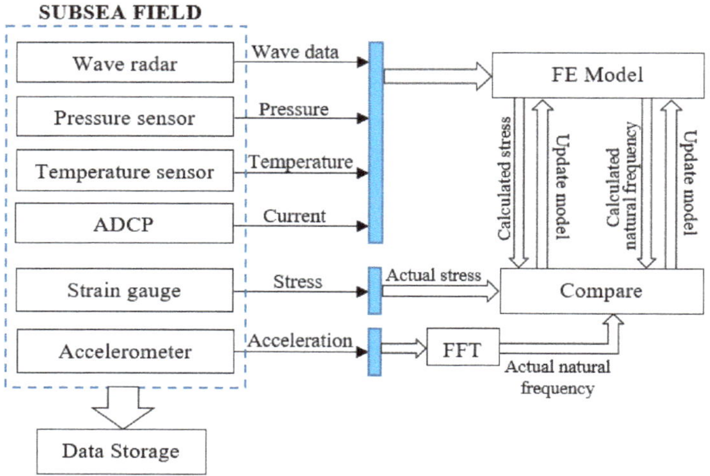

Figure 8. A field sensor system and the data-driven model [4].

Another component of the digital twin is the IoT/sensor system installed in the physical asset. IoT brings together low-cost sensors, cloud computing, and BDA in subsea pipeline systems where robustness, reliability, and security are highly desired. The third component is the data analytics to find the insight between the measured sensor data and apply a machine-learning algorithm to find the RFL based on the measured strain gauge data. The hydrodynamic load can be measured in real-time using the field sensor system and fed into the digital model. Wireless data loggers are connected to the IoT-based systems which transmit the data to store in the cloud for online analytics, visualization and reporting. The data stored in the cloud is accessible onshore or onboard for data analytics.

Figure 9 illustrates the predictive maintenance model based on sensor data analytics and ML algorithms. The predictive maintenance schedule is estimated by using a system of artificial neural network (ANN). The memory blocks of a long short-term memory (LSTM) are used for the layers of a recurrent neural network (RNN).

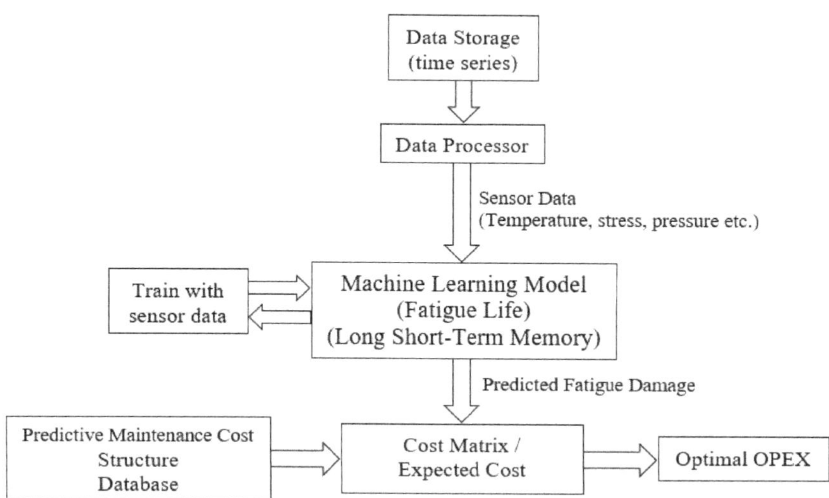

Figure 9. A predictive maintenance model based on sensor data analytics [4].

6. Fast Assessment of Failures and Inspection Planning

In the technically demanding submarine environment, subsea pipelines are susceptible to various damage threats, which may lead to catastrophic failures of the disastrous economic and environmental variety. For instance, the waves and currents can lead to a scouring of the soil underneath the pipelines and free span problems where a segment of the pipeline becomes unsupported except at the two ends of the free span length. Free span lengths above the allowable design limit can result in fatigue damage through vortex-induced vibrations (VIV) [65,85–87]. Corrosion and erosion can occur from chemical attacks and abrasion due to the internal fluids containing abrasive sand particles travelling at high velocities. Corrosions, when combined with tensile stresses, can lead to stress cracking and leakage in subsea pipelines [46–52,65,88].

To detect and correct malfunctions of the physical assent, the digital twin applies ML, DL, and AI algorithms. It was revealed in a survey that asset monitoring and maintenance is the most anticipated application area for the digital twin [5]. The fitness-for-service (FFS) of the physical asset is continuously monitored by the digital twin to identify potential failures. The big-data analytics capabilities of the digital twin can monitor the asset and send warnings to the responsible parties.

Digital twins provide high-fidelity accurate models and keep updating through the lifecycle of the pipelines with gathered data from sensors and inspections. Thus, they can reproduce the current state of the pipelines in the virtual space. Comparisons between digital twin simulated data and collected data can help determine the failure mode. One advantage of the state monitoring by the digital twin is that users can monitor the product state from any remote location through the unique identifier incorporated by the digital twin [31].

A computer vision-based digital twin model for real-time corrosion inspection was proposed in [89]. The CNN algorithm was used for the automated corrosion identification and classification from the remotely operated vehicles (ROV) images and ILI data. Based on the corrosion assessment by the digital twin, predictive and prescriptive maintenance strategies are recommended.

During the service life of subsea pipelines, failure mechanisms such as external/internal sheath damage, fatigue damage, or corrosion may arise. Therefore, high OPEX is consumed to confirm the fitness of the system [90].

Based on the reliability assessments [91–93], the risk-based inspection planning methodology has been used for the integrity management of subsea pipelines. The steps of the methodology include (1) data gathering, (2) development of risk criteria, (3) probability of failure (PoF) and consequence of failure (CoF), and (4) risk evaluation and enhanced inspection strategy [94–99]. Bayesian network (BN) and genetic algorithm (GA) were used to develop a framework for the inspection decision-making for subsea pipelines [100]. Using digital twins, risk target data analysis and risk estimation by prognostic and ML techniques were performed [101,102].

7. Life Extension Assessments

With the increasing maturity of the oil and gas industry, the requirement to operate a subsea pipeline beyond the design life is becoming commonplace [103]. The life extension of the subsea pipelines opens up many development opportunities. If the pipelines can be reused for future developments, significant capital expenditure reduction can be achieved [104,105].

The process of a life extension assessment considering consolidated guidelines [106–109] is summarized as follows:

(1)   Definition of the operational context and premises for an extended operational period, and identification of new threats.
(2)   Assessment of current condition, functionality and integrity of the system (Diagnostic).
(3)   Reassessment of the technical lifetime (Prognostic).
(4)   Identification of Life Extension measures (Prescriptive).
(5)   Development of a Life Extension program.

The basic premise of the life-extension process is similar regardless of which guideline is applied. Figure 10 displays a typical life extension process. The detailed descriptions of the key steps in the process can be found in [104].

A digital twin can be of great value to support ALCM and hence support optimal management of the asset lifetime [110]. The life extension assessment process clearly maps into the classification and implementation levels of the digital twin.

A digital twin supported by the system of systems concepts was proposed in [110] to represent an aid for model-based condition assessment, and the estimation of RUL contributes with prescriptive capabilities to the identification of measures related to condition-based or predictive maintenance policies, optimal operation, and control to extend equipment lifetime. In this regard, the life cycle losses and costs can be reduced. Following the standards and guidelines for life extension evaluations, a risk picture can be automated and integrated within levels 2 and 3 of the digital twin to guarantee the visualization of current and future risks. This can then serve as inputs into level 4 to identify the risk-based extension measures and to visualize the mitigated risk picture.

Another digital twin concept was proposed to provide an accurate estimation of the true fatigue life of assets to unlock potential fatigue life and ultimately extend the life of assets in the oil and gas industry [111]. The digital twin concept was divided into four tiers that allow for unlocking the RFL of the subsea asset:

(1)   High-resolution modelling of the asset (RB-FEA).
(2)   Update the model to reflect real-world conditions (Digital Twin).
(3)   Fatigue calculations based on continuous monitoring.
(4)   Statistical correlation between sea states and fatigue damage. Retrospective fatigue calculation.

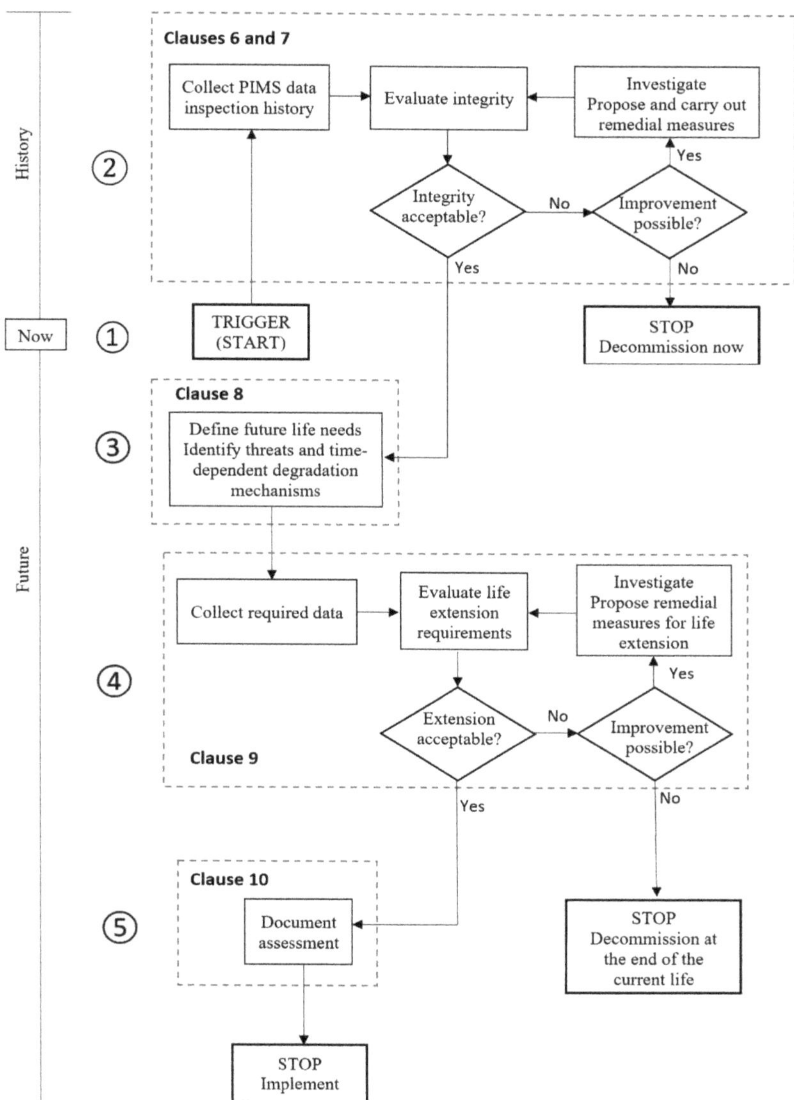

Figure 10. The life extension process in [107].

The reduced basis finite element analysis (RB-FEA) technology by Akselos is faster than conventional FEA in which higher accuracy is ensured by using posterior accuracy indicators and automated model enrichment [112]. Among the existing case studies [111,113], the Akselos digital twin is used for the life extension of subsea assets.

Fluid flow assurance issues including hydrates, wax deposition, asphaltenes and naphthenates, slugging, scales, corrosion, and erosion, bring a critical operational challenge to the production and transportation of pipelines [114]. Computational fluid dynamics (CFD) can be applied, combined with different models (e.g., CSMHyK rheological model, Eulerian–Eulerian CFD-model, population balance model), to provide accurate analysis in terms of flow assurance [115]. The models can be integrated into commercial CFD packages such as FLUENT, STAR-CD, TransAT, and STAR-CCM+ for such analyses. DNV developed hydraulic modelling software Synergi Gas for the optimization and simulation of gas distribution and transmission networks. In the further development of digital twins in subsea pipelines, flow assurance and fluid composition tracking need to be taken into consideration.

Table 1 lists the reviewed papers on the application of digital twins on different areas of subsea pipelines. It reveals that the maintenance and manufacturing of the pipelines are two top-ranked applications of digital twins. It can also be seen that half of the publications are journal papers. Given that the investigations and discussions on digital twins were mainly published in international conferences in the early 2010s, as reported in [5,34], more researchers intend to publish their research in journals in the coming years.

Table 1. A categorical review of the applications of digital twins on subsea pipelines.

Application	Reference Number	Year of Publication	Type of Document
Construction	[63]	2018	Conference
Design	[4]	2019	Conference
Design	[18]	2019	Conference
Life extension	[110]	2019	Conference
Life extension	[111]	2019	Conference
Maintenance	[11]	2017	Conference
Maintenance	[74]	2020	Journal
Maintenance	[78]	2019	Journal
Maintenance	[79]	2018	Journal
Maintenance	[80]	2019	Journal
Maintenance	[81]	2018	Magazine
Maintenance	[82]	2019	Journal
Manufacturing	[6]	2015	Journal
Manufacturing	[10]	2019	Journal
Manufacturing	[17]	2017	Conference
Manufacturing	[32]	2021	Journal
Monitoring	[89]	2021	Conference
Risk assessment	[101]	2022	Journal
Risk assessment	[102]	2022	Journal
Risk assessment	[113]	2018	Conference

8. Conclusions

With the recent wave of digitalization, the digital twin has been discussed as a powerful technology in a variety of industries including the oil and gas industry. The emergence of the digital twin provides an efficient way to realize remote monitoring and control, downtime prediction, and risk reduction of oil and gas subsea pipeline systems. This paper provides detailed coverage of recent publications on various applications of digital twins for subsea pipelines in terms of design, construction, service life, and assessments of life extension.

The key opportunities identified for improved integrity management enabled by digital twin applications are data contextualization, standardization, automated anomaly detection, and learning through sharing. The information of RFL provided by twin models will be valuable for the assessment of the extension of the service life of the subsea assets. With the calibrated twin model by considering actual environmental conditions, fatigue damage can be evaluated in real-time during the service life of the pipeline systems. Thus, the owners and authorities will be able to know the RFL and issue actions to optimize fatigue life or provide improvements or reinforcements for the lifetime extension of the assets.

On the other hand, the following main challenges of the use of digital twins have also been identified. Data related to the information on the conditions and risks are, in most cases, stored in different systems. Limited access to the data on servers is another issue in the use of digital twins. The physical and virtual facilities need to be protected against cyber-attacks by advanced cyber-security protocols [3,5,34,110]. Regarding the social impact, it was revealed that digital twin technologies can result in the redistribution of the workplace without much impact on employment [112].

The influence of digital twins also relies on the quantitative changes in new technologies. The following steps are suggested to make the research and development of digital twins more coherent: unify data and model standards; create a public database for sharing data and models; develop products and services to help digital twins become easier to build and use; develop universal platforms and tools for digital twin applications; and establish forums for practitioners and researchers.

Author Contributions: Conceptualization, B.-Q.C. and P.M.V.; writing—original draft preparation, B.-Q.C.; writing—review and editing, P.M.V. and C.G.S.; visualization, B.-Q.C.; supervision, C.G.S. All authors have read and agreed to the published version of the manuscript.

Funding: This work was developed in the scope of the project "Cementitious cork composites for improved thermal performance of pipelines for ultradeep waters—SUBSEAPIPE", with reference no. POCI-01-0145-FEDER-031011 funded by European Regional Development Fund (FEDER) through COMPETE2020—Operational Program Competitive-ness and Internationalization (POCI) and with financial support from the Portuguese Foundation for Science and Technology (Fundação para a Ciência e Tecnologia—FCT), under contract 02/SAICT/032108/2017. This study contributes to the Strategic Research Plan of the Centre for Marine Technology and Ocean Engineering, which is financed by FCT, under contract UIDB/UIDP/00134/2020.

Conflicts of Interest: The authors declare no conflict of interest.

References

1. Silva, L.M.R.; Guedes Soares, C. An integrated optimization of the floating and subsea layouts. *Ocean Eng.* **2019**, *191*, 106557. [CrossRef]
2. Silva, L.M.R.; Guedes Soares, C. Oilfield development system optimization under reservoir production uncertainty. *Ocean Eng.* **2021**, *225*, 108758. [CrossRef]
3. DNV GL. *2020/1022: How Digital Tools and Solutions Can Improve Subsea Integrity Management*; DNV GL: Oslo, Norway, 2020.
4. Bhowmik, S. Digital twin of subsea pipelines: Conceptual design integrating IoT, machine learning and data analytics. In Proceedings of the Annual Offshore Technology Conference 2019, Houston, TX, USA, 6–9 May 2019. [CrossRef]
5. Wanasinghe, T.R.; Wroblewski, L.; Petersen, B.K.; Gosine, R.G.; James, L.A.; de Silva, O.; Mann, G.K.I.; Warrian, P.J. Digital Twin for the Oil and Gas Industry: Overview, Research Trends, Opportunities, and Challenges. *IEEE Access* **2020**, *8*, 104175–104197. [CrossRef]
6. Rosen, R.; von Wichert, G.; Lo, G.; Bettenhausen, K.D. About the importance of autonomy and digital twins for the future of manufacturing. *IFAC-PapersOnLine* **2015**, *48*, 567–572. [CrossRef]
7. Gelernter, D.H. *Mirror Worlds: Or the Day Software Puts the Universe in a Shoebox—How It Will Happen and What It Will Mean*; Oxford University Press: New York, NY, USA, 1993; OCLC 23868481; ISBN 978-0195079067. [CrossRef]
8. Grieves, M. Virtually intelligent product systems: Digital and physical twins. In *Complex Systems Engineering: Theory and Practice*; Flumerfelt, S., Schwartz, K.G., Mavris, D., Briceno, S., Eds.; American Institute of Aeronautics and Astronautics: Reston, VA, USA, 2019; pp. 175–200. [CrossRef]
9. Li, C.; Mahadevan, S.; Ling, Y.; Choze, S.; Wang, L. Dynamic Bayesian network for aircraft wing health. *AIAA J.* **2017**, *55*, 930–941. [CrossRef]
10. Mandolla, C.; Messeni, A.; Percoco, G.; Urbinati, A. Computers in Industry Building a digital twin for additive manufacturing through the exploitation of blockchain: A case analysis of the aircraft industry. *Comput. Ind.* **2019**, *109*, 134–152. [CrossRef]
11. Magargle, R.; Johnson, L.; Mandloi, P.; Davoudabadi, P.; Kesarkar, O.; Krishnaswamy, S.; Batteh, J.; Pitchaikani, A. A simulation based digital twin for model-driven health monitoring and predictive maintenance of an automotive braking system. In Proceedings of the 12th International Modelica Conference, Prague, Czech Republic, 15–17 May 2017; pp. 35–46. [CrossRef]
12. Damjanovic-Behrendt, V. A digital twin-based privacy enhancement mechanism for the automotive industry. In Proceedings of the International Conference on Intelligent Systems (IS), Madeira, Portugal, 25–27 September 2018; pp. 272–279. [CrossRef]
13. Torkamani, A.; Andersen, K.G.; Steinhubl, S.R.; Topol, E.J. High-definition medicine. *Cell* **2017**, *170*, 828–843. [CrossRef]
14. Liu, Y.; Zhang, L.; Yang, Y.; Zhou, L.; Ren, L.; Wang, F.; Liu, R.; Pang, Z.; Deen, M.J. A novel cloud-based framework for the elderly healthcare services using digital twin. *IEEE Access* **2019**, *7*, 49088–49101. [CrossRef]
15. Laaki, H.; Miche, Y.; Tammi, K. Prototyping a digital twin for real time remote control over mobile networks: Application of remote surgery. *IEEE Access* **2019**, *7*, 20325–20336. [CrossRef]
16. Jimenez, J.I.; Jahankhani, H.; Kendzierskyj, S. Health care in the cyberspace: Medical cyber-physical system and digital twin challenges. In *Digital Twin Technologies and Smart Cities*; Springer: Cham, Switzerland, 2020; pp. 79–92. [CrossRef]

17. Post, J.; Groen, M.; Klaseboer, G. Physical model based digital twins in manufacturing processes. In Proceedings of the 11th Forming Technology Forum, Enschede, The Netherlands, 12–13 October 2017.

18. Howard, D. The digital twin: Virtual validation in electronics development and design. In Proceedings of the 2019 Pan Pacific Microelectronics Symposium, Kauai, HI, USA, 11–14 February 2019; pp. 1–9. [CrossRef]

19. Bilberg, A.; Malik, A.A. Digital twin driven human-robot collaborative assembly. *CIRP Ann.* **2019**, *68*, 499–502. [CrossRef]

20. Mohammadi, N.; Taylor, J.E. Smart city digital twins. In Proceedings of the 2017 IEEE Symposium Series on Computational Intelligence (SSCI), Honolulu, HI, USA, 27 November–1 December 2017; pp. 1–5. [CrossRef]

21. Ruohomäki, T.; Airaksinen, E.; Huuska, P.; Kesäniemi, O.; Martikka, M.; Suomisto, J. Smart city platform enabling digital twin. In Proceedings of the International Conference on Intelligent Systems (IS), Madeira, Portugal, 25–27 September 2018; pp. 155–161. [CrossRef]

22. Piascik, R.; Vickers, J.; Lowry, D.; Scotti, S.; Stewart, J.; Calomino, A. *Technology Area 12: Materials, Structures, Mechanical Systems, and Manufacturing Road Map*; NASA Office of Chief Technologist: Washington, DC, USA, 2010.

23. Caruso, P.; Dumbacher, D.; Grieves, M. Product lifecycle management and the quest for sustainable space exploration. In Proceedings of the AIAA SPACE 2010 Conference & Exposition, Anaheim, CA, USA, 30 August–2 September 2010. [CrossRef]

24. Glaessgen, E.; Stargel, D. The digital twin paradigm for future NASA and US Air Force vehicles. In Proceedings of the 53rd AI—AA/ASME/ASCE/AHS/ASC Structures, Structural Dynamics and Materials Conference, Honolulu, HI, USA, 23–26 April 2012. [CrossRef]

25. Tuegel, E. The airframe digital twin: Some challenges to realization. In Proceedings of the 53rd AIAA/ASME/ASCE/AHS/ASC Structures, Structural Dynamics and Materials Conference, Honolulu, HI, USA, 23–26 April 2012. [CrossRef]

26. Maher, D.P. On software standards and solutions for a trusted Internet of Things. In Proceedings of the 51st Hawaii International Conference on System Sciences (HICSS-51), Waikoloa Village, HI, USA, 3–6 January 2018. [CrossRef]

27. Chen, B.Q.; Videiro, P.M.; Guedes Soares, C. Review of digital twin of ships and offshore structures. In *Developments in Maritime Technology and Engineering*; Guedes Soares, C., Santos, T.A., Eds.; Taylor and Francis: London, UK, 2021; pp. 445–451. [CrossRef]

28. Panetta, K. Gartners Top 10 Technology Trends for 2017. Available online: https://www.gartner.com/smarterwithgartner/gartners-top-10-technology-trends-2017/ (accessed on 27 January 2022).

29. Cearley, D.W.; Walker, M.J.; Blosch, M. *Top 10 Strategic Technology Trends for 2018*; Gartner: Stamford, CT, USA, 2017.

30. Panetta, K. Gartner Top 10 Strategic Technology Trends for 2019. Available online: https://www.gartner.com/smarterwithgartner/gartners-top-10-technology-trends-2019/ (accessed on 27 January 2022).

31. Liu, M.; Fang, S.; Dong, H.; Xu, C. Review of digital twin about concepts, technologies, and industrial applications. *J. Manuf. Syst.* **2021**, *58*, 346–361. [CrossRef]

32. Son, Y.H.; Kim, G.-Y.; Kim, H.C.; Jun, C.; Noh, S. Do Past, present, and future research of digital twin for smart manufacturing. *J. Comput. Des. Eng.* **2021**, *9*, 1–23. [CrossRef]

33. Wang, J.; Li, X.; Wang, P.; Liu, Q. Bibliometric analysis of digital twin literature: A review of influencing factors and conceptual structure. *Technol. Anal. Strateg. Manag.* **2022**, 1–15. [CrossRef]

34. Atalay, M.; Murat, U.; Oksuz, B.; Parlaktuna, A.M.; Pisirir, E.; Testik, M.C. Digital twins in manufacturing: Systematic literature review for physical-digital layer categorization and future research directions. *Int. J. Comput. Integr. Manuf.* **2022**, 1–27. [CrossRef]

35. Tao, F.; Zhang, M.; Nee, A.Y.C. Five-dimension digital twin modeling and its key technologies. *Digit. Twin Driven Smart Manuf.* **2019**, 63–81. [CrossRef]

36. Qi, Q.; Tao, F.; Hu, T.; Anwer, N.; Liu, A.; Wei, Y.; Wang, L.; Nee, A.Y.C. Enabling technologies and tools for digital twin. *J. Manuf. Syst.* **2021**, *58*, 3–21. [CrossRef]

37. Tao, F.; Qi, Q. Make more digital twins. *Nature* **2019**, *573*, 490–491. [CrossRef]

38. Guo, B.; Song, S.; Ghalambor, A.; Lin, T.R. *Offshore Pipelines: Design, Installation, and Maintenance*; Gulf Professional Publishing: Houston, TX, USA, 2013. [CrossRef]

39. Cai, J.; Jiang, X.; Lodewijks, G. Residual ultimate strength of offshore metallic pipelines with structural damage–a literature review. *Ships Offshore Struct.* **2017**, *12*, 1037–1055. [CrossRef]

40. Bai, Q.; Bai, Y. *Subsea Pipeline Design, Analysis, and Installation*; Gulf Professional Publishing: Waltham, MA, USA, 2014.

41. Yazdi, M.; Khan, F.; Abbassi, R. Operational subsea pipeline assessment affected by multiple defects of microbiologically influenced corrosion. *Process Saf. Environ. Prot.* **2022**, *158*, 159–171. [CrossRef]

42. Wang, Z.; Tang, Y.; Guedes Soares, C. Imperfection study on lateral thermal buckling of subsea pipeline triggered by a distributed buoyancy section. *Mar. Struct.* **2021**, *76*, 102916. [CrossRef]

43. Wang, Z.; Tang, Y.; Yang, J.; Guedes Soares, C. Analytical study of thermal upheaval buckling for free spanning pipelines. *Ocean Eng.* **2020**, *218*, 108220. [CrossRef]

44. Bruton, D.A.; Bolton, M.; Carr, M.; White, D. Pipe-soil interaction with flowlines during lateral buckling and pipeline walking—The SAFEBUCK JIP. In Proceedings of the Offshore Technology Conference, Houston, TX, USA, 5–8 May 2008. [CrossRef]

45. Bruton, D.A.; Carr, M.; Sinclair, F.; MacRae, I. Lessons learned from observing walking of pipelines with lateral buckles, including new driving mechanisms and updated analysis models. In Proceedings of the Offshore Technology Conference, Houston, TX, USA, 3–6 May 2010. [CrossRef]

46. Chen, B.; Zhang, X.; Guedes Soares, C. The effect of general and localized corrosions on the collapse pressure of subsea pipelines. *Ocean Eng.* **2022**, *247*, 110719. [CrossRef]

47. Zhang, X.; Chen, B.; Guedes Soares, C. Effect of non-symmetrical corrosion imperfection on the collapse pressure of subsea pipelines. *Mar. Struct.* **2020**, *73*, 102806. [CrossRef]

48. Teixeira, A.P.; Guedes Soares, C.; Netto, T.A.; Estefen, S.F. Reliability of pipelines with corrosion defects. *Int. J. Press. Vessel. Pip.* **2008**, *85*, 228–237. [CrossRef]

49. Teixeira, A.P.; Palencia, O.G.; Guedes Soares, C. Reliability analysis of pipelines with local corrosion defects under external pressure. *J. Offshore Mech. Arct. Eng.* **2019**, *141*, 1–10. [CrossRef]

50. Netto, T.A. On the effect of narrow and long corrosion defects on the collapse pressure of pipelines. *Appl. Ocean Res.* **2009**, *31*, 75–81. [CrossRef]

51. ASME. *B31G Manual for Determining the Remaining Strength of Corroded Pipelines (Supplement to ASME B31 Code for Pressure Piping)*; The American Society of Mechanical Engineers: New York, NY, USA, 2012.

52. DNV. *Recommended Practice DNV-RP-F101: Corroded Pipelines*; DNV: Oslo, Norway, 2019.

53. DNV. *Standard DNV-ST-F101: Submarine Pipeline Systems*; DNV: Oslo, Norway, 2021.

54. American Petroleum Institute. *API Specification 5L—Line Pipe*, 46th ed.; American Petroleum Institute: Washington, DC, USA, 2018.

55. Li, R.; Chen, B.; Guedes Soares, C. Design equation for the effect of ovality on the collapse strength of sandwich pipes. *Ocean Eng.* **2021**, *235*, 109367. [CrossRef]

56. An, C.; Duan, M.; Estefen, S.F.; Su, J. *Structural and Thermal Analyses of Deepwater Pipes*; Springer: Cham, Switzerland, 2021; ISBN 9783030535391. [CrossRef]

57. Xia, M.; Kemmochi, K.; Takayanagi, H. Analysis of filament-wound fiber-reinforced sandwich pipe under combined internal pressure and thermomechanical loading. *Compos. Struct.* **2001**, *51*, 273–283. [CrossRef]

58. Estefen, S.F.; Netto, T.A.; Pasqualino, I.P. Strength analyses of sandwich pipes for ultra deepwaters. *Trans. ASME J. Appl. Mech.* **2005**, *72*, 599–608. [CrossRef]

59. Yang, J.; Paz, C.M.; Estefen, S.F.; Fu, G.; Lourenço, M.I. Collapse pressure of sandwich pipes with strain-hardening cementitious composite—Part 1: Experiments and parametric study. *Thin-Walled Struct.* **2020**, *148*, 106605. [CrossRef]

60. Li, R.; Chen, B.Q.; Guedes Soares, C. Effect of ovality length on collapse strength of imperfect sandwich pipes due to local buckling. *J. Marit. Sci. Eng.* **2022**, *10*, 12. [CrossRef]

61. Bhardwaj, U.; Teixeira, A.P.; Guedes Soares, C. Uncertainty in collapse strength prediction of sandwich pipelines. *J. Offshore Mech. Arct. Eng.* **2022**, *144*, 041702. [CrossRef]

62. DNV. *Standard DNV-RP-A204: Qualification and Assurance of Digital Twins*; DNV: Oslo, Norway, 2021.

63. Hlady, J.; Glanzer, M.; Fugate, L. Automated creation of the pipeline digital twin during construction—Improvement to construction quality and pipeline integrity. In Proceedings of the 2018 12th International Pipeline Conference, Calgary, AB, Canada, 24–28 September 2018; Volume 2, pp. 1–12. [CrossRef]

64. Bhowmik, S.; Noiray, G.; Naik, H. Subsea pipeline design automation using digital field twin. In Proceedings of the Abu Dhabi International Petroleum Exhibition & Conference ADIP 2019, Abu Dhabi, United Arab Emirates, 11–14 November 2019. [CrossRef]

65. Ho, M.; El-Borgi, S.; Patil, D.; Song, G. Inspection and monitoring systems subsea pipelines: A review paper. *Struct. Health Monit.* **2020**, *19*, 606–645. [CrossRef]

66. Kaiser, M.J. The global offshore pipeline construction service market 2017–Part I. *Ships Offshore Struct.* **2018**, *13*, 65–95. [CrossRef]

67. Karuks, E.; Rohn, M. Mechanical Protection Coating for Coated Metal Substrate. U.S. Patent No. 4,395,159, 5 February 1985.

68. Sumner, G.R. Offshore Pipeline Insulated with a Cementitious Coating. U.S. Patent No. 5,476,343, 19 December 1995.

69. Bass, R.M.; Newberry, B.L.; Langner, C.G. Annulus for Electrically Heated Pipe-in-Pipe Subsea Pipeline. U.S. Patent No. 6,814,146, 9 November 2004.

70. Zhang, X.; Duan, M.; Guedes Soares, C. Lateral buckling critical force for submarine pipe-in-pipe pipelines. *Appl. Ocean. Res.* **2018**, *78*, 99–109. [CrossRef]

71. Bhardwaj, U.; Teixeira, A.P.; Guedes Soares, C. Reliability assessment of a subsea pipe-in-pipe system for major failure modes. *Int. J. Press. Vessel. Pip.* **2020**, *188*, 104177. [CrossRef]

72. Bertram, S.J.; Fan, Y.; Raffelt, D.; Michalak, P. An applied machine learning approach to subsea asset inspection. In Proceedings of the Abu Dhabi International Petroleum Exhibition & Conference, ADIPEC 2018, Abu Dhabi, United Arab Emirates, 12–15 November 2018. [CrossRef]

73. Daily, J.; Peterson, J. Predictive maintenance: How big data analysis can improve maintenance. In *Supply Chain Integration Challenges in Commercial Aerospace*; Springer: Cham, Switzerland, 2017; pp. 267–278. [CrossRef]

74. Errandonea, I.; Beltrán, S.; Arrizabalaga, S. Digital Twin for maintenance: A literature review. *Comput. Ind.* **2020**, *123*, 103316. [CrossRef]

75. Swanson, L. Linking maintenance strategies to performance. *Int. J. Prod. Econ.* **2001**, *70*, 237–244. [CrossRef]

76. Shafiee, M. Maintenance strategy selection problem: An MCDM overview. *J. Qual. Maint. Eng.* **2015**, *21*, 378–402. [CrossRef]

77. Ansari, F.; Glawar, R.; Nemeth, T. PriMa: A prescriptive maintenance model for cyber-physical production systems. *Int. J. Comput. Integr. Manuf.* **2019**, *32*, 482–503. [CrossRef]

78. Werner, A.; Zimmermann, N.; Lentes, J. Approach for a holistic predictive maintenance strategy by incorporating a digital twin. *Procedia Manuf.* **2019**, *39*, 1743–1751. [CrossRef]

79. Vathoopan, M.; Johny, M.; Zoitl, A.; Knoll, A. Modular fault ascription and corrective maintenance using a digital twin. *IFAC-PapersOnLine* **2018**, *51*, 1041–1046. [CrossRef]

80. Rajesh, P.K.; Manikandan, N.; Ramshankar, C.S.; Vishwanathan, T.; Sathishkumar, C. Digital twin of an automotive brake pad for predictive maintenance. *Procedia Comput. Sci.* **2019**, *165*, 18–24. [CrossRef]

81. Strohmeier, F.; Schranz, C.; Guentner, G. i-maintenance: A digital twin for smart maintenance. *ERCIM News* **2018**, *115*, 12–14.

82. Liu, Z.; Chen, W.; Zhang, C.; Yang, C.; Chu, H. Data super-network fault prediction model and maintenance strategy for mechanical product based on digital twin. *IEEE Access* **2019**, *7*, 177284–177296. [CrossRef]

83. Bashiri, M.; Badri, H.; Hejazi, T.H. Selecting optimum maintenance strategy by fuzzy interactive linear assignment method. *Appl. Math. Model.* **2011**, *35*, 152–164. [CrossRef]

84. Arzaghi, E.; Abaei, M.M.; Abbassi, R.; Garaniya, V.; Chin, C.; Khan, F. Risk-based maintenance planning of subsea pipelines through fatigue crack growth monitoring. *Eng. Fail. Anal.* **2017**, *79*, 928–939. [CrossRef]

85. Nikoo, H.M.; Bi, K.; Hao, H. Effectiveness of using pipe-in-pipe (PIP) concept to reduce vortex-induced vibrations (VIV): Three-dimensional two-way FSI analysis. *Ocean Eng.* **2018**, *148*, 263–276. [CrossRef]

86. Shabani, M.M.; Shabani, H.; Goudarzi, N.; Taravati, R. Probabilistic modelling of free spanning pipelines considering multiple failure modes. *Eng. Fail. Anal.* **2019**, *106*, 104169. [CrossRef]

87. Li, X.; Zhang, Y.; Abbassi, R.; Khan, F.; Chen, G. Probabilistic fatigue failure assessment of free spanning subsea pipeline using dynamic Bayesian network. *Ocean Eng.* **2021**, *234*, 109323. [CrossRef]

88. Ferreira, A.D.M.; Afonso, S.M.B.; Willmersdorf, R.B.; Lyra, P.R.M. Multiresolution analysis and deep learning for corroded pipeline failure assessment. *Adv. Eng. Softw.* **2021**, *162–163*, 103066. [CrossRef]

89. Bhowmik, S. Digital twin for offshore pipeline corrosion monitoring: A deep learning approach. In Proceedings of the Offshore Technology Conference, Houston, TX, USA, 16–19 August 2021. OTC-31296-MS. [CrossRef]

90. Hameed, H.; Bai, Y.; Ali, L. A risk-based inspection planning methodology for integrity management of subsea oil and gas pipelines. *Ships Offshore Struct.* **2021**, *16*, 687–699. [CrossRef]

91. Bhardwaj, U.; Teixeira, A.P.; Guedes Soares, C. Reliability assessment of corroded pipelines with different burst strength models. In *Developments in Maritime Technology and Engineering*; Guedes Soares, C., Santos, T.A., Eds.; Taylor and Francis: London, UK, 2021; Volume 1, pp. 687–696. [CrossRef]

92. Bhardwaj, U.; Teixeira, A.P.; Guedes Soares, C.; Samdani Azad, M.; Punurai, W.; Asavadorndeja, P. Reliability assessment of thick high strength pipelines with corrosion defects. *Int. J. Press. Vessel. Pip.* **2019**, *177*, 103982. [CrossRef]

93. Bhardwaj, U.; Teixeira, A.P.; Guedes Soares, C. Uncertainty in reliability of thick high strength pipelines with corrosion defects subjected to internal pressure. *Int. J. Press. Vessel. Pip.* **2020**, *188*, 104170. [CrossRef]

94. Seo, J.K.; Cui, Y.; Mohd, M.H.; Ha, Y.C.; Kim, B.J.; Paik, J.K. A risk-based inspection planning method for corroded subsea pipelines. *Ocean Eng.* **2015**, *109*, 539–552. [CrossRef]

95. Stadie-Frohbös, G.; Lampe, J. Risk based inspection for aged offshore pipelines. In Proceedings of the ASME 32nd International Conference on Ocean, Offshore and Arctic Engineering, Nantes, France, 9–14 June 2013; American Society of Mechanical Engineers: New York, NY, USA, 2013; p. V02BT02A024. [CrossRef]

96. ABS. *Guide for Surveys Using Risk-Based Inspection for the Offshore Industry*; American Bureau of Shipping: Houston, TX, USA, 2003.

97. API. *Risk-Based Inspection. API-RP-580*, 3rd ed.; American Petroleum Institute: Washington, DC, USA, 2016.

98. ASME. *B31G—Manual for Determining the Remaining Strength of Corroded Pipelines*; Technical Report; ASME: New York, NY, USA, 1991.

99. Singh, M.; Markeset, T. A methodology for risk-based inspection planning of oil and gas pipes based on fuzzy logic framework. *Eng. Fail. Anal.* **2009**, *16*, 2098–2113. [CrossRef]

100. Liu, X.; Zheng, J.; Fu, J.; Nie, Z.; Chen, G. Optimal inspection planning of corroded pipelines using BN and GA. *J. Pet. Sci. Eng.* **2018**, *163*, 546–555. [CrossRef]

101. Priyanka, E.B.; Thangavel, S.; Gao, X.Z.; Sivakumar, N.S. Digital twin for oil pipeline risk estimation using prognostic and machine learning techniques. *J. Ind. Inf. Integr.* **2022**, *26*, 100272. [CrossRef]

102. Priyanka, E.B.; Thangavel, S.; Prabhakaran, P. Rank-based risk target data analysis using digital twin on oil pipeline network based on manifold learning. *Proc. Inst. Mech. Eng. Part E J. Process Mech. Eng.* **2022**. [CrossRef]

103. Franklin, J.G.; Stowell, B.E.; Jee, T.P.; Hawkins, M. Standardizing the approach for offshore pipeline lifetime extension. In Proceedings of the Abu Dhabi International Petroleum Exhibition and Conference ADIPEC 2008, Abu Dhabi, United Arab Emirates, 3–6 November 2008; Volume 3, pp. 1354–1360. [CrossRef]

104. Selman, A.; Hubbard, R. How to age gracefully—Pipeline life extension. In Proceedings of the SPE Asia Pacific Oil & Gas Conference and Exhibition, Perth, Australia, 25–27 October 2016. [CrossRef]

105. Ferreira, N.N.; Martins, M.R.; de Figueiredo, M.A.G.; Gagno, V.H. Guidelines for life extension process management in oil and gas facilities. *J. Loss Prev. Process Ind.* **2020**, *68*, 104290. [CrossRef]

106. Norwegian Oil and Gas Association. *Guideline 122—Norwegian Oil and Gas Recommended Guidelines for the Management of Life Extension*; Norwegian Oil and Gas Association: Forus, Norway, 2017.

107. *ISO/TS 12747*; Petroleum and Natural Gas Industries—Pipeline Transportation Systems—Recommended Practice for Pipeline Life Extension. ISO: Geneva, Switzerland, 2011.

108. *NORSOK Y-002*; Life Extension for Transportation System. NORSOK: Oslo, Norway, 2010.

109. *NORSOK Y-009*; Life Extension for Subsea Systems. NORSOK: Oslo, Norway, 2011.

110. Altamiranda, E.; Colina, E. A system of systems digital twin to support life time management and life extension of subsea production systems. In Proceedings of the OCEANS 2019, Marseille, France, 17–20 June 2019. [CrossRef]

111. Knezevic, D.; Fakas, E.; Riber, H.J. Predictive digital twins for structural integrity management and asset life extension—JIP concept and results. In Proceedings of the SPE Offshore Europe Conference and Exhibition, OE 2019, Aberdeen, UK, 3–6 September 2019; pp. 1–6. [CrossRef]

112. Rasheed, A.; San, O.; Kvamsdal, T. Digital twin: Values, challenges and enablers from a modeling perspective. *IEEE Access* **2020**, *8*, 21980–22012. [CrossRef]

113. Sharma, P.; Knezevic, D.; Huynh, P.; Malinowski, G. RB-FEA based digital twin for structural integrity assessment of Offshore structures. In Proceedings of the Annual Offshore Technology Conference, Houston, TX, USA, 30 April–3 May 2018; Volume 4, pp. 2942–2947. [CrossRef]

114. Theyab, M.A. Fluid flow assurance issues: Literature review. *SciFed J. Pet.* **2018**, *2*, 1–11.

115. Balakin, B.V.; Lo, S.; Kosinski, P.; Hoffmann, A.C. Modelling agglomeration and deposition of gas hydrates in industrial pipelines with combined CFD-PBM technique. *Chem. Eng. Sci.* **2016**, *153*, 45–57. [CrossRef]

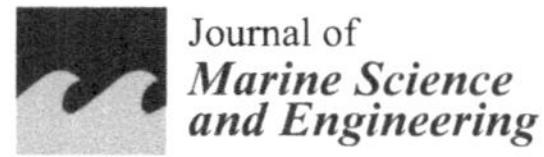

Journal of
*Marine Science
and Engineering*

Article

Effect of Ovality Length on Collapse Strength of Imperfect Sandwich Pipes Due to Local Buckling

Ruoxuan Li, Bai-Qiao Chen * and C. Guedes Soares

Centre for Marine Technology and Ocean Engineering (CENTEC), Instituto Superior Técnico, Universidade de Lisboa, 1049-001 Lisbon, Portugal; ruoxuan.li@centec.tecnico.ulisboa.pt (R.L.); c.guedes.soares@centec.tecnico.ulisboa.pt (C.G.S.)
* Correspondence: baiqiao.chen@centec.tecnico.ulisboa.pt

Abstract: The effect of ovality length on imperfect sandwich pipes is investigated using the finite element method in the scenario of local buckling under external pressure. First, the finite element model of the imperfect sandwich pipelines is established in ANSYS and is validated by comparing the results from numerical simulation with those from experiments. Then, the effect of ovality features on the collapse strength of the sandwich pipes is studied. At last, based on the calculation results from 1200 cases, a prediction equation is proposed to represent the relationship between collapse strength and ovality length of imperfect sandwich pipes. Good agreement is achieved between the proposed equation and the calculation results, leading to the conclusion that the proposed simplified model can be an efficient tool in the evaluation of the local collapse strength of subsea sandwich pipes under external pressure.

Keywords: subsea pipeline; sandwich pipe; collapse strength; initial ovality

Citation: Li, R.; Chen, B.-Q.; Guedes Soares, C. Effect of Ovality Length on Collapse Strength of Imperfect Sandwich Pipes Due to Local Buckling. *J. Mar. Sci. Eng.* **2022**, *10*, 12. https://doi.org/10.3390/jmse10010012

Academic Editor: Bruno Brunone

Received: 20 November 2021
Accepted: 21 December 2021
Published: 24 December 2021

Publisher's Note: MDPI stays neutral with regard to jurisdictional claims in published maps and institutional affiliations.

1. Introduction

Subsea pipelines play an important role in the offshore oil and gas industry, as the fastest, safest, and most economical and reliable means of transporting oil and gas continuously [1,2]. Among all mechanical properties of pipelines during the service life, the carrying capacity of external pressure is the first concern. In addition, it is necessary to maintain a high temperature of the liquid to keep the smooth flow. Due to the outstanding thermal protection, sandwich pipes (SPs) are applied in many subsea applications [3]. The SPs are made of three layers, two metal pipes as the inner and outer pipe, and a core layer between the two pipes. The core layer is usually light weighted and has low heat conductivity with low cost [4–6].

During the whole procedures of manufacture, transportation and installation, imperfections are unavoidable to be introduced onto the structure. The level of the imperfection of the cross-section is described by the ovality, Δ,

$$\Delta = \frac{D_{\max} - D_{\min}}{D_{\max} + D_{\min}} \tag{1}$$

The collapse strength of circular pipe is highly dependent on the ovality [4,6,7]. After the concept of SPs was proposed by Netto et al. [8] and Xia et al. [9], the collapse strength of SPs under external pressure was investigated [10,11]. The post-buckling behaviour and stability of SPs were studied by Arjomandi and Taheri [12], in which the importance of core material in SPs was figured out.

In addition to experiments, numerical simulation is also an active method [13–15] for the analysis of offshore structures such as subsea pipelines. An et al. [6] studied the collapse strength of SPs experimentally and numerically. By increasing the hydrostatic pressure in a water chamber, the collapse strength of the specimens was obtained, then finite element method (FEM) calculations were performed in ABAQUS in 2D cases.

Following the study of the improvement of the introduction of the core layer in SPs [5], the influence of steel grades was investigated by Yang et al. [16]. After obtaining experiment results, numerical simulation was performed. Then, Yang et al. [17] proposed an equation to fit the data. The accuracy of eight fitted equations was shown, more specifically, the average error is about 8% to 20% with the maximum error of 100% approximately, indicating the difficulties to capture the complex behaviour of the collapse pressure of an SP without the guidance of the physical background.

More recently, Li et al. [18] proposed an equation to describe the relationship between the collapse strength and geometrical parameters of SPs based on a series of numerical calculations.

It was found in Li and Guedes Soares [19] that the ovality length has a significant influence on the collapse strength of the pipeline under external pressure. Hence, the objective of this work is to investigate the effect of the ovality length on the collapse strength of SPs. At first, validation of the FEM calculations in ANSYS is performed by comparing with the experimental results from [6]. Then, the effect of the shape of ovality is examined by a series of FEM cases changing the geometrical parameters of SPs and the ovality. Finally, an equation is fitted to the results, to describe the relationship between the collapse strength of SPs and ovality.

2. Validation of the Finite Element Calculations

The results of the numerical calculations of this research are compared with those from the experiments of An et al. [6]. In the experiment, the collapse strength of SPs under external pressure is determined in a hyperbaric chamber. The length of SPs is 1750 mm. The geometrical parameters of the specimens' cross-section are illustrated in Table 1, in which three cases are calculated to compare with SP1~3, respectively. The radii of the outer pipe (R_o) and that of the inner pipe (R_i) of the three cases are very similar. The thickness of the outer pipe (t_o) is set to be the same in three cases, as well as that of the inner pipe (t_i). The difference between the three cases is the ovality of the outer (Δ_o) and inner (Δ_i) pipe is changeable.

Table 1. Geometrical parameters of cases performed by An et al. [6] in experiments.

Case	R_o (mm)	R_i (mm)	t_o (mm)	t_i (mm)	Δ_o (%)	Δ_i (%)
SP1	101.4	76.2	2.0	1.8	0.41	0.32
SP2	101.5	76.3	2.0	1.8	0.47	0.22
SP3	101.5	76.3	2.0	1.8	0.39	0.23

The collapse strength test results are, 30.5 MPa for SP1, 30.6 MPa for SP2, and 29.7 MPa for SP3.

The physical properties of the core layer and pipes are shown in Figure 1. By measuring the curves in Figure 1, for the core layer, the yield stress is 27 MPa, Young's modulus is 9 GPa. In FEA, the perfect elastic-plastic material model is applied for the core layer. For the inner pipe, the yield stress and ultimate stress are 215 MPa and 474 MPa, Young's modulus is 82.1 GPa. For the outer pipe, the yield stress and ultimate stress are 147 MPa and 381 MPa, Young's modulus is 59.9 GPa. In FEA, the flexibility-linear hardening model is applied for pipes.

The numerical calculations of this research are performed in ANSYS with an implicit computation procedure. The element type is SOLID185. The finite element mesh and load condition are shown in Figure 2. The length of the pipe is 1200 mm. Only a quarter of the cross-section of a half-pipe is modelled because of the symmetry. In the circumferential direction, the number of elements is 20, with an element size of about 7 mm on average. In the radius direction, the number of elements is 1, 4, and 1, respectively, for the inner pipe, core layer, and outer pipe. The element size is 5.8 mm for the core layer. In the axial direction, the number of elements is 120, corresponding to an element size of 5 mm. A rigid end is created in the position of $z = 0$ to simulate the plug from the experiment. Symmetrical boundary conditions are set in the plane *XOZ* and *YOZ*.

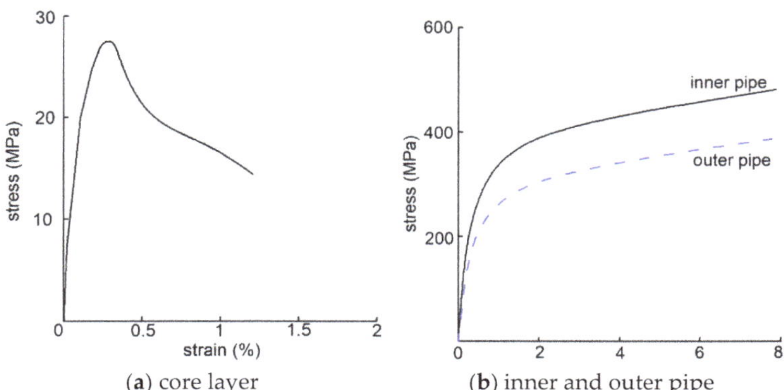

(**a**) core layer   (**b**) inner and outer pipe

Figure 1. Strain–stress relationship of SPs materials in the experiment by An et al. [6].

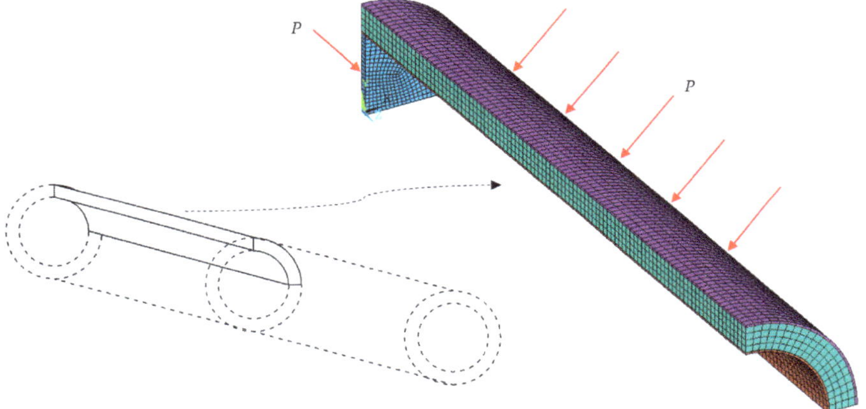

Figure 2. Mesh division and loading condition of numerical calculation case.

A fully bonded adhesion condition is adopted between the interfaces of the core layer and metal pipes. In ANSYS, this is realized by coupling the displacement of the nodes from both sides of the interfaces which is the same as that used in Li et al. [18].

The uniform pressure acts on both the outer surface of the outer pipe and the end of the pipe at the position of $z = 0$. Thus, the loading condition of this case is the coupling of external pressure and axial compression. Two kinds of load increase simultaneously which is the same as that in the experiment.

At this step, the half-wave number in the circumferential direction is selected to be 2, which will be studied later in Section 3. The length of the ovality λ is 50 mm, whose meaning is shown in Figure 3.

The comparison results between calculations and experiments are shown in Table 2. Good agreement is confirmed between the collapse strength (P_e) measured in experiments from [6] and the collapse strength (P_{co}) from FEM calculation in this research.

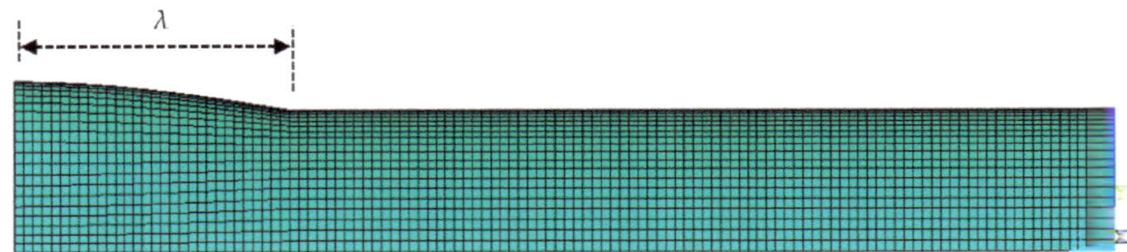

Figure 3. Sample of initial imperfection in the axial direction (factor = 100).

Table 2. Comparison of FEM calculation and experiment results from [6].

Case	Results in [6], P_e (MPa)	Results from FEM, P_{co} (MPa)	Error (%) *
SP1	30.5	29.9	−1.97
SP2	30.6	30.0	−1.96
SP3	29.7	30.0	1.01

* Error = $(P_{co} - P_e)/P_e \times 100\%$.

Then, the mesh sensitivity is performed. As shown in Table 3, five cases with different element sizes are calculated for every pipe configuration, SP1 SP2, and SP3. The elements of the core layer from each case are almost cubic. The comparison results of the calculation and experiment are shown in Table 4. Despite the small errors in all five cases, a relatively larger error appears with finer mesh in Table 4. The reason is that the ovality measured in experiments could not be as perfect as that used in FEM, and the collapse strength is influenced by many factors as shown in the conclusions of this research. From the results, the mesh division of N20 has acceptable accuracy and high computation efficiency.

Table 3. Detail mesh division for N16~N60.

Case	Axial Direction (Half-Length)		Circular Direction (1/4 Circle)		Thickness Direction for Core Layer		Thickness Direction for Metal Pipe	
	Number of Elements	Element Size (mm)	Number of Elements	Element Size (°)	Number of Elements	Element Size (mm)	Number of Elements	Element Size, o (mm)
N16	100	6	16	5.625	3	7.73	1	2/1.8
N20	120	5	20	4.5	4	5.8	1	2/1.8
N30	120	5	30	3	6	3.87	1	2/1.8
N40	200	3	40	2.25	8	2.9	2	1/0.9
N60	300	2	60	1.5	12	1.93	3	0.67/0.

Table 4. Comparison results between calculation and experiment of N16~N60.

Case	Results in [6], P_e (MPa)	N16		N20		N30		N40		N60	
		P_{co} (MPa)	Error (%) *	P_{co} (MPa)	Error (%) *	P_{co} (MPa)	Error (%) *	P_{co} (MPa)	Error (%) *	P_{co} (MPa)	Error (%) *
SP1	30.5	30.2	−0.98	29.9	−1.97	29.8	−2.30	29.1	−4.59	28.8	−5.
SP2	30.6	30.2	−1.31	30.0	−1.96	29.8	−2.61	29.0	−5.23	28.8	−5.
SP3	29.7	30.3	2.02	30.0	1.01	29.9	0.67	29.0	−2.36	28.8	−3.

* Error = $(P_{co} - P_e)/P_e \times 100\%$.

3. Effect of Ovality Shape in the Circumferential Direction

The circular pipes are usually manufactured by rolling flat plates. During this procedure, ovality in pipe circumferential direction is usually introduced into the pipe, according to the variable,

$$\omega = \omega_0 \mathrm{Cos}(n_c \theta) \tag{2}$$

where

w is the additional variable in the radius direction;

w_0 is the maximum amplitude;

n_c is the number of half-waves in the circumferential direction;

θ is the central angle.

Different types of ovality are compared to each other. For the overall ovality, the value of n_c is selected to be 2, 3, and 4. In this section, half of the whole circle is modelled in each case instead of one quarter in the previous because of the case Nc3. The cross-sections of three patterns are shown in Figure 4, with a scale factor of 100 for a clear illustration. The comparison result of the collapse pressure of the sandwich pipes is shown in Table 5.

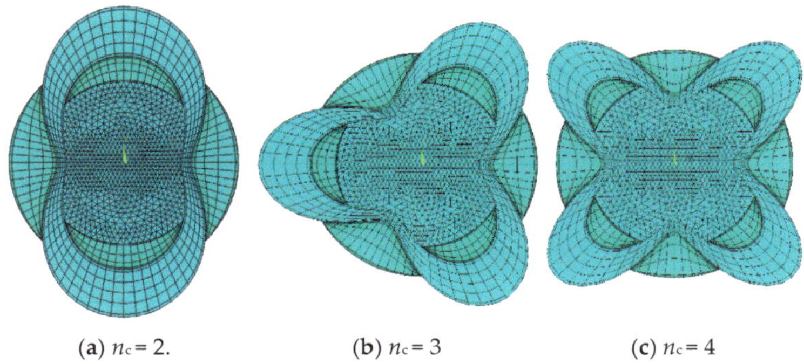

(a) n_c = 2.   (b) n_c = 3   (c) n_c = 4

Figure 4. Half-wave numbers in the circumferential direction with the value of Δ is 0.2% (scale factor: 100).

Table 5. Comparison results between calculation and experiment of Nc2~Nc4.

Case	Results in [6], P_e (MPa)	Nc2		Nc3		Nc4	
		P_{co} (MPa)	Error (%) *	P_{co} (MPa)	Error (%) *	P_{co} (MPa)	Error (%) *
SP1	30.5	30.0	−1.74	30.3	−0.66	32.2	5.42
SP2	30.6	30.0	−1.95	30.3	−0.87	32.2	5.08
SP3	29.7	30.0	1.12	30.4	2.30	32.2	8.26

* Error = $(P_{co} - P_e)/P_e \times 100\%$.

From Table 5, the case Nc2 with n_c = 2 is the most severe situation. Therefore, in this paper, the half wave number in the circumferential direction is set to be 2 for the parametrical study.

4. Parametric Study

In the previous calculation, the specimens are under loads of both axial compression and external pressure. The purpose is to compare with the results from the experiment. After the validation, in parametrical study cases, the load is set to be only the uniform external pressure. The length of SPs is 1200 mm, half of the length and one-quarter of the cross-section is modelled. The physical property of the core layer is the same as shown in Figure 1a. The physical property of the metal pipe is selected as the same as that of the outer pipe in Figure 1b.

Five groups of pipes with various radii of outer and inner pipes are considered. As shown in Table 6, the value of R_o/R_i is almost evenly distributed between 1.2 and 2.0. In this section, the ovality of the outer and inner pipe is selected to be of the same value, unlike that in [18], since the effect of ovality is not the principal task in this research.

Table 6. Outer and inner radius in each group of calculation.

Group	R_o (mm)	R_i (mm)	R_o/R_i
1	60	50	1.2
2	100	70	1.43
3	80	50	1.6
4	90	50	1.8
5	100	50	2.0

In each group, four ratios of radius over thickness are set, R/t = 20, 25, 40, and 50, for both inner and outer pipe. The value of ovality (Δ) is set to be 0.01, 0.05, 0.2, and 1%. The length of ovality area (λ) is 50, 80, 120, 160, 200, 300, 400, 600, and 1200 mm. To perform dimensionless, a ratio of λ/R_o is selected, this parameter means the gradient of an ovality in SPs axial direction. In this regard, 240 cases are calculated in each group. The case name is represented by a five-number '*abcde*', as shown in Table 7. In detail, '*a*' means the group, '*b*' means the ratio of radius and thickness of the outer pipe, '*c*' means the ratio of radius and thickness of the inner pipe, '*d*' means the ovality length, and '*e*' means the ovality.

Table 7. Explanation of case names.

a	Group	*b,c* *	R/t	*d*	λ (mm)	*d*	λ (mm)	*e*	Δ (%)
1	1	1	20	1	50	5	200	1	0.01
2	2	2	25	2	80	6	300	2	0.05
3	3	3	40	3	120	7	400	3	0.2
4	4	4	50	4	160	8	600	4	1
5	5					9	1200		

* *b*: R_o/t_o; *c*: R_i/t_i.

For example, in the case series of '313*de*', the relationships between collapse strength (P_{co}) and ovality (Δ) and ovality length (λ) are shown in Figures 5 and 6, respectively.

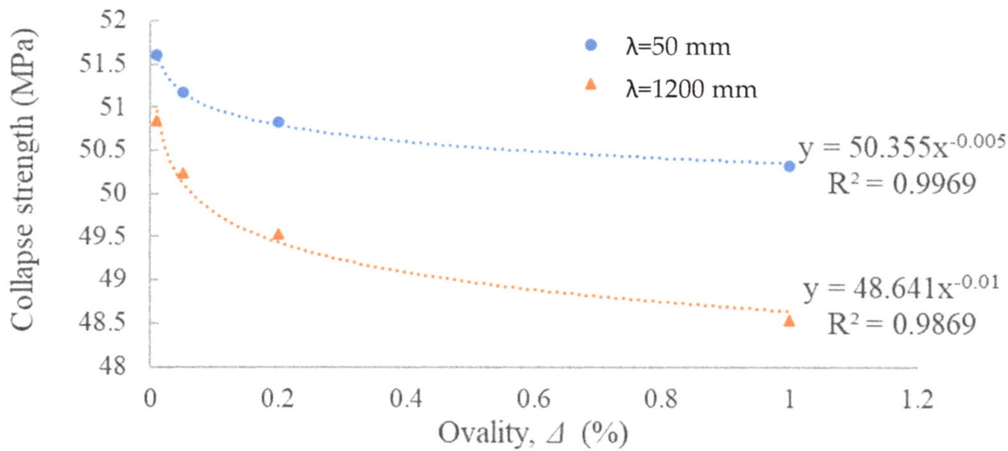

Figure 5. Relationship between collapse strength and ovality of case series of '313*de*'.

The relationships between collapse strength and ovality/ovality length follow a power function. The good fit is obtained as illustrated in Figures 5 and 6. Moreover, the physical meaning of power function is suitable. When the values of Δ and λ are set to be zero, it means no imperfection is introduced to the SPs, so the collapse strength of intact SPs is only decided by the material properties and geometrical parameters. As a result, in the final fitted equation, the power function is selected. Five variables are included, where Δ and λ/R_o describe the features of ovality, R_o/t_o, R_i/t_i, and R_o/R_i describe the collapse strength of intact SPs.

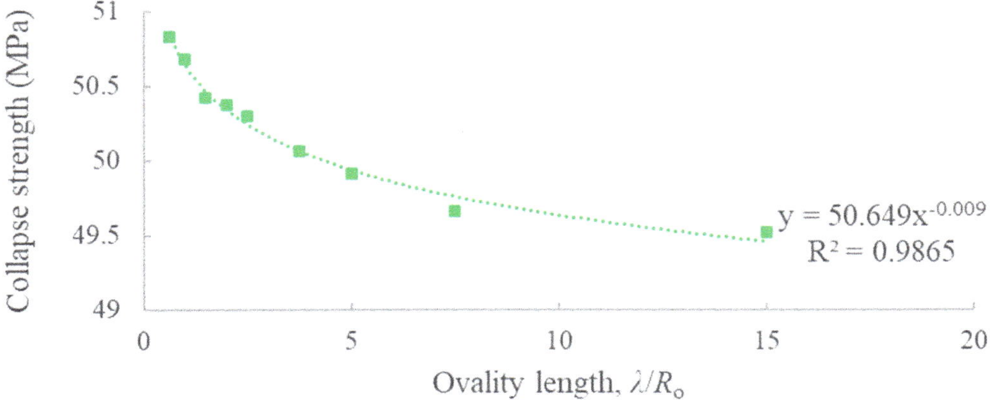

Figure 6. Relationship between collapse strength and ovality length of case series of '313*de*'.

After summarizing all 1200 calculation cases, a fitted equation is derived,

$$P_{\text{fit}} = Ce^{md\cdot\Delta + nl\cdot\lambda/R_o} \tag{3}$$

where

$$md = -0.2816\left(\frac{R_o}{R_i}\right)^2 + 0.9743\left(\frac{R_o}{R_i}\right) - 0.8525$$

$$nl = -0.1516\left(\frac{R_o}{R_i}\right)^2 + 0.5229\left(\frac{R_o}{R_i}\right) - 0.4617$$

$$C = k_c\left(\frac{R}{t}\right) + b_c, \quad \frac{R}{t} = \left(\frac{R_o}{t_o}\right)^{1/3}\left(\frac{R_i}{t_i}\right)^{2/3}$$

$$k_c = -0.5867\left(\frac{R_o}{R_i}\right) - 0.1918, \quad b_c = 65.94\left(\frac{R_o}{R_i}\right) - 11.62$$

In Equation (3), the expression of fitted collapse strength (P_{fit}) contains several constants. These constants are regarded as the effect of material properties of SPs.

The accuracy of Equation (3) is checked. Some cases from Group 1 have large errors. In these cases, the value of λ/R_o is too large, for example, when λ = 1200 mm, the value of λ/R_o is 20. A large value of λ/R_o means the ovality area occurred to a large part of SPs. The issue discussed in this research focuses on the local buckling so that the cases with λ/R_o ratios large than 16 are removed. Therefore, 1136 cases are summarized. The comparison result of the other cases between fitted equation and calculation is shown in Figure 7, in which good agreement is achieved.

In Equation (3), Δ and λ/R_o are the features to describe the ovality. Except for these two parameters, the other two variables, R_o/R_i and R/t, describe the geometrical property of the SP It is concluded that the R/t ratio is a dominant factor not only in the collapse strength of the single-walled pipes but also in the collapse strength of SPs. The influence of the core layer is described by R_o/R_i, a higher value of it means a thicker layer of the core material. There are several constant values included in Equation (3) controlled by the physical properties of SP, since the material has not been changed in this research. The influence of the physical properties is worth to be investigated in the future.

In terms of the geometrical property of SPs, the increase in R_o/R_i and decrease in R/t can both lead to the improvement of the collapse strength of imperfect SPs.

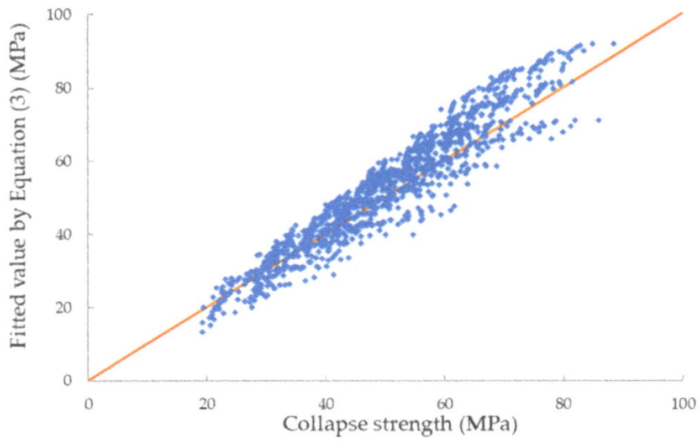

Figure 7. Comparison between the results obtained by using the fitted equation and from the calculation.

5. Discussion

In Equation (3), the relationship between collapse strength and ovality length and ovality follows a power function. The influence of ratios of radius to thickness (R/t) of inner and outer pipe refers to Equation (2) in [18]. It is confirmed that the influence of the R/t of the inner pipe is greater than that of the outer pipe.

The value of md in Equation (3) is almost twice of nl when the ratio of R_o/R_i is set, that means for a certain SP, the effect of ovality is a more dominant factor of ovality to reduce the loading capacity when the SP is acted by external pressure.

The relationship between collapse strength of imperfect SPs and R/t is shown in Figure 8, and the relationship between collapse strength and R_o/R_i is shown in Figure 9. In both figures, the ovality is the same with the parameter $\lambda/R_o = 2.0$ and $\Delta = 0.2\%$. It is clear that the collapse strength is greater of SPs with a higher value of R_o/R_i or a lower value of R/t. The changing rates of the curves are almost the same in both figures. However, it is concluded from the expression of C in Equation (3) that the influence of R/t is greater than that of R_o/R_i by checking the partial derivatives of both parameters. Due to the material strength of the metal pipe that is much higher than that of the core layer, the improvement of strength of both outer and inner pipes results in significant promotion of the collapse strength of SPs. Thus, in order to obtain SPs with a higher carrying capacity of external pressure, it is recommended to increase the strength of metal pipes, especially the inner pipe.

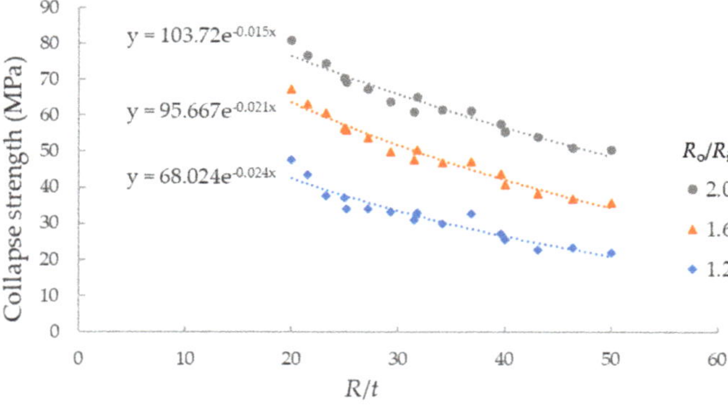

Figure 8. The relationship between collapse strength and parameter R/t ($\lambda/R_o = 2.0$, $\Delta = 0.2\%$).

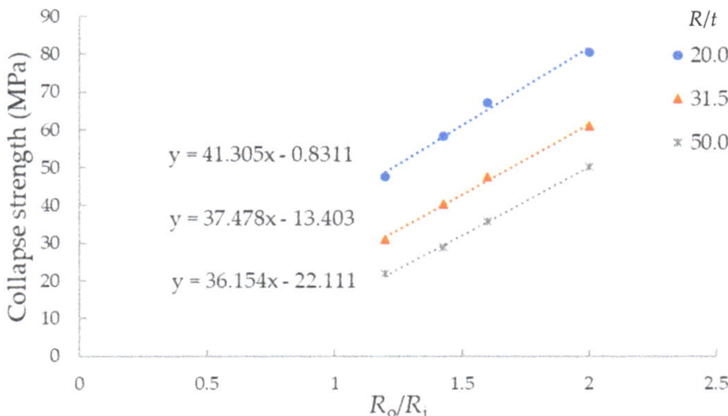

Figure 9. The relationship between collapse strength and parameter R_o/R_i ($\lambda/R_o = 2.0$, $\Delta = 0.2\%$).

Considering local buckling as the main concern, 64 cases with the ratio of λ/R_o greater than 16 are taken out of consideration. From the comparison result of calculation and Equation (3), the error of most cases is smaller than 20%, except for 51 cases out of 1136. The average value of error is only 4.4%, which indicates a good fit.

6. Conclusions

In this paper, the effect of ovality length on the collapse strength of SPs is investigated, with an emphasis on the local buckling scenarios. After the numerical calculation is validated against experiment results, the effect of half wavenumbers in the circumferential direction is checked. Then, 1200 cases are calculated, considering the ratio of R_o/R_i in the range of 1.2 to 2.0, ratios of R_o/t_o and R_i/t_i in the ranges of 20 to 50, and the ratio of λ/R_o in the range of 0.5 to 15. The relationship between the collapse strength and ovality length and ovality follows a power function. Fitting 1200 cases of numerical calculations, Equation (3) is derived. Good agreement is achieved between Equation (3) and calculation results, leading to the conclusion that the proposed simplified model can be efficiently used in the evaluation of the collapse pressure of subsea pipelines.

Author Contributions: Conceptualization, C.G.S.; methodology, R.L.; formal analysis, R.L. and B.-Q.C.; writing—original draft preparation, R.L. and B.-Q.C.; writing—review and editing, C.G.S.; visualization, R.L.; supervision, C.G.S. All authors have read and agreed to the published version of the manuscript.

Funding: This work was developed in the scope of the project "Cementitious cork composites for improved thermal performance of pipelines for ultradeep waters—SUBSEAPIPE, with the reference no. POCI-01-0145-FEDER-031011 funded by European Regional Development Fund (FEDER) through COMPETE2020—Operational Program Competitive-ness and Internationalization (POCI) and with financial support from the Portuguese Foundation for Science and Technology (Fundação para a Ciência e Tecnologia—FCT). This study contributes to the Strategic Research Plan of the Centre for Marine Technology and Ocean Engineering, which is financed by FCT, under contract UIDB/UIDP/00134/2020.

Acknowledgments: This work was developed in the scope of the project "Cementitious cork composites for improved thermal performance of pipelines for ultradeep waters—SUBSEAPIPE, with the reference no. POCI-01-0145-FEDER-031011 funded by European Regional Development Fund (FEDER) through COMPETE2020—Operational Program Competitiveness and Internationalization (POCI) and with financial support from the Portuguese Foundation for Science and Technology (Fundação para a Ciência e Tecnologia—FCT). This study contributes to the Strategic Research Plan of the Centre for Marine Technology and Ocean Engineering, which is financed by FCT, under contract UIDB/UIDP/00134/2020.

Conflicts of Interest: The authors declare no conflict of interest.

References

1. Silva, L.M.R.; Teixeira, A.P.; Guedes Soares, C. A methodology to quantify the risk of subsea pipeline systems at the oilfield development selection phase. *Ocean Eng.* **2019**, *179*, 213–225. [CrossRef]
2. Silva, L.M.R.; Guedes Soares, C. Oilfield development system optimization under reservoir production uncertainty. *Ocean Eng.* **2021**, *225*, 108758. [CrossRef]
3. Wang, Z.K.; Guedes Soares, C. Upheaval thermal buckling of functionally graded subsea pipelines. *Appl. Ocean Res.* **2021**, *116*, 102871. [CrossRef]
4. Castello, X.; Estefen, S.F. Limit strength and reeling effects of sandwich pipes with bonded layers. *Int. J. Mech. Sci.* **2007**, *49*, 577–588. [CrossRef]
5. An, C.; Castello, X.; Duan, M.; Toledo Filho, R.D.; Estefen, S.F. Ultimate strength behaviour of sandwich pipes filled with steel fiber reinforced concrete. *Ocean Eng.* **2012**, *55*, 125–135. [CrossRef]
6. An, C.; Duan, M.; Toledo Filho, R.D.; Estefen, S.F. Collapse of Sandwich Pipes with PVA Fiber Reinforced Cementitious Composites Core under External Pressure. *Ocean Eng.* **2014**, *82*, 1–13. [CrossRef]
7. Gong, S.; Wang, X.; Zhang, T.; Liu, C. Buckle propagation of sandwich pipes under external pressure. *Eng. Struct.* **2018**, *175*, 339–354. [CrossRef]
8. Netto, T.A.; Santos, J.M.C.; Estefen, S.F. Sandwich pipes for ultra-deep waters. In Proceedings of the 4th International Pipeline Conference, Calgary, AB, Canada, 29 September–3 October 2002; Volume 36207, pp. 2093–2101. [CrossRef]
9. Xia, M.; Kemmochi, K.; Takayanagi, H. Analysis of filament-wound fiber-reinforced sandwich pipe under combined internal pressure and thermomechanical loading. *Compos. Struct.* **2001**, *51*, 273–283. [CrossRef]
10. Paz, C.M.; Fu, G.; Estefen, S.F.; Lourenço, M.I.; Chujutalli, J.A.H. Sandwich pipe: Reel-lay installation effects. In Proceedings of the ASME 34th International Conference on Ocean, Offshore and Arctic Engineering, St. John's, NL, Canada, 31 May–5 June 2015. OMAE2015-41089. [CrossRef]
11. Xu, Q.; Gong, S.; Hu, Q. Collapse analyses of sandwich pipes under external pressure considering inter-layer adhesion behaviour. *Mar. Struct.* **2016**, *50*, 72–94. [CrossRef]
12. Arjomandi, K.; Taheri, F. Stability and post-buckling response of sandwich pipes under hydrostatic external pressure. *Int. J. Press. Vessel. Pip.* **2011**, *88*, 138–148. [CrossRef]
13. Jin, Z.; Shen, X.; Yan, S.; Ye, H.; Gao, Z.; Chen, Z. A three-dimensional analytical solution for sandwich pipe systems under linearly varying external pressures. *Ocean Eng.* **2016**, *124*, 298–305. [CrossRef]
14. Hastie, J.C.; Kashtalyan, M.; Guz, I.A. Analysis of filament-wound sandwich pipe under combined internal pressure and thermal load considering restrained and closed ends. *Int. J. Press. Vessel. Pip.* **2021**, *191*, 104350. [CrossRef]
15. Chen, B.Q.; Guedes Soares, C. Experimental and numerical investigation on welding simulation of long stiffened steel plate specimen. *Mar. Struct.* **2021**, *75*, 102824. [CrossRef]
16. Yang, J.; Paz, C.M.; Estefen, S.F.; Fu, G.; Lourenço, M.I. Collapse pressure of sandwich pipes with strain-hardening cementitious composite-Part 1, Experiments and parametric study. *Thin-Walled Struct.* **2020**, *148*, 106605. [CrossRef]
17. Yang, J.; Estefen, S.F.; Fu, G.; Paz, C.M.; Lourenço, M.I. Collapse pressure of sandwich pipes with strain-hardening cementitious composite-Part 2, A suitable prediction equation. *Thin-Walled Struct.* **2020**, *148*, 106606. [CrossRef]
18. Li, R.; Chen, B.; Guedes Soares, C. Design equation for the effect of ovality on the collapse strength of sandwich pipes. *Ocean Eng.* **2021**, *235*, 109367. [CrossRef]
19. Li, R.; Guedes Soares, C. Numerical study on the effects of multiple initial defects on the collapse strength of pipelines under external pressure. *Int. J. Press. Vessel. Pip.* **2021**, *194*, 104484. [CrossRef]

Journal of
Marine Science and Engineering

Article

CFRP Strengthening and Rehabilitation of Inner Corroded Steel Pipelines under External Pressure

Jianxing Yu [1,2], Weipeng Xu [1,3,*], Yang Yu [1,3], Fei Fu [2], Huakun Wang [4], Shengbo Xu [1,3] and Shibo Wu [1,3]

[1] State Key Laboratory of Hydraulic Engineering Simulation and Safety, Tianjin University, Tianjin 300072, China; yjx2000@tju.edu.cn (J.Y.); yang.yu@tju.edu.cn (Y.Y.); edwinb@tju.edu.cn (S.X.); wushibo@tju.edu.cn (S.W.)
[2] College of Mechanical and Marine Engineering, Beibu Gulf University, Qinzhou 535000, China; fufei201916@163.com
[3] Tianjin Key Laboratory of Port and Ocean Engineering, Tianjin University, Tianjin 300072, China
[4] School of Architecture and Civil Engineering, Xiamen University, Xiamen 361005, China; hkwang@xmu.edu.cn
* Correspondence: xuweipeng@tju.edu.cn

Abstract: The objective of this study was to investigate the performance of pipelines repaired with carbon-fibre-reinforced polymer (CFRP) under external pressure. The repaired pipeline experienced defects in terms of thinning of its local inner wall. The three-dimensional finite element method was used to analyse the collapse pressure of repaired pipes with internal corrosion defects. The traction–separation law and interlaminar damage criterion were applied to simulate the collapse process of repaired pipes. The results show that the collapse pressure of the composite-repaired pipe increased and the CFRP significantly reduced the strain in the defect region. It was observed that the ovality of the corrosion defect region was reduced and that the repair effectiveness mainly depended on the length, thickness, and interlayer cohesion.

Keywords: repair; carbon-fibre-reinforced polymer (CFRP); pipeline; collapse; inner corroded

Citation: Yu, J.; Xu, W.; Yu, Y.; Fu, F.; Wang, H.; Xu, S.; Wu, S. CFRP Strengthening and Rehabilitation of Inner Corroded Steel Pipelines under External Pressure. *J. Mar. Sci. Eng.* **2022**, *10*, 589. https://doi.org/10.3390/jmse10050589

Academic Editors: Bai-Qiao Chen and Carlos Guedes Soares

Received: 10 April 2022
Accepted: 25 April 2022
Published: 26 April 2022

Publisher's Note: MDPI stays neutral with regard to jurisdictional claims in published maps and institutional affiliations.

1. Introduction

In recent years, with the reduction in resources inland and increasing difficulties in mining, submarine energy has been developing. Deep-sea energy mainly includes oil, gas, and minerals, which are transported through pipelines. Pipelines are widely used to transport deep-sea resources because of their safety and efficiency. However, the submarine environment is complex, so pipelines can be worn thin and corroded with the transportation of ore, which could eventually lead to the collapse of the pipeline in the deep-sea high-pressure environment, causing serious leakage problems.

Researchers have analysed the collapse of submarine pipelines in deep-sea high-pressure environments [1–3]. The local buckling of the pipeline occurs under high external pressure, and this buckling would propagate along the pipeline at a high speed, resulting in catastrophic accidents [4]. Dyau and Kyriakides and Sakakibara et al. [5,6] studied the pipe buckling caused by local collapse and proposed an empirical analysis method to evaluate the residual collapse pressure of an internally corroded pipeline. Netto [7] proposed an empirical formula of pipe collapse pressure through experiments and numerical simulation data. Meanwhile, researchers have investigated the influence of different defects. Ramasamy and Tuan Ya [8] studied the effect of dent defects on the collapse pressure. Wang et al. [9–11] studied the influence of pipelines' external surface corrosion on their external-pressure-bearing capacity. Through tests and program development and calculation, it was found that random pitting corrosion increases the risk of a pipeline's collapse and changes its buckling shape. Moreira Junior et al. [12] studied the effects of internal local corrosion and wall thickness eccentricity on pipeline collapse pressure. Zhang and Pan [13] analysed the effects of thickness eccentricity and overall ovality on pipe re crushing pressure. In [14,15], a local corrosion model is developed, and it is found that the defects caused by internal wear and corrosion have a great impact on the ability of a pipeline to withstand external water pressure and axial load.

In view of the above-mentioned situations that lead to local instability of pipelines, several researchers have put forward disaster-reduction measures for the arrestor, including steel and composite materials [16]. In the early stage, Johns et al. [17] and Hahn et al. [18] carried out experimental research on the influence of pipe arrestors on buckling propagation, and their research proved that the arrestor can enhance the pipe's local collapse pressure. Kyriakides's team has performed many experiments and analyses on the performance of steel buckling arrestors, finding that steel ring buckling arrestors can effectively prevent pipe buckling propagation and large area damage along pipelines. They also found that the effect of integral buckling arrestors is significantly better than that of slip-on arrestors. The interaction between buckling arrestors and pipes was also found to have a significant influence on propagation pressure [19–23]. However, the arrestors can only prevent the propagation of buckling and do not stop the collapse process. Once the pipeline collapses, it needs to be replaced to restore it to its working capacity but doing so is expensive. If effective measures can be taken in time to repair the pipeline, large area damage and high maintenance costs can be prevented. The study of deep-water pipeline repair is of great significance. Common repair methods of deep-water pipes include welding, grouting, sleeves, and composite materials [24–28].

At present, methods of repairing defective pipelines mainly involve cutting the damaged pipe section and replacing it, which is expensive and unsafe. Several experiments have found that the effect of the CFRP is better than that of GFRP when repairing pipe of the same size, but the tensile strength of GFRP is lower than that of the CFRP [29,30]. Several studies have been carried out to provide a reliable experimental and theoretical basis for CFRP pipeline repairs. It has been found that the CFRP has advantages in terms of high specific strength and stiffness and performance weight ratio [31,32]. Meanwhile, the bonding of CFRP and pipelines could improve pipelines' pressure capacity [33], increase their ability to resist the lateral impact [34,35], strengthen the pipeline weld [25], and enhance the stability of its tubular structure [36] and the bearing capacity of the pipeline axial force and bending moment. Using their developed numerical model, Mokhtari and Alavi Nia [37] found that the repaired pipeline can reduce the influence of the staggered soil layer on the pipeline. Shamsuddoha et al. [38] proposed a CFRP manufacturing method and found that the bearing capacity of the repaired pipeline was improved. Meriem-Benziane et al. [39] studied the repair of the longitudinal crack of an API X65 pipeline under internal pressure with CFRP and found that CFRP could effectively enhance its bearing capacity under internal pressure, and they proposed a suitable safety evaluation method for CFRP crack repair. The bonding mechanical properties between CFRP and steel have also been widely investigated, with an overspread use in many fields [40] thanks to the good mechanical properties obtained in combination with steel plates. It was found that the surface roughness of the steel plate and repair environment have an impact on the bonding properties of CFRP [41]. The steel plate repaired by CFRP has strong durability in seawater [42], the CFRP could repair the defected steel plate and still maintain a good repair effect in the marine environment [43], and the ultimate strength of cracked structures can be restored by CFRP [43]. In conclusion, CFRP is suitable for repairing deep-water pipelines.

Nevertheless, the research on improving the structural strength using CFRP has not been fully optimized. Based on the test data [44], suitable numerical models for simulating pipelines repaired by CFRP have been proposed in the literature. In this study, the finite element method is used to determine the collapse pressure and collapse mode of internally corroded pipelines repaired using CFRP. In addition, the mechanism of CFRP repairing internally corroded pipelines under external water pressure is studied. The collapse pressure, stress, ovality, and plastic deformation of the repaired pipeline are calculated with a numerical method. Finally, the effects of thickness and length of the

CFRP, interlayer bonding, and internal corrosion metal loss rate on the collapse pressure are systematically analysed.

2. Description of the Repaired Model

A three-dimensional elastic–plastic finite element model is used to simulate the collapse and buckling propagation of pipelines under external pressure. This study is divided into two parts. In the first part, a model suitable for analysing the CFRP buckling arrestor is developed, and the feasibility and accuracy of the model are verified in relation to the literature [44]. The second part investigates the effectiveness of CFRP on the buckling behaviour of an annular internally corroded steel pipe.

2.1. Numerical Model

The aim of this paper is to study the repair effect of CFRP on an inner corroded pipeline under external pressure, and the methodology followed is a simulation experiment. ABAQUS/Standard code [45] was applied for investigation; steel pipe and CFRP were modelled using brick elements C3D8R (8-node linear brick).

For the loading, based on the requirements of the experiment, the continuous water injection in the pressure vessel was simulated using numerical simulation technology. As the internal water pressure increases, the pipeline collapses, and the water pressure in the pressure vessel drops. Loading continued until the water pressure was steady. The keyword "FLUID FLUX" was used to simulate the process of continuous and uniform water injection and pressurization in the pressure vessel.

For boundary conditions, the pressure vessel was fixed, and the sides of the steel pipe were set to be symmetrical along the X-axis (XSYMM: U1 = UR2 = UR3 = 0) and along the Y-axis (YSYMM: U2 = UR1 = UR3 = 0). The end of the steel pipe was set to be hinged, allowing axial movement along the pipe. The displacement in the remaining directions was 0. The pipe section at the defect and the section of the CFRP were set to be symmetrical along the Z-axis (ZSYMM = U3 = UR1 = UR2 = 0).

The steel is elastic–plastic material. When the element stress is less than the yield stress, the pipeline is in the elastic stage, and the stress–strain slope is Young's modulus E. When the element stress is greater than the yield strength, the steel is in the plastic stage. The element will incur plastic deformation. The pressure vessel adopts rigid body.

First, to verify the accuracy of the model, we used the same material property parameters as the test group 28D 2. As shown in Figure 1, the CFRP was located at 1/3 and 2/3 of the stainless-steel pipes, with the pipe diameter D = 25.4 mm and the wall thickness t = 0.9 mm, respectively. An initial defect in the shape of the dent was generated at the position 6D away from the end so that the collapse position of the pipe would start from the end. To apply dents, the pipe was clamped in the jaws of a universal testing machine and a rigid semi-circular rod of the same diameter was gradually pressed to produce an ovality ratio of less than 1% Δ_0. The defects were quantified by measuring the maximum and minimum diameters (D_{max} and D_{min}) of the dented pipe and measured by the ovality that Δ_0 indicates: $\Delta_0 = (D_{max} - D_{min})/(D_{max} + D_{min})$.

A dent with an ovality Δ_0 less than 1% was also introduced into the numerical model, as shown in Figure 1. The effect of CFRP on the propagation pressure is the goal of this study. The initial defect was only to ensure that the collapse position started from one end, so the residual stress effect of the dent defect in the test can be ignored. When the accuracy of the model was verified, the symmetrical model was used to save the calculation cost, as shown in Figure 2.

Then, a model was established to investigate the mechanism of CFRP in repairing corroded pipelines. Pipelines are corroded internally, and local pipe walls are thinned. As shown in Figure 2, the CFRP wrapped around the local defect. The outside of the pipe was evenly subjected to external pressure. The repair effect and mechanism of the CFRP arrestor on internally corroded pipes were studied by numerical simulation. Figure 2 illustrates the geometric features, where the pipe length is 1 m, D is the pipe diameter (D = 25.4 mm), t is the thickness of the pipe wall ($t = 0.9$ mm), L is the longitudinal length of the CFRP, t is the wall thickness at the pipe defect, w is the longitudinal length of the defect, and h is the thickness of the CFRP. Pipelines were not subjected to internal pressure.

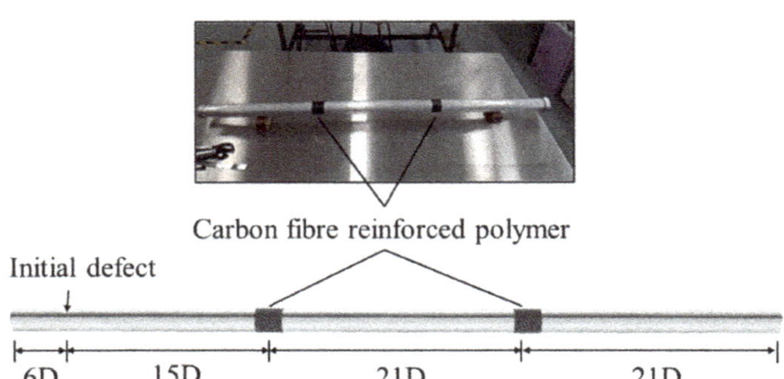

Figure 1. Comparison of experimental (**up**) [44] and numerical models (**down**).

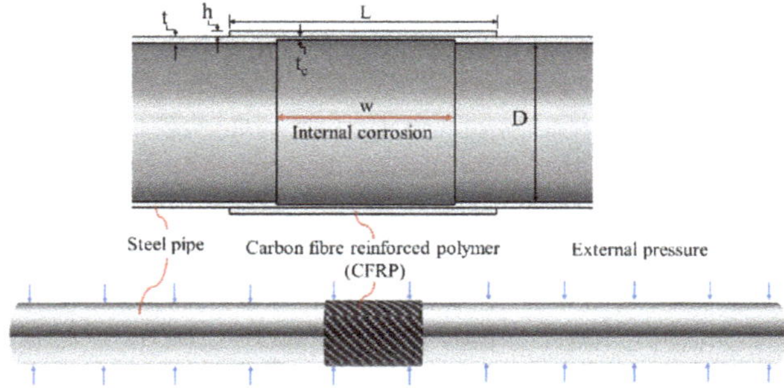

Figure 2. Schematic diagram of internal defects and repaired pipeline.

2.2. Adhesive Model

In this study, the CFRP was first subjected to external water pressure and then separated from the pipe when buckling propagation occurred, as shown in Figure 3a. As can be seen from Figure 3b, when the pipe buckling propagates across the CFRP, the pipe cross-section is U-shaped. There is a large gap between the pipe and the CFRP; thus it can be inferred that the failure mode of the interlaminar bonding is the steel–adhesive interface failure.

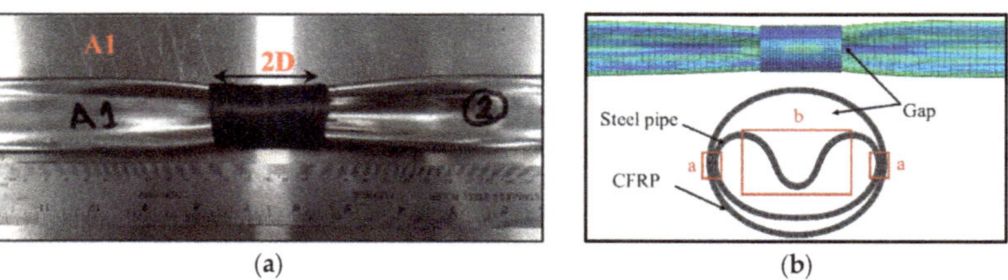

Figure 3. U-shape buckling propagation through arrestor (**a**) Experiment result, (**b**) Simulation result [44].

The interaction of pipelines with CFRP was simulated using the bond layer model in ABAQUS®. The material parameters described in the literature were adopted for the interlaminar bonding of the pipe [45]. Since the CFRP is subjected to external water pressure in this study, if the finite thickness method in the literature is used to simulate the adhesive layer, the calculation will not converge during the hydraulic loading process. Therefore, the simulation method of the no-thickness adhesive layer was adopted in this study. As shown in Figure 4, the advantage of this method is that when the CFRP layer is subjected to external water pressure, a hard contact can be set between the CFRP and the steel pipe, and the pressure exerted on the CFRP can be transmitted to the outer surface of the pipe, making it more realistic to restore the process of experiments.

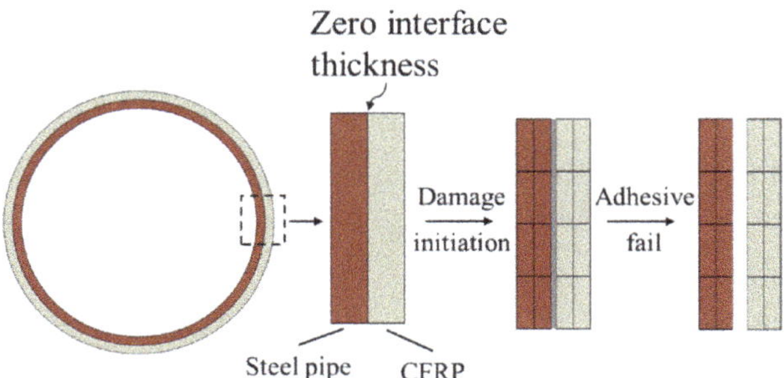

Figure 4. Schematic diagram of zero-thickness adhesive layer.

The behaviour of the adhesive layer is governed by the traction–separation law, and the following parameters were introduced into the model: interface stiffness (behaviour before damage), maximum stress before damage (damage initiation criterion), and the damage evolution law in the cohesive zone. Stress was taken as the main damage initiation criterion.

Initiation of damage is assumed to begin when a quadratic interaction function involving the separation ratios and stress ratios reaches a value of one. This criterion can be represented as [46]:

$$\left\{ \frac{\langle t_n \rangle}{t_n^0} \right\}^2 + \left\{ \frac{\langle t_s \rangle}{t_s^0} \right\}^2 + \left\{ \frac{\langle t_t \rangle}{t_t^0} \right\}^2 = 1 \tag{1}$$

$$\left\{ \frac{\langle \delta_n \rangle}{\delta_n^0} \right\}^2 + \left\{ \frac{\langle \delta_s \rangle}{\delta_s^0} \right\}^2 + \left\{ \frac{\langle \delta_t \rangle}{\delta_t^0} \right\}^2 = 1 \tag{2}$$

where t_n, t_s, and t_t represent the pure mode nominal stresses (mode-I, mode-II, and mode-III, respectively), and t_n^0, t_s^0, and t_t^0 represent the corresponding pure mode nominal strengths. δ_n^0, δ_s^0, δ_t^0 represent the peak value of the separation distance when the normal direction or the first and second shear directions are separated.

For uncoupled behaviour, each traction component depends only on its conjugate nominal strain. In the local element directions, the stress–strain relations for uncoupled behaviour are as follows [47]:

$$\left\{ \begin{array}{c} t_n \\ t_s \\ t_t \end{array} \right\} = \left(\begin{array}{ccc} K_{nn} & & \\ & K_{ss} & \\ & & K_{tt} \end{array} \right) \left\{ \begin{array}{c} \varepsilon_n \\ \varepsilon_s \\ \varepsilon_t \end{array} \right\} \tag{3}$$

The quantities ε_n, ε_s, and ε_t represent the corresponding nominal strains, while $K_{nn} = E$, $K_{ss} = G$, $K_{tt} = G$, E, and G are the longitudinal and transverse elastic moduli of the adhesive, respectively.

The critical fracture energy input in the numerical model is the value inferred from the currently known parameters. This value is the area of G_{II} in Figure 5. The relationship of K_{ss}, t_s^0, δ_s^0, and G_{II} is shown in Figure 5. Table 1 shows the values introduced in ABAQUS®.

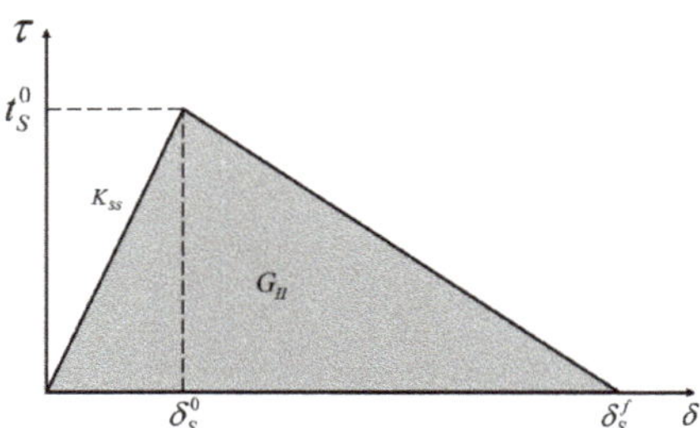

Figure 5. Traction–separation law with linear softening available in ABAQUS® [45].

Table 1. Parameters of the traction–separation behaviour for the cohesive elements to model the adhesive layer [45].

	K_{nn}	1451
Interfacial stiffness (N/mm^3)	K_{ss}	537
	K_{tt}	537
	t_n^0	19.0
Damage initiation (N/mm^2)	t_s^0	19.0
	t_t^0	19.0
Damage evolution (N/mm)	G_{II}	1.40

To simulate the mutual contact of the inner surfaces of the steel pipe after buckling, a penalized friction formula model was adopted. In addition, a surface-to-surface master–slave contact interaction was employed between the outer face of the pipe and the inner face of the CFRP.

2.3. Boundary Condition

Firstly, to verify the accuracy of the model, a full-scale pipeline model was established that was the same as the test in the literature [44], with boundary conditions that were the same as the test, and the external pressure of the pipeline was simulated using fluid-chamber technology. As shown in Figure 6, the one end of the model is only allowed to move along the axis of the pipe, and the fluid chamber is restricted by all degrees of freedom.

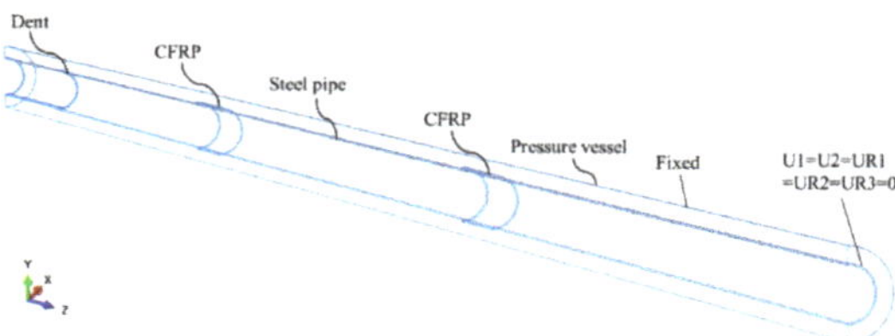

Figure 6. Schematic diagram of the boundary conditions of the pipeline model.

In the study of corrosion repair by CFRP, a one-eighth model was established by the method of symmetrical boundary to reduce the computational cost. As shown in Figure 7, the pressure vessel was fixed, and the sides of the steel pipe were set to be symmetrical along the X-axis (XSYMM: U1 = UR2 = UR3 = 0) and along the Y-axis (YSYMM: U2 = UR1 = UR3 = 0). The end of the steel pipe was set to be hinged, allowing axial movement along the pipe. The displacement in the remaining directions was 0. The pipe section at the defect and the section of the CFRP were set to be symmetrical along the Z-axis (ZSYMM = U3 = UR1 = UR2 = 0).

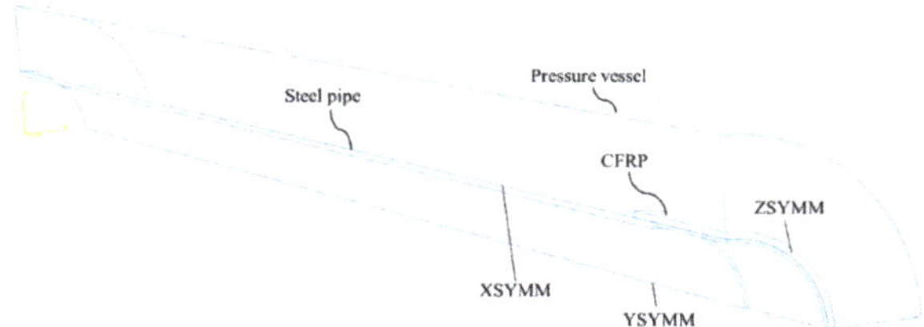

Figure 7. The 1/8 Simplified Model Boundary Condition Diagram.

2.4. CFRP and Steel Pipes' Properties

The material properties described in [44] were used for numerical simulation studies. As shown in Table 2, the material of the CFRP was manufactured by the Prepreg (PP) curing method in this paper. The steel pipe material was selected as the steel pipe property with Coupon ID 28D2 in the text. The tests were paused at the vicinity of the yield and ultimate points in order to capture the upper and lower yield and ultimate stresses, respectively. σ^L_y is the lower yield stress, σ^U_y is the upper yield stress, σ^L_u is the lower ultimate stress, and σ^U_u is the upper ultimate stress, V_f is the corresponding volume fractions.

Table 2. Steel pipe and CFRP material parameters [44].

	E (GPa)	198.3
	$\sigma_{y0.2\%}$ (MPa)	338.4
	σ^L_y (MPa)	386.9
SS-304 pipeline (28D2)	σ^U_y (MPa)	391.6
	σ^L_u (MPa)	658.4
	σ^U_u (MPa)	678.3
	Elongation at Rupture (%)	67.20%
	$E1$	132.7
	$E2$	7.4
CFRP	σ^U_1 (MPa)	1967.5
	σ^U_2 (MPa)	13.6
	V_f	0.547

3. Numerical Method for Collapse of Repaired Pipe

3.1. Loading Regime

Based the requirements of the experiment, the continuous water injection in the pressure vessel was simulated using numerical simulation technology. As the internal water pressure increases, the pipeline collapses, and the water pressure in the pressure vessel drops. Loading continued until the water pressure was steady. The keyword "FLUID FLUX" was used to simulate the process of continuous and uniform water injection and pressurization in the pressure vessel.

3.2. Methodology of the Present Work

This work is divided into two parts. Firstly, the pressure vessel model simulated the collapse behaviour of the corroded pipeline before and after repair and analysed the mechanism of CFRP repairing the pipeline. Secondly, the influence of different lengths and thicknesses of CFRP on the collapse pressure of steel pipes was determined. Additionally, for pipelines with different defect lengths, CFRP with different lengths was used, and the influence of CFRP corresponding to defects of different lengths was determined.

3.3. Pipeline Failure Criteria

The failure of the pipeline structure was judged based on the pressure change in the pressure vessel. This judgment was performed in the same way as in the literature [8]. As the water was injected into the pressure vessel, the surface pressure of the pipeline increased, and the deformation of the pipeline before the instability was still due to elastic strain. When the pressure reached a critical value, such as at position 1 in Figure 8, the pipeline was locally deformed, resulting in an increase in the ovality of the pipeline defect position. Global buckling occurred in pipes due to local deformation. The pressure in the vessel no longer increases, and the peak pressure is defined as the position of the pipeline failure, which is also the maximum bearing capacity of the pipeline.

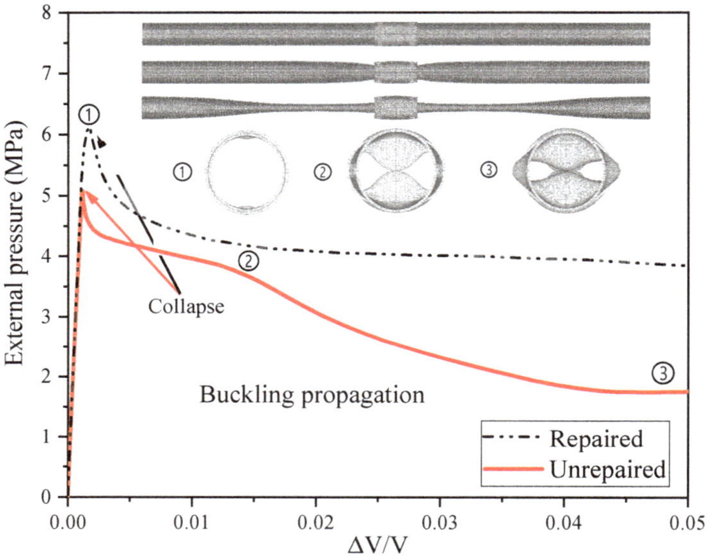

Figure 8. Diagram of external pressure and interface change in the pipeline before and after repair.

With the continuous injection of water, the pipeline at the defect is the first to contact, as shown in the cross-sectional view 2 in Figure 8. Then, the range of the pipeline instability expands along the axial direction of the pipeline, and the pressure no longer changes significantly, as shown in position 3. This process was referred to as buckling propagation in previous studies. The X-axis represents the change in pipe volume. ΔV represents the volume of water injected into the high-pressure vessel. V represents the internal volume of the pipe.

3.4. Model Verification and Mesh-Size Sensitivity

First, to verify the accuracy of the model, the same full-scale pipeline model as the one in [44] was established, and the buckling propagation and buckling crossing phenomena of the pipeline under the action of external pressure were observed. Figure 9 shows the comparison between the experimental results and the numerical simulation results. The numerical simulation reproduces the experimental situation well. The buckling of the pipeline passes through A1 and A2 successively, and the section of the pipeline presents a dumbbell shape during the propagation process and a U-shape when it passes through the CFRP.

Based on the numerical model described above, the collapse pressure was compared with Mahmoud's experimental results [44]. It can be observed from Table 3 that the predicted collapse pressure is in good agreement with the experimental results, and most of the errors are within $\pm 5\%$. In the following sections, the numerical models are applied for parametric studies. σ^L_y is the lower yield stress, and σ^U_u is the upper ultimate stress.

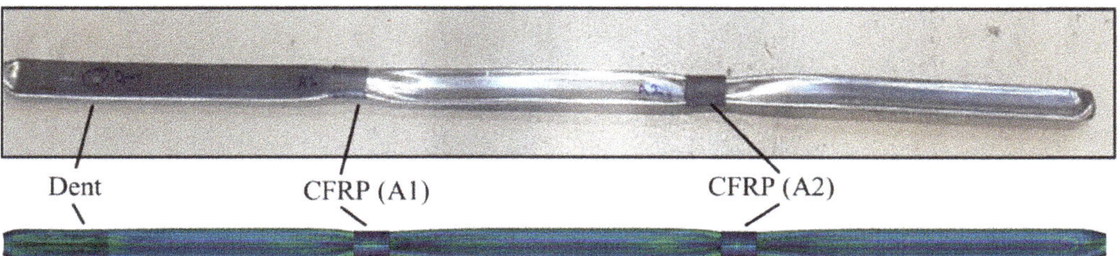

Figure 9. Comparison of experimental results [44] and numerical simulations.

Table 3. Comparison between the numerical simulation and experiments in the literature [44].

Exp. ID	E (GPa)	σ_y 0.2% (MPa)	σ^L_y (MPa)	σ^U_u (MPa)	CFRP ID	h/t	L/D	P_{Xi} (MPa)	P_{XS} (MPa)	Error (%)
28D1PPF1	196.2	341.2	387.7	672.2	1_A1	2	2	5.30	5.53	4.34
					1_A2	2	2	7.00	7.21	3.00
					1_A3	2	2	6.50	6.68	2.77
28D2PPF2	198.3	338.4	386.9	678.3	2_A1	1	2	6.20	6.33	2.10
					2_A2	1	2	7.00	7.02	0.29

h is the thickness of CFRP, t is the thickness of the pipe wall, L is the longitudinal length of the CFRP, D is the pipe diameter, P_{Xi} is the result of experiment, P_{XS} is the result of numerical simulation, *Error* is the deviation between numerical simulation and test results, and *Error* = $(P_{XS} - P_{Xi}) * 100\%/P_{Xi}$.

A meshing method with high accuracy and low computational cost was found through the different meshing of the model. Based on the geometry and loading characteristics, a 1/8 model was established to investigate the collapse pressure of the repaired pipes under external pressure. Mesh independence was determined before model verification, and the collapse pressure was used as the index of convergence. The results are summarized in Table 4, and the meanings of each parameter are shown in Figure 10: N_h is the mesh number in the hoop direction of the concrete jacket, N_t is the mesh number in the thickness direction of the CFRP and the steel pipe, S_a is the element size in the axial direction, and P_{co} is the collapse pressure.

Table 4. Mesh convergence test.

No.	N_h	N_t	S_a	Element Number of Steel Pipe	Element Number of CFRP	Element Number Total	P_{co}/MPa	Error/%
1	10	1	2	2900	130	3030	5.33	−7.79
2	10	1	4	1450	60	1510	4.98	−13.84
3	10	1	6	970	40	1010	6.11	5.71
4	20	2	2	11,600	520	12,120	5.76	−0.35
5	20	2	4	5800	240	6040	5.92	2.42
6	20	1	6	1940	80	2020	5.75	−0.52
7	30	2	2	17,400	780	18,180	5.77	−0.17
8	30	2	4	8700	360	9060	5.75	−0.52
9	30	1	6	2910	120	3030	6.03	4.33
10	40	2	2	23,200	1040	24,240	5.75	−0.52
11	40	4	2	46,400	2080	48,480	5.70	−1.38
12	40	4	1	92,800	4000	96,800	5.78	0.00

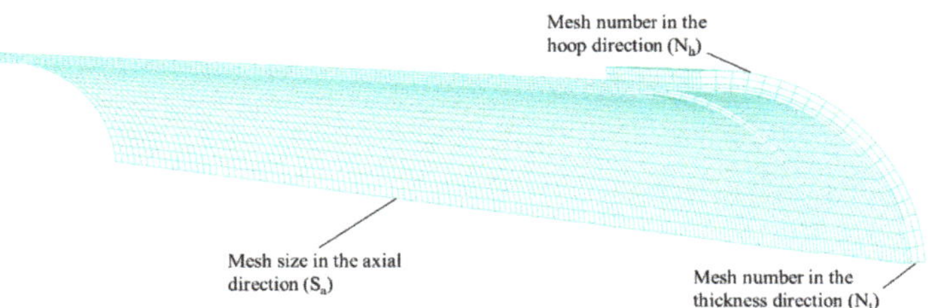

Figure 10. Schematic diagram of meshing.

It was observed that when the number of elements was increased from 9060 to 96800, the collapse pressure deviation remained at 1%. Considering the accuracy and computational cost, No. 4 was used for further analysis.

4. Results and Discussion

4.1. Stress Response under Ultimate External Pressure

CFRP significantly improves the pipe stress distribution and increases the maximum stress. In most cases, the occurrence of pipeline collapse is due to excessive stress in one or more locations. When the local stress of the pipeline exceeds the yield strength of the material, large deformation occurs in this part of the region. The deformation is generally caused by a small range of deformation, which drives the surrounding steel pipe to collapse in a larger area. The pipeline is subjected to external water pressure. Initially, due to its radial stiffness, the pipe can resist a part of the unstable force caused by the external water pressure due to the imperfect shape of the steel pipe.

However, when a large area of deformation occurs, the radial stiffness of the pipeline decreases and is not enough to support the external pressure; therefore, the pipeline becomes unstable, resulting in collapses. As shown in Figure 11, when the pipe is not repaired, the maximum stress of the steel pipe is located at the centre of the defect, reaching 390 MPa, which is similar to the yield strength of the steel pipe material.

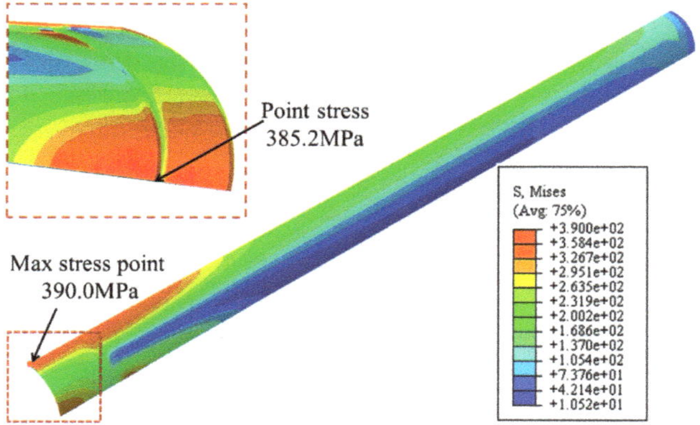

Figure 11. Stress distribution of the pipeline at the collapse pressure (unrepaired pipe).

The defects on the junction of the inner side of the steel pipe and the pipeline also have a large stress concentration, reaching 385.2 MPa. It can be seen that the collapse of the steel pipe is closely related to the yield strength of the steel pipe, and the authors of [5] have provided similar explanations. Figure 12 shows the stress distribution of the repaired pipeline at the maximum external water pressure. It can be seen from the highlighted areas in Figure 12 that the maximum stress on the pipeline reaches 408.5 MPa, which exceeds the yield strength of the steel. Similarly, there is stress concentration at the inner junction of the steel pipe, but the stress is smaller than it is for the unpaired pipeline, and the range of stress concentration is significantly larger than it is for the unpaired pipeline. The CFRP can effectively limit the local deformation of the pipeline and disperse the stress inside the pipeline.

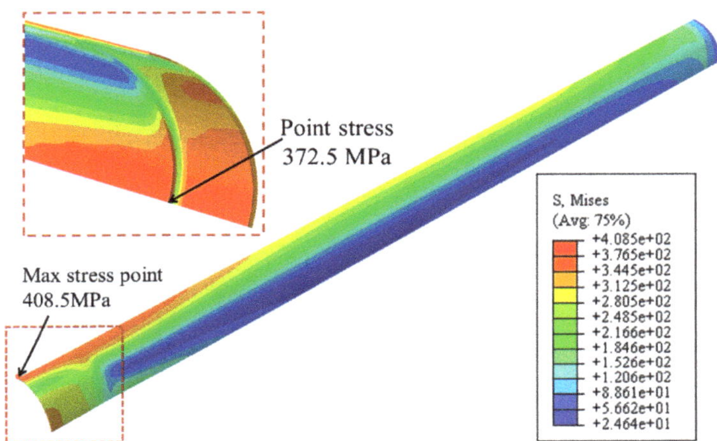

Figure 12. Stress distribution of the pipeline at the collapse pressure (repaired pipe).

To further study the repairing effect of CFRP on steel pipes, the stress distribution of carbon fibres was independently analysed. It can be seen from Figure 13 that when the external water pressure reaches 8.12 MPa, the stress above the middle of the defect and at the junction are larger than the unpaired pipelines, but both are significantly smaller than the failure strength of CFRP, while the interaction force between the steel pipe and the CFRP is uniform. As the pressure increases to 10.14 MPa, the maximum stress of the CFRP reaches 1650 MPa (less than the ultimate stress 1967.5 MPa), and the CFRP is damaged and deformed. The local failure of the CFRP is an important reason for the instability of the pipeline.

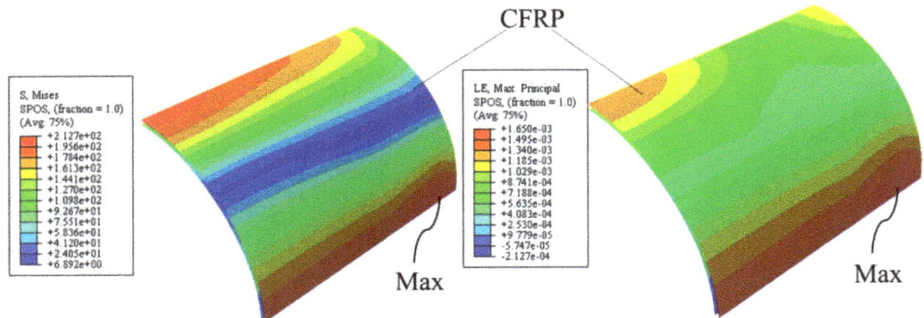

Figure 13. CFRP stress distribution diagram at limit state.

The influence of the interaction between the CFRP and steel pipe on the crushing pressure was further analysed through interlayer bonding. It is shown in Figure 14 that when the external pressure is 8.12 MPa, although the interlayer bonding is damaged, the stiffness of the bonding layer does not decrease, as shown in Figure 15.

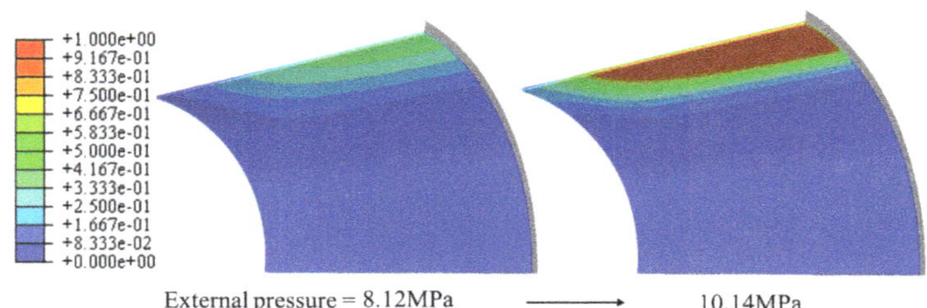

Figure 14. Quadratic displacement damage initiation criterion for cohesive surfaces.

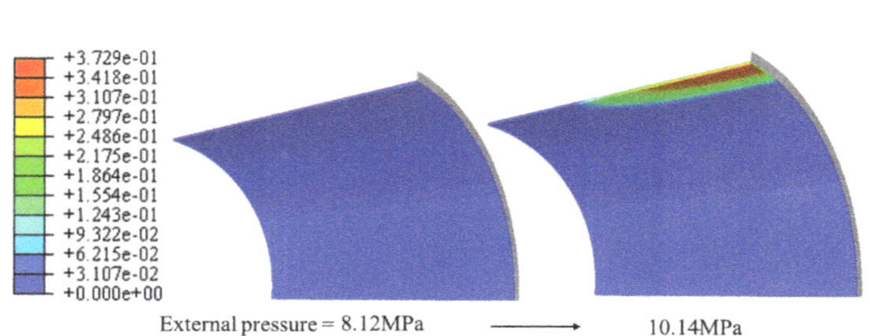

Figure 15. Scalar stiffness degradation for cohesive surfaces.

When the external water pressure reaches the collapse pressure, a large area of bonding failure occurs in the centre of the defect, and the failure rate reaches 100%, while the stiffness damage only occurs in the middle position, and the damage degree reaches 37.2%. Therefore, the decrease in the stiffness of the interlayer bonding layer is an important factor that causes the local instability of the pipeline, and the ultimate bearing capacity of the repaired pipeline can be improved by improving the ductility and the maximum-failure tensile strength of the interlayer bonding.

The effect of the interlayer adhesion on the steel pipe was analysed above, and the interlayer contact was analysed to further determine the mechanism of the CFRP on the steel pipe. When the pressure was 5.68 MPa, the deformation of the steel pipe was small, and the contact stress with the CFRP was concentrated in the red box in Figure 16, which is the junction between the internal corrosion defect and the complete pipe section.

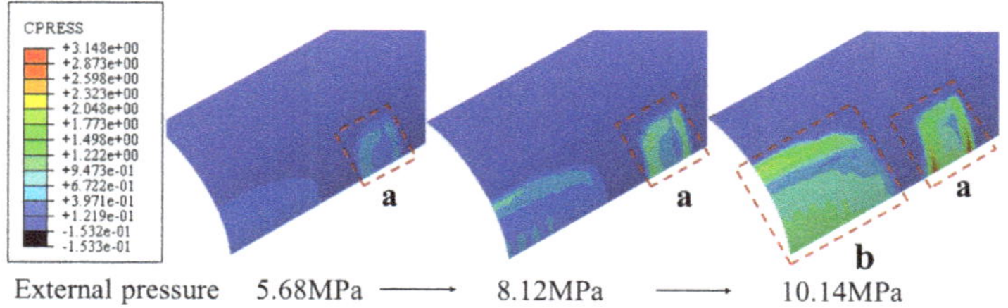

Figure 16. Interaction stress diagram at the limit state.

This stress concentration is mainly due to the deformation in the middle of the defect, and the steel pipes on both sides are in contact with the CFRP first. Since the elastic modulus of the CFRP is larger than that of the steel pipes, there is a large contact force at the junction. Figure 16 shows that when the pressure is increased to 10.14 MPa, the central part of the defect is obviously concave and deformed under collapse pressure, the defect is protruded outward as a whole, and more than half of the large area contact force appears. The CFRP limits the deformation of the pipeline very well.

The position of maximum contact force is located at the junction of the defect and the pipe section. Therefore, the effect of the CFRP on the pipeline is mainly to limit the deformation at the junction of the defect and the complete pipe segment through the contact force, increasing the local radial stiffness of the pipeline. Defective pipes with internal thinning are restrained from deformation by contact between the CFRP and the steel pipe.

4.2. Stress–Time History Response Analysis

In this section, the repairing effect of CFRP on steel pipes is investigated by analysing the stress changes in the pipes during the test. Figure 17 shows the 1/4 section of the steel pipe. In Figure 18, the relationship between the two and the effect on the collapse pressure of the pipe are explored through an analysis of the maximum stress and maximum ovality of the pipe. The ovality is defined by the following formula:

$$\Delta = \frac{R_{\max} - R_{\min}}{R_{\max} + R_{\min}} \times 100\% \tag{4}$$

where $R_{\max}$ represents the maximum radius of the pipeline, and $R_{\min}$ represents the minimum diameter of the pipeline. The initial ovality at the pipe section is 1%.

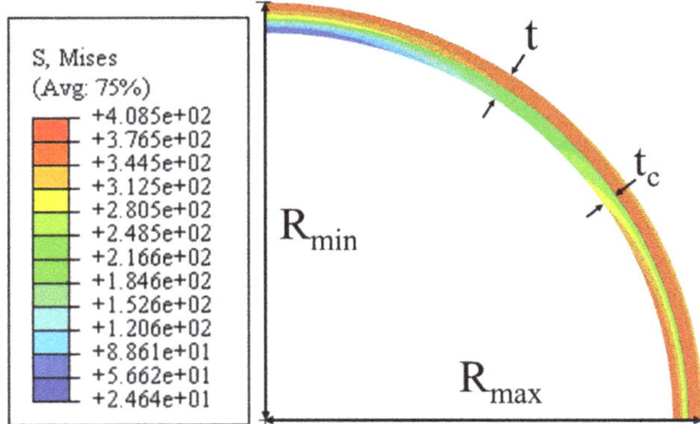

Figure 17. Schematic diagram of ovality. Where t is the thickness of the pipe without defect, and t_c is the pipe wall thickness at the local defect location.

CFRP significantly enhances the external pressure resistance of the pipeline and reduces the ovality when the pipeline is crushed. As shown in Figure 18, the maximum stress and ovality of the steel pipe before and after repair both increase with the increase in external pressure. Before the repair, when the external pressure reached 7.84 MPa, the ovality in the middle of the pipeline defect was 4.08%, and the maximum stress of the repaired pipeline reached 404.67 MPa, which was higher than the maximum stress of 385.18 MPa before the repair. Therefore, when the external pressure reached 10.15 MPa, the local stress of the repaired pipeline exceeded the yield strength of the steel pipe material. However, there is no large-scale collapse due to the limitation in CFRP. According to previous studies, the collapse pressure of the pipeline is positively correlated with the ovality. Therefore, CFRP increases the radial stiffness of the pipeline and the stability of the defect parts, limits the deformation of the steel pipe, and improves the overall stability of the pipeline.

In the above analysis, it was found that the maximum stress of the repaired steel pipe does not increase uniformly with the increase in the external pressure. To explore this phenomenon and make corrections to the future CFRP manufacturing process, this section analyses the change in the maximum stress point of the steel pipe.

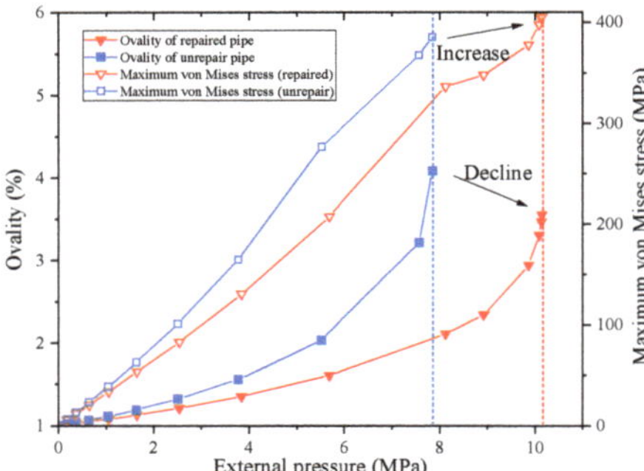

Figure 18. Variation of maximum stress and ovality with external pressure.

As shown in Figure 19, the external pressure–stress curves for points one, two, and three are extracted, respectively. It can be found from Figure 20 that the initial stress concentration of the steel tube occurs at point one. This shows that when the external water pressure is small, the middle position of the pipeline defect is most prone to deformation. Currently, the interaction force between the CFRP and the steel pipe is still very small. When the external pressure is greater than 2.54 MPa, the maximum stress of the steel pipe appears at point two. This shows that the CFRP has begun to limit the deformation of the defective pipe section. At this time, the overall strength of the defect is increased, the defect pipeline will not have local large deformation, and the stress concentration position will appear at the junction of the defect and the complete pipe section. When the external water pressure is 8.9–9.9 MPa, the maximum stress of the steel pipe changes from point two to point three. It can be seen from Figure 20 that point three maintains a linear increase before crushing, while the stress at point two increases at a slower rate of 8.12 MPa. This is because the bond between the CFRP and the steel pipe is damaged, the bond force at the centre of the pipe defect decreases, and the force on the compressed parts on both sides continues to increase. When the external water pressure is greater than 9.9 MPa, the maximum stress of the pipeline is concentrated in the depression in the centre of the defect position, and the stability of the steel pipe is greatly reduced by this external pressure. The central area undergoes large deformation due to the decrease in the bond strength. CFRP can effectively disperse the stress concentration when the water pressure is small, and the stress at the concave position of the pipeline is no longer a dangerous point in the pipeline. CFRP can significantly improve the stress concentration of steel pipes and delay the occurrence of pipeline collapse.

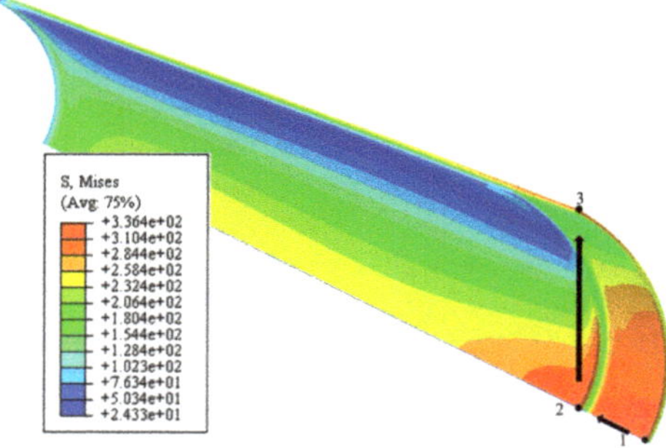

Figure 19. Schematic diagram of the maximum stress transfer path. (The direction of the arrow indicates the direction of movement of the point of maximum stress).

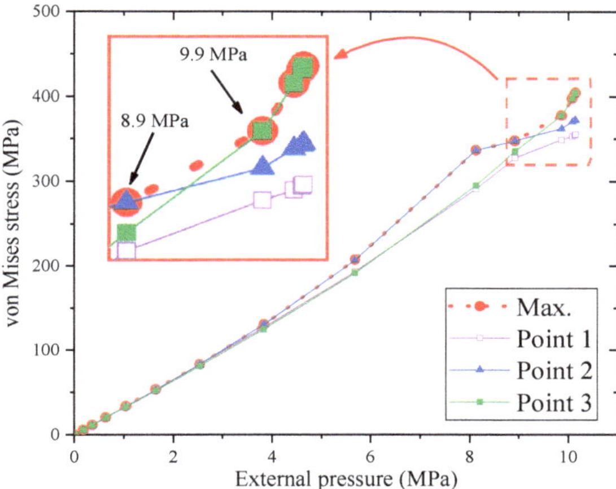

Figure 20. Stress curves of three points in the pipeline.

5. Influential Factors

5.1. Longitudinal Length of Defect and CFRP

The length of the defect influences the repair effect of the pipeline. As shown in Figure 21, the repaired pipe does not fully restore the strength of the pipe. The curve of $L/w = 0$ indicates that the collapse pressure of unrepaired pipe decreases with the increase in defect length. When the ratio of the pipeline defect length to diameter $w/D > 4$, the change in pipeline collapse pressure decreases. The same phenomenon is observed when CFRP fully covers the pipeline defect location. However, if the CFRP does not fully cover the defect position, the collapse pressure of the pipeline will still decrease with the increase in the defect length.

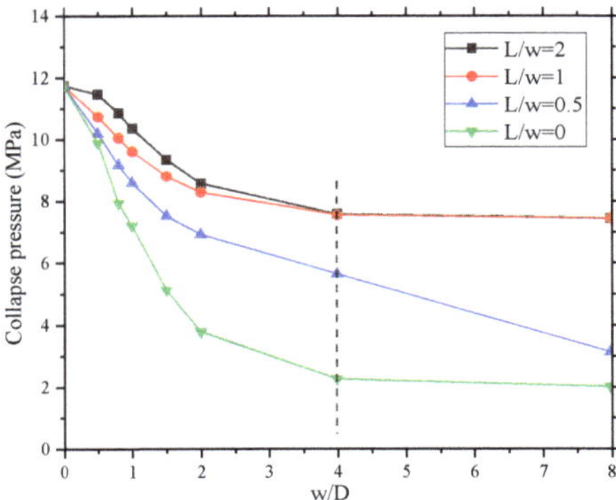

Figure 21. Repairing effect of CFRP on defects of different lengths.

The repair effect of CFRP is not linearly enhanced. As shown in Figure 22, when the defect length/diameter $w/D < 2$, the increase in the CFRP length can significantly increase the collapse pressure of the pipeline. However, when $L/w > 2$, the effect of CFRP is small, and the effect of increasing the length of the CFRP is not significant. It can be seen from the data analysis that the collapse pressure of the repaired pipeline is 3.1–274% higher than that of the pipeline before repair. When L/w is less than two, the collapse pressure of the repaired pipe decreases with the increase in defect width. The repair effect increases gradually from 3.1% and reaches the highest when $w/D = 4$. When w/D is greater than four, the collapse pressure can be increased to 65% of the pipe without defect, which is

274% higher than the unpaired pipe. When the CFRP length is greater than the defect length, the repair effect is similar. When the defect length is less than the diameter, the collapse pressure of the repaired pipeline can reach 72.98–97.53% of the initial external pressure capacity. However, as shown in Figure 22, when the defect length of the pipeline is greater than the CFRP, and the defect length is less than four times the diameter, the repair effect increases with the increase in the defect length; when the defect length is greater than four times the diameter, the repair effect increases. The effect decreases when the defect length increases.

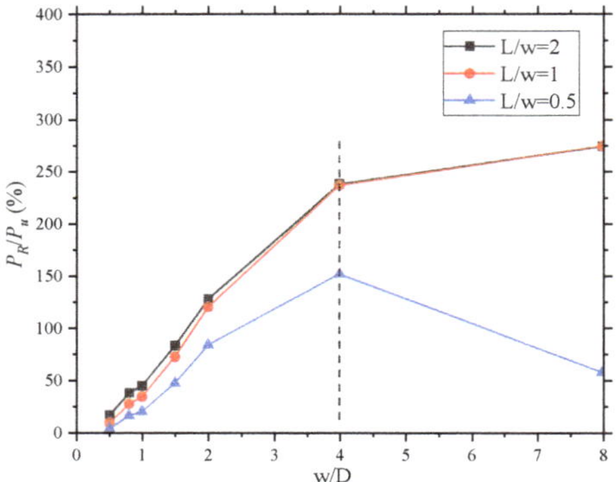

Figure 22. Comparison of the collapse pressure of the pipeline before (P_u) and after repair (P_R).

5.2. Metal Loss Rate

In this section, the influence of different degrees of defects on the repair effect is analysed. t_c/t refers to the defect wall thickness/wall thickness. The model used is $L/w = 1$, $w/D = 4$. The thickness of the CFRP is 0.5–3 times the wall thickness of the steel pipe, and the wall thickness at the defect is 0.3–0.8 times the wall thickness before corrosion. As shown in Figure 23, when the CFRP thickness is greater than the pipe wall thickness, the repair effect for defects with $t_c/t < 0.3$ differs by 3.09%. When the CFRP thickness is twice the pipe wall thickness, the repair effect is almost the same. For pipes with large wall thickness defects, increasing the thickness of the CFRP has no obvious effect. For steel pipes with small defects, if the thickness t_c/t of the defected part is greater than 0.3, increasing the thickness of the CFRP can effectively improve the repair efficiency. It can be seen from the bottom two curves in Figure 23 that when $h/t = 0.5$, the ultimate bearing capacity of the pipeline is increased by less than 1%.

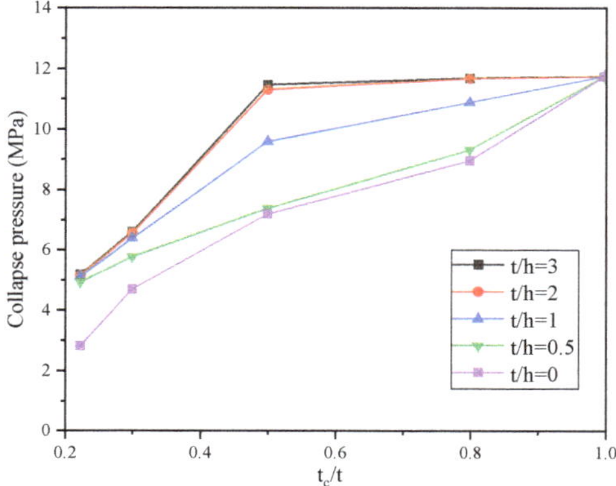

Figure 23. Comparison of the effect of metal loss rate.

When the t_c/t defect wall thickness/wall thickness is less than 0.3, increasing the CFRP thickness cannot significantly improve the repair effect; when the ratio of the t_c/t defect wall thickness and the wall thickness is greater than 0.3, increasing the CFRP wall thickness can improve the ultimate bearing capacity of the repaired pipeline. When the CFRP wall thickness is two times the steel pipe wall thickness, the repair efficiency is the highest, t_c/t is greater than 0.5, and the ultimate bearing capacity of the pipeline after repair is the same as before the repair.

5.3. Thickness of CFRP

In this section, the effects of different sizes of CFRP are analysed. The adopted model is $L/w = 0.5-2$, $h/t = 0.5-3$. The thickness of the CFRP is 0.5–3 times the wall thickness of the steel pipe, and the length is 0.5–2 times the length of the defect. As shown in Figure 24, the collapse pressure of the repaired pipeline gradually increases as the thickness of the CFRP increases. Therefore, the greater the thickness, the better the effect of the repair. The repair efficiency has a nonlinear relationship with the length of the CFRP, as shown in Figure 25: when the length of the CFRP is the same as the pipeline defect, the repair efficiency is the highest. When the CFRP length is greater than the defect length, the repair effect decreases with the increase in thickness. When the CFRP thickness is the same as the pipe thickness, increasing the CFRP length can significantly improve the repair efficiency. Therefore, when the thickness of the CFRP is larger than the thickness of the steel pipe wall, the repair effect is not significantly improved.

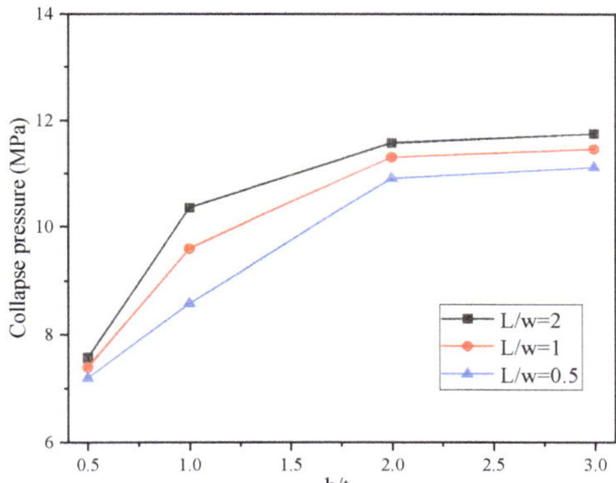

Figure 24. Influence of CFRP thickness on the repair effect.

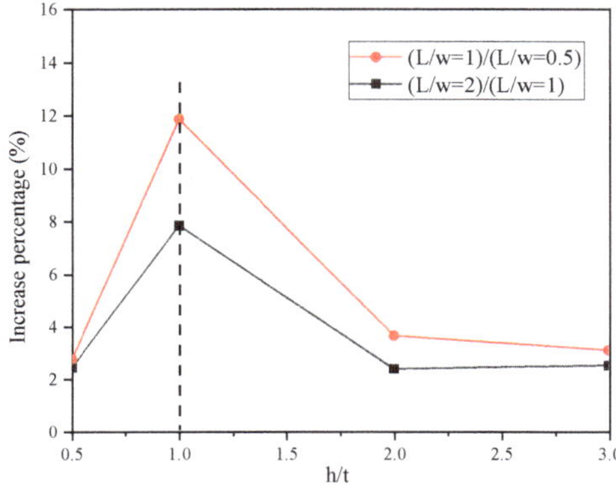

Figure 25. Comparison of repairing effects of different lengths of CFRP.

6. Conclusions

This study investigated the buckling failure of internally corroded steel pipes repaired using the CFRP. The repaired internally corroded pipeline was analysed using the three-dimensional elastic–plastic finite element method. The collapse phenomenon of the repaired steel pipe under external hydrostatic pressure was simulated using the cohesive element and energy criterion. The results show that the CFRP thickness and length can change the collapse pattern and the pipeline's stress distribution. A decrease in the ovality with corrosion defects was observed, and it was found that the repair of CFRP mainly depends on the length, thickness, and interlayer cohesion. From the above results, the following conclusions can be drawn:

1. CFRP limits the deformation at the junction of defects and complete pipe segments, thereby increasing the local radial stiffness of the pipe and the stability of the defect parts. The interaction of CFRP with the steel pipe causes the deformation at the defects of the steel pipe to be limited by the uncorroded pipe. CFRP can effectively limit the local deformation of the pipeline and disperse the stress inside it, which improves the pipeline's overall stability.
2. CFRP disperses the stress concentration well when the water pressure is small and the stress at the concave position of the pipeline is no longer a dangerous point of the pipeline. CFRP can significantly improve the stress concentration of steel pipes and delay the occurrence of pipeline collapse. Local CFRP failure is an important cause of pipeline instability.
3. The decrease in stiffness of the interlayer bonding layer is an important factor that causes the local instability of the pipeline, and the ultimate bearing capacity of the repaired pipeline can be improved by improving the ductility and the maximum-failure tensile force of the interlayer bonding.
4. When the defect length is less than the diameter, the collapse pressure of the repaired pipeline can reach 72.98–97.53%, and the repair effect increases with the increase in CFRP length. When the defect length is less than the CFRP length and the defect length is less than four times the diameter, the crushing pressure of the repaired pipeline is increased by 3.1–274%, and the repair effect increases with the increase in the defect length. When the defect length is greater than four times the diameter, the repair effect decreases with the increase in the defect length.
5. When the defect wall thickness/steel wall thickness is less than 0.3, increasing the CFRP thickness cannot significantly improve the repair effect; when the ratio of the t_c/t defect wall thickness and the wall thickness is greater than 0.3, increasing the CFRP wall thickness can improve the ultimate bearing capacity of the repaired pipeline. When the CFRP wall thickness is two times the steel pipe wall's thickness, the repair efficiency is the highest, and the bearing limit of the pipeline after repair can reach 100%. When it is more than two times the steel pipe wall's, it has no obvious effect on the bearing limit after the repair. If the ratio of the wall thickness of the defect and the wall thickness of the steel pipe is greater than 0.5, the ultimate bearing capacity of the pipeline after repair is the same as before the repair.

Author Contributions: Conceptualization, J.Y., W.X. and Y.Y.; methodology, J.Y., W.X. and Y.Y.; software, W.X., Y.Y., F.F. and H.W.; validation, W.X., Y.Y., F.F. and H.W.; formal analysis, S.X. and W.W.; investigation, S.X. and S.W.; resources, F.F.; data curation, F.F.; writing—original draft preparation, W.X. and Y.Y.; writing—review and editing, F.F., H.W., S.X. and S.W.; visualization, J.Y.; supervision, J.Y.; project administration, J.Y.; funding acquisition, J.Y., Y.Y. and F.F. All authors have read and agreed to the published version of the manuscript.

Funding: This research is supported by Guangxi Zhuang Autonomous Region Bagui Scholars Program and the National Natural Science Foundation of China (Grant No. 52071234, Grant No. 51879189).

Institutional Review Board Statement: Not applicable.

Informed Consent Statement: Not applicable.

Conflicts of Interest: The authors declare that they have no known competing financial interests or personal relationships that could have appeared to influence the work reported in this paper.

References

1. Yun, H.D.; Peek, R.R.; Paslay, P.R.; Kopp, F.F. Loading History Effects for Deep-Water S-Lay of Pipelines. *J. Offshore Mech. Arct. Eng.* **2004**, *126*, 156–163. [CrossRef]
2. Xie, P.; Zhao, Y.; Yue, Q.; Palmer, A.C. Dynamic loading history and collapse analysis of the pipe during deepwater S-lay operation. *Mar. Struct.* **2015**, *40*, 183–192. [CrossRef]
3. Yuan, L.; Kyriakides, S. Liner wrinkling and collapse of bi-material pipe under axial compression. *Int. J. Solids Struct.* **2015**, *60–61*, 48–59. [CrossRef]
4. Kyriakides, S. Effects of Reeling on Pipe Structural Performance—Part I: Experiments. *J. Offshore Mech. Arct. Eng.* **2017**, *139*, 051706. [CrossRef]
5. Dyau, J.Y.; Kyriakides, S. On the localization of collapse in cylindrical shells under external pressure. *Int. J. Solids Struct.* **1993**, *30*, 463–482. [CrossRef]
6. Sakakibara, N.; Kyriakides, S.; Corona, E. Collapse of partially corroded or worn pipe under external pressure. *Int. J. Mech. Sci.* **2008**, *50*, 1586–1597. [CrossRef]
7. Netto, T.A. A simple procedure for the prediction of the collapse pressure of pipelines with narrow and long corrosion defects—Correlation with new experimental data. *Appl. Ocean Res.* **2010**, *32*, 132–134. [CrossRef]
8. Ramasamy, R.; Ya, T.T. Nonlinear finite element analysis of collapse and post-collapse behaviour in dented submarine pipelines. *Appl. Ocean Res.* **2014**, *46*, 116–123. [CrossRef]
9. Wang, H.; Yu, Y.; Yu, J.; Jin, C.; Zhao, Y.; Fan, Z.; Zhang, Y. Effect of 3D random pitting defects on the collapse pressure of pipe—Part II: Numerical analysis. *Thin-Walled Struct.* **2018**, *129*, 527–541. [CrossRef]
10. Wang, H.; Yu, Y.; Yu, J.; Duan, J.; Zhang, Y.; Li, Z.; Wang, C. Effect of 3D random pitting defects on the collapse pressure of pipe—Part I: Experiment. *Thin-Walled Struct.* **2018**, *129*, 512–526. [CrossRef]
11. Wang, H.; Yu, Y.; Xu, W.; Li, Z.; Yu, S. Time-variant burst strength of pipe with corrosion defects considering mechano-electrochemical interaction. *Thin-Walled Struct.* **2021**, *169*, 108479. [CrossRef]
12. Junior, N.M.; Carrasquila, A.; Figueiredo, A.A.; da Fonseca, C. Worn pipes collapse strength: Experimental and numerical study. *J. Pet. Sci. Eng.* **2015**, *133*, 328–334. [CrossRef]
13. Zhang, X.; Pan, G. Collapse of thick-walled subsea pipelines with imperfections subjected to external pressure. *Ocean Eng.* **2020**, *213*, 107705. [CrossRef]
14. Chen, B.-Q.; Zhang, X.; Soares, C.G. The effect of general and localized corrosions on the collapse pressure of subsea pipelines. *Ocean Eng.* **2022**, *247*, 110719. [CrossRef]
15. Chen, Y.; Dong, S.; Zang, Z.; Ao, C.; Liu, H.; Gao, M.; Ma, S.; Zhang, E.; Cao, J. Buckling analysis of subsea pipeline with idealized corrosion defects using homotopy analysis method. *Ocean Eng.* **2021**, *234*, 108865. [CrossRef]
16. Yu, J.; Xu, W.; Yu, Y.; Wang, H.; Li, H.; Xu, S.; Han, M. Effectiveness of concrete grouting method for deep-sea pipeline repairs. *Thin-Walled Struct.* **2021**, *169*, 108336. [CrossRef]
17. Johns, T.G.; Mesloh, R.E.; Sorenson, J.E.; Laboratories, B.C. Propagating Buckle Arrestors for Offshore Pipelines. *J. Press. Vessel Technol.* **1978**, *100*, 206–214. [CrossRef]
18. Hahn, G.D.; She, M.; Carney, J.F. Buckle propagation and arresting in offshore pipelines. *Thin-Walled Struct.* **1994**, *18*, 247–260. [CrossRef]
19. Lee, L.-H.; Kyriakides, S. On the arresting efficiency of slip-on buckle arrestors for offshore pipelines. *Int. J. Mech. Sci.* **2004**, *46*, 1035–1055. [CrossRef]
20. Netto, T.A.; Kyriakides, S. Dynamic performance of integral buckle arrestors for offshore pipelines. Part I: Experiments. *Int. J. Mech. Sci.* **2000**, *42*, 1405–1423. [CrossRef]
21. Netto, T.A.; Kyriakides, S. Dynamic performance of integral buckle arrestors for offshore pipelines. Part II: Analysis. *Int. J. Mech. Sci.* **2000**, *42*, 1425–1452. [CrossRef]
22. Lee, L.-H.; Kyriakides, S.; Netto, T.A. Integral buckle arrestors for offshore pipelines: Enhanced design criteria. *Int. J. Mech. Sci.* **2008**, *50*, 1058–1064. [CrossRef]
23. Bardi, F.; Kyriakides, S. Plastic buckling of circular tubes under axial compression—Part I: Experiments. *Int. J. Mech. Sci.* **2006**, *48*, 830–841. [CrossRef]
24. Bruce, W.A. *Comparison of Fiber-Reinforced Polymer Wrapping Versus Steel Sleeves for Repair of Pipelines*; Elsevier: Amsterdam, The Netherlands, 2015; ISBN 9780857096920.
25. Farag, M.H.; Mahdi, E. New approach of pipelines joining using fiber reinforced plastics composites. *Compos. Struct.* **2019**, *228*, 111341. [CrossRef]
26. Lim, K.S.; Azraai, S.N.A.; Yahaya, N.; Noor, N.M.; Zardasti, L.; Kim, J.-H.J. Behaviour of steel pipelines with composite repairs analysed using experimental and numerical approaches. *Thin-Walled Struct.* **2019**, *139*, 321–333. [CrossRef]
27. Yu, J.; Xu, W.; Chen, N.-Z.; Jiang, S.; Xu, S.; Li, H.; Han, M. Effect of dent defects on the collapse pressure of sandwich pipes. *Thin-Walled Struct.* **2022**, *170*, 108608. [CrossRef]

28. Sun, C.; Mao, D.; Zhao, T.; Shang, X.; Wang, Y.; Duan, M. Investigate Deepwater Pipeline Oil Spill Emergency Repair Methods. *Aquat. Procedia* **2015**, *3*, 191–196. [CrossRef]
29. Keller, M.W.; Jellison, B.D.; Ellison, T. Moisture effects on the thermal and creep performance of carbon fiber/epoxy composites for structural pipeline repair. *Compos. Part B Eng.* **2013**, *45*, 1173–1180. [CrossRef]
30. Wonderly, C.; Grenestedt, J.; Fernlund, G.; Cĕpus, E. Comparison of mechanical properties of glass fiber/vinyl ester and carbon fiber/vinyl ester composites. *Compos. Part B Eng.* **2005**, *36*, 417–426. [CrossRef]
31. Lukács, J.; Nagy, G.; Török, I.; Égert, J.; Pere, B. Experimental and numerical investigations of external reinforced damaged pipelines. *Procedia Eng.* **2010**, *2*, 1191–1200. [CrossRef]
32. Xu, X.; Shao, Y.; Gao, X.; Mohamed, H.S. Stress concentration factor (SCF) of CHS gap TT-joints reinforced with CFRP. *Ocean Eng.* **2022**, *247*, 110722. [CrossRef]
33. Kong, D.; Huang, X.; Xin, M.; Xian, G. Effects of defect dimensions and putty properties on the burst performances of steel pipe wrapped with CFRP composites. *Int. J. Press. Vessel. Pip.* **2020**, *186*, 104139. [CrossRef]
34. Elchalakani, M. Rehabilitation of corroded steel CHS under combined bending and bearing using CFRP. *J. Constr. Steel Res.* **2016**, *125*, 26–42. [CrossRef]
35. Elchalakani, M.; Karrech, A.; Basarir, H.; Hassanein, M.F.; Fawzia, S. CFRP strengthening and rehabilitation of corroded steel pipelines under direct indentation. *Thin-Walled Struct.* **2017**, *119*, 510–521. [CrossRef]
36. Mohamed, H.S.; Shao, Y.; Chen, C.; Shi, M. Static strength of CFRP-strengthened tubular TT-joints containing initial local corrosion defect. *Ocean Eng.* **2021**, *236*, 109484. [CrossRef]
37. Mokhtari, M.; Nia, A.A. The influence of using CFRP wraps on performance of buried steel pipelines under permanent ground deformations. *Soil Dyn. Earthq. Eng.* **2015**, *73*, 29–41. [CrossRef]
38. Shamsuddoha; Manalo, A.; Aravinthan, T.; Islam, M.; Djukic, L. Failure analysis and design of grouted fibre-composite repair system for corroded steel pipes. *Eng. Fail. Anal.* **2021**, *119*, 104979. [CrossRef]
39. Meriem-Benziane, M.; Abdul-Wahab, S.A.; Zahloul, H.; Babaziane, B.; Hadj-Meliani, M.; Pluvinage, G. Finite element analysis of the integrity of an API X65 pipeline with a longitudinal crack repaired with single- and double-bonded composites. *Compos. Part B Eng.* **2015**, *77*, 431–439. [CrossRef]
40. Kupski, J.; Teixeira de Freitas, S. Design of adhesively bonded lap joints with laminated CFRP adherends: Review, challenges and new opportunities for aerospace structures. *Compos. Struct.* **2021**, *268*, 113923. [CrossRef]
41. Borrie, D.; Liu, H.B.; Zhao, X.L.; Raman, R.K.S.; Bai, Y. Bond durability of fatigued CFRP-steel double-lap joints pre-exposed to marine environment. *Compos. Struct.* **2015**, *131*, 799–809. [CrossRef]
42. Ramirez, F.A.; Carlsson, L.A.; Acha, B.A. Evaluation of water degradation of vinylester and epoxy matrix composites by single fiber and composite tests. *J. Mater. Sci.* **2008**, *43*, 5230–5242. [CrossRef]
43. Hu, L.; Li, M.; Yiliyaer, T.; Gao, W.; Wang, H. Strengthening of cracked DH36 steel plates by CFRP sheets under fatigue loading at low temperatures. *Ocean Eng.* **2022**, *243*, 110203. [CrossRef]
44. Alrsai, M.; Karampour, H.; Hall, W.; Lindon, A.K.; Albermani, F. Carbon fibre buckle arrestors for offshore pipelines. *Appl. Ocean Res.* **2021**, *111*, 102633. [CrossRef]
45. Jimenez-Vicaria, J.D.; Pulido, M.D.G.; Castro-Fresno, D. Influence of carbon fibre stiffness and adhesive ductility on CFRP-steel adhesive joints with short bond lengths. *Constr. Build. Mater.* **2020**, *260*, 119758. [CrossRef]
46. Campilho, R.D.S.G.; de Moura, M.F.S.F.; Pinto, A.M.G.; Morais, J.J.L.; Domingues, J.J.M.S. Modelling the tensile fracture behaviour of CFRP scarf repairs. *Compos. Part B Eng.* **2009**, *40*, 149–157. [CrossRef]
47. Zhang, Y.; Liu, Z.; Xin, J.; Wang, Y.; Zhang, C.; Zhang, Y. The attenuation mechanism of CFRP repaired corroded marine pipelines based on experiments and FEM. *Thin-Walled Struct.* **2021**, *169*, 108469. [CrossRef]

Journal of
Marine Science and Engineering

Article

Probabilistic Collapse Design and Safety Assessment of Sandwich Pipelines

Utkarsh Bhardwaj, Angelo Palos Teixeira * and C. Guedes Soares

Centre for Marine Technology and Ocean Engineering (CENTEC), Instituto Superior Técnico,
Universidade de Lisboa, 1049-001 Lisbon, Portugal
* Correspondence: teixeira@centec.tecnico.ulisboa.pt

Abstract: This paper presents an approach for probabilistic design and safety assessment of sandwich pipelines under external pressure. The methodology consists of the categorisation of sandwich pipeline collapse strength models based on interlayer adhesion conditions. The models are validated by comparing their predictions against collapse test data of sandwich pipelines. The accuracy of the strength models and their prediction uncertainty are used to select the best model in each category. Regarding interlayer adhesion categories, uncertainty propagation of models' predictions over a wide range is assessed by the Monte Carlo simulation method. The proposed methodology is demonstrated using a case study of a sandwich pipeline with adequate probabilistic modelling of the basic random variables. Different limit states are defined for three categories of sandwich pipelines, based on which structural reliability indices are estimated. In employing the First Order Reliability Method for sensitivity analysis, the importance of basic variables of the limit states is evaluated. Later, a parametric analysis is conducted, presenting reliability variations for several design and operational scenarios of sandwich pipelines. Finally, to achieve a uniform level of structural reliability of sandwich pipelines, a few suggestions are provided, and practical partial safety factors are calculated. The results of the present analysis can provide guidance on the probabilistic design and operational safety assessment of sandwich pipelines.

Keywords: sandwich pipes; reliability assessment; uncertainty propagation; FORM; collapse strength; partial safety factors

Citation: Bhardwaj, U.; Teixeira, A.P.; Guedes Soares, C. Probabilistic Collapse Design and Safety Assessment of Sandwich Pipelines. *J. Mar. Sci. Eng.* **2022**, *10*, 1435. https://doi.org/10.3390/jmse10101435

Academic Editor: Fuping Gao

Received: 26 August 2022
Accepted: 29 September 2022
Published: 5 October 2022

Publisher's Note: MDPI stays neutral with regard to jurisdictional claims in published maps and institutional affiliations.

1. Introduction

As the oil and gas industries move to ultra-deep waters and arctic conditions for hydrocarbon exploitation, they face challenges in designing well-insulated pipelines and risers proficient at enduring extreme operating conditions. In such cases, the pipelines are subjected to extreme hydrostatic pressure and are designed mainly to resist to collapse. The Pipe-in-Pipe (PiP) system has been seen as the most feasible solution to such problems [1]. A PiP system consists of two concentric pipes, where the annulus either is filled with non-structural insulating material, or it may carry water or an umbilical cable, among others. Some significant design guidance for PiPs can be tracked from DNV–ST–F101 [2] as the failure modes described are also applicable to them. The focus of the PiP system is to increase the insulation capacity to prevent any blockage of the pipe due to temperature falling below the temperature of paraffin or hydrate forming. Sandwich pipelines (SP) are a special case of PiP in which the cavity between pipes is filled with a structural core. In modern SPs, the thermal and structural requirements are dealt with from an overall perspective implying that the core material provides thermal insulation as well as structural resistance against the burst of the inner pipe and collapse of the outer pipe [3]. Thereby, greater strength with adequate flow assurance can be obtained in the case of SPs.

An SP may also be constituted by multiple layers of material bonded together over the main pipe, contributing to their single property as a global structural response. However, the present study is limited to two layered SPs (outer and inner metallic pipe). Figure 1 presents a sample of the SP specimen and geometry considered in this study. As far as the operating conditions of SPs are concerned, local buckling or simply collapse under external pressure is presumably the most dominating failure mode among all possible structural failures. Therefore, the collapse behaviour of SPs (two-layered) has been widely investigated through analytical, numerical, and experimental works [4–7]. In particular, the factors and circumstances governing collapse and buckle propagation of PiPs [8] and SPs [9,10] have been extensively studied.

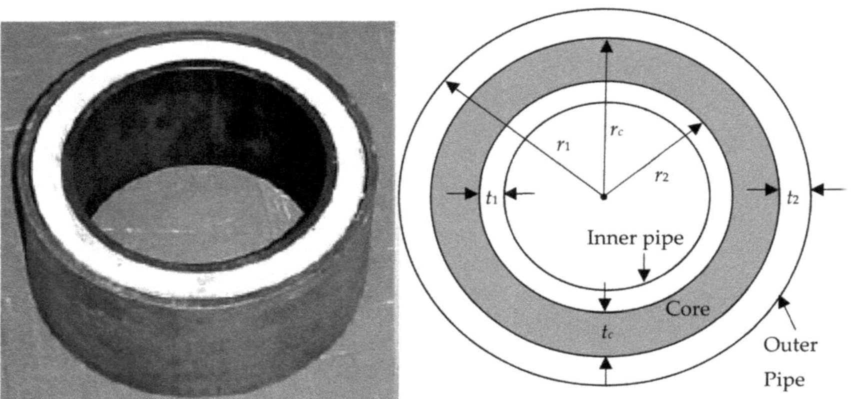

Figure 1. Illustration of SP specimen and geometry [11].

The structural safety of SPs with an objective to support maintenance decisions can be assessed through a probabilistic analysis with duly defined uncertainty associated with the design variables. For conventional single-walled pipelines, probabilistic analyses using different burst failure models [12–14] are commonly conducted, whereas reliability assessment under external pressure is less frequent [15]. As per the literature review [4–7], collapse failure is the main concern of SP, and hence, an accurate collapse strength prediction is of utmost importance to the conceptual design process and for subsequent maintenance planning during the operational life of the SP system. Nonetheless, challenges exist regarding the prediction of the collapse pressure of SPs as it depends on the material and geometric properties of each layer (inner and outer pipe) and the extent of inter-layer adhesion [16]. A considerable amount of research work has been carried out analytically and numerically to deduce the collapse strength models for SPs. These studies have comprehended a broad variety of SP design configurations with different core materials and bonding conditions of the layers. However, the differences lie in theories, assumptions and idealisations of the failure behaviour adopted in developing these models. In the absence of dedicated codes for SPs, a few relevant models available in the literature can be adopted for design and reliability assessments.

Among the pioneer models, Sato and Patel [17] have developed a model for elastic buckling of SPs with a perfect bonded inter-layer relation. Their model can also be used for unbonded relation of SP layers. Early development in the last decade has indicated the greater significance of interlayer adhesion conditions on the collapse strength. In particular, Arjomandi and Taheri [16] have advocated four conditions of inter-layer relation: (i) the core layer is fully bonded to the inner and outer layer; (ii) the outer layer is bonded to the core, while the inner layer is free to slide in the tangential direction; (iii) the inner layer is bonded to the core, while the outer layer is free to slide; and (iv) the core is unbounded to both layers. They utilised the shell theory and developed kinematic relations for inner and outer pipes. Furthermore, this kinematic relation was applied linearly to the core layer. These simplifications of the collapse phenomenon, however, neglect radial and shear strain in the layers and nonlinear strain in the core layer. Hashemian and Mohareb [18,19] have used this nonlinearity in the formation of a complex kinematic relation to all the layers and developed an eigenvalue solution for SP. The models developed by them are rather complicated and arduous to use for quick prediction of collapse strength.

Due to the inherent complexity of the collapse phenomenon of SPs, there is always a compromise between the accuracy and simplicity of the models. Some of the complexities in the above-mentioned models arise due to the assumption of perfect geometry, elasticity, shell theory, and inter-layer adhesion, among others. Obviously, because of certain idealisation of the problem at hand in model formulations, their predictions deviate from reality in a wider range of SP configuration. As SP is a novel concept in pipeline technology, its mechanical behaviour is still not fully understood, not to mention the inherent reliability level during operation. As per the best of the authors' knowledge, there is a scarcity of such an approach for SP that contributes to its structural safety and subsequent design. When the structural reliability analysis is required to be carried out, a target safety level should be deliberated to ensure that a certain safety level is always achieved. Moreover, to avoid unnecessary conservatism, accurate partial safety factors are to be calculated.

In accounting for the above facts, the present study assesses the structural safety of SP for collapse failure and reliability-based design probabilistically. The remainder of the paper is organised as follows. Section 2 describes the methodology comprising a discussion about the adopted collapse strength models and the approach for deterministic and probabilistic assessment of the collapse of SPs. In Section 3, a case study is developed to demonstrate the application of the methodology and the results are discussed. Following the results, important conclusions are drawn.

2. Methodology

This section describes the methodology adopted in this paper for holistic probabilistic analysis for safety assessment of SPs against collapse failure and reliability-based design. The present work follows up on the authors' previous work [20], where the focus was to perceive and quantify the uncertainty among the collapse strength predictions of SPs. However, the present work concentrates on a methodology on how to assess the safety level in SP with uncertainty-associated and reliability-based design, as shown in Figure 2. The main steps and sub-steps of the proposed methodology are also presented in Figure 2. In the absence of a dedicated design code for SP, available strength models are identified. As collapse is the most potential failure mode, the collapse strength models are the appropriate choice for assessment. Further, these models are classified as per the interlayer adhesion conditions. The next step focuses on the assessment of uncertainty in the model's predictions. The results of these analyses would help in selecting the best model in each category for ultimate limit state formulation. In using appropriate probabilistic modelling of limit state variables, reliability against collapse is assessed. Finally, for a target reliability level, a design case is conceptualised, and safety factors are estimated for safer operations. The steps are detailed in the following sub-sections.

2.1. Collapse Strength Models for SP

The collapse failure of SPs has become the focus of recent studies in which various aspects have been dealt with [9,21]. Especially, Castello et al. [11,22–24] have investigated the strength of SPs in different circumstances. These numerical studies are very useful yet specific to circumstances. So, they may not be used in general for collapse strength prediction. For the design and maintenance activities, the accurate and quick prediction of collapse strength is paramount. With such an objective, researchers have proposed several equations (also called strength models) with analytical or numerical methods with empirical techniques. Appendix A details the various collapse models of SP used in this study. Following a comprehensive analytical study, Sato and Patel [17] have proposed a model assuming that the core occupies the entire cavity. Using finite element simulations, Arjomandi and Taheri [16,25–27] have proposed a number of models depending on SP inter-layer adhesion conditions. In their recent study with nearly 12,000 finite element simulations [27], they recommended 12 models for three kinds of core material (soft, moderate, and hardcore). He et al. [28] have numerically investigated the effect of inter-layer adhesion strength, the relative angle between the main axis of the inner and outer pipes with geometrical and material properties on the collapse strength of polypropylene core SP. Recently, using cementitious composite, Yang et al. [4,5] conducted an experimental study followed by comprehensive numerical simulations for SP configurations resulting in a novel model. Additionally, they have used a model formed through automatic machine learning software EUREQA [29]. It should be noted that some of the above-mentioned models are very specific to the core material; however, for the sake of comparison, they are used in the present study. Moreover, it would also be interesting to know the importance of the basic variables in such models through sensitivity analysis.

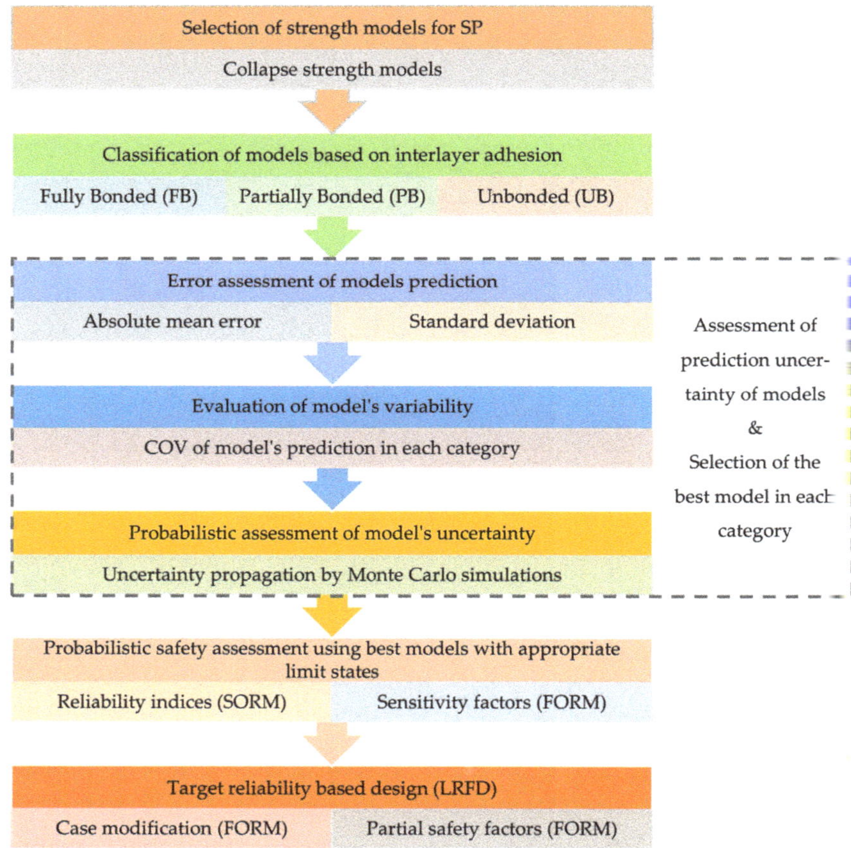

Figure 2. Outline of the main steps and sub steps in the proposed methodology.

2.2. Classification of Models

After a deep study and review of the collapse strength models, the authors propose to group the models into three categories, based on the inter-layer adhesion condition, i.e., (1) fully bonded core (FB) with the inner and outer layer, (2) partially bonded (PB) core that comprises core bonded with the inner and outer layer and core bonded with certain shear strength among layers and (3) unbonded core (UB) that can freely slide among layers (Appendix A). The category PB further has a few sub-classes as inner layer bonded to the core (IB), outer layer bonded to the core (OB), with a given shear stress of 1.5 (1.5) or 15 (15) MPa as evident with He et al. models [28].

The collapse strength models for SP are presented with their notations in Appendix A, showing few deviations in their structure. It should be noted that for the sake of consistency and simplification, the representation of the models in Appendix A may differ from that of the corresponding reference. The present study simplifies and modifies transcription, typesetting, and other issues.

The models' symbols are improvised in the present study based on the model developer and interlayer adhesion condition specified under bracket—(). For example, models developed by Sato and Patel [17] are denoted by SP(FB) and SP(UB) for fully bonded and unbonded core conditions. Arjomandi and Taheri [16,25–27] have proposed a set of models starting with T1(FB) for fully bonded conditions, T2 and T3(FB,IB,OB,UB) based on different interlayer adhesion conditions and finally, 12 models according to adhesion and core elasticity. He et al. models [28] are denoted as HT(1.5) and HT(15). Finally, the two models developed by Yang et al. [4,5] for unbonded core are denoted as Yg(UB) and EQ(UB)—developed by EUREQA.

Due to their complexity and computational difficulty in further reliability formulation, the Hashemian and Mohareb [18,19] models are not incorporated here. It should also be noted that the present study only considers models for two layered SPs.

For the sake of consistency, all analyses are performed in the three categories of interlayer adhesion conditions to distinguish the results clearly.

2.3. Error Assessment

After selecting the collapse strength models (Appendix A), they are compared with collapse test data obtained from the literature (shown in Appendix B). All analyses are categorised into three sets of inter-layer adhesion conditions, as described above. Two important statistical parameters used in this paper to investigate the collapse strength models performance against test data are the absolute mean error (i.e., relative error on average) and its standard deviation, as shown below:

$$\text{Absolute mean error } (\mu) = \frac{\sum \text{Abs}\left(P_c / P_{\text{exp}} - 1 \right)}{N} \tag{1}$$

$$\text{Standard deviation } (s) = \sqrt{\frac{\sum \left(P_c / P_{\text{exp}} - 1 - \mu \right)^2}{N - 1}} \tag{2}$$

where P_c denotes the calculated collapse pressure from a model, P_{exp} is experimental collapse pressure, and N is the total number of experiments.

In the previous work [20], it was revealed that a very high level of uncertainty is propagated while using model uncertainty factors. Therefore, they are not derived and used here. Moreover, more experimental data specific to each bonding condition is required to account for models' uncertainty properly.

2.4. Deterministic Evaluation of Uncertainty

The deterministic collapse strength predictions from the collapse models are compared in a range of different SP design configurations. The variation in collapse strength is illustrated by the variation in these design variables. Uncertainty in terms of coefficient of variation (COV) among the different prediction models in each category is evaluated.

2.5. Probabilistic Assessment

Next, a practical SP case study is formulated, and stochastic properties are given to the basic design variables. The uncertainty propagation from the models is conducted using Monte Carlo simulation that generates a vector of independent random variables **X** to characterise the uncertainty on the mathematical modelling, loading, dimensions, and material properties of pipelines. This allows assessing the uncertainty in the model output (collapse pressure) due to the uncertainty in input variables (material and geometrical variables). The present study also assumes all variables to be mutually independent. If two or more basic variables are dependent, transformation methods such as Nataf or Rosenblatt may be used [30].

Later, the collapse strength models are used to formulate the limit state functions, $g(X)$, which are used to evaluate the structural reliability of SPs. Based on the generalised form $g(X)$ of limit state function formulated, the probability of a limit state violation, typically referred to as probability of failure (P_f), can be calculated as:

$$P_f = \int_{g(x) \leq 0} f_X(x) dx \tag{3}$$

where $f_X(x)$ is the joint probability density function of the vector X of random variables. Failure occurs when $g(X)$ is less than or equal to zero. This region is called a failure region, whereas $g(X) > 0$ represents the safe region. Multidimensional integration is required to solve Equation (3), where the dimension of the integral is equal to the number of variables. The direct calculation of P_f from the integral is rather complicated, and computational techniques are often unfeasible. Therefore, approximate reliability methods such as the Monte Carlo Simulation (MCS) method or the First/Second-Order Reliability Methods (FORM/SORM) are frequently used.

The First Order Reliability Method (FORM) uses a first-order Taylor series expansion (linearisation in a standard normal space) of the limit state function about a point known as the most likely failure point or design point (u*). The first phase in this method consists of transforming non-normal variables into standard normal variables (see, e.g., Ditlevsen & Madsen [31] for further details on the transformation methods). The design point is obtained by solving a constrained optimisation problem that involves finding the point located on the limit-state surface, which has the minimum distance to the origin in the standard normal space. The smallest distance between the origin and this specific point on the limit state surface is known as the reliability index (β). For linear and close-to-linear limit state functions, this solution is judiciously accurate. The Second Order Reliability Method (SORM) is mainly beneficial when the limit state function is nonlinear. The normal way is to approximate the failure surface by a quadratic hyper-surface rather than by a hyper-plane. If the limit state function is not too highly nonlinear, the failure possibility can be assessed from the limit state surface curvatures (δ_i) at the FORM design point.

$$\beta^{\text{SORM}} = -\Phi^{-1} \left\{ \Phi(-\beta) \sum_{j=1}^{k} \left[\prod_{i=1}^{n-1} (1 - \beta.\delta_i) \right]^{-1/2} \right\} \tag{4}$$

where $\delta_i = -\left[\frac{\partial^2 u_n}{\partial u_i^2} \right]$ is the ith principal curvature of the limit state surface $g(u^*) = 0$ at u*. The above formulation is recognised as an asymptotic approximation which is accurate when $\beta \to \infty$. In other cases, some improvement terms are also added to the probability of failure. More formal discussion and derivations are accessible at [31]. Further details about the basic concepts and their applications of these approaches can be referred to in [30]. In the present study, FORM and SORM are adopted for reliability evaluation using the formulated limit state functions.

Furthermore, the present study examines the significance of the basic limit state variables on the reliability indices through a sensitivity analysis. The sensitivity measure or factor (α_i) can be calculated using the FORM approach [32], as:

$$\alpha_i = \frac{\left(\frac{\partial g(u)}{\partial u_i} \right)_{u=u^*}}{\sqrt{\sum_{i=1}^{m} \left(\frac{\partial g(u)}{\partial u_i} \right)^2_{u=u^*}}} \tag{5}$$

with i = 1,2,3 … … m being the index related to each random variable involved in reliability analysis. The partial derivatives are estimated at the FORM design point u*, which represents the most probable failure point of the design random variables in the standard Gaussian space. The positive α_i value indicates that the random variable (i) contributes positively (i.e., increases safety level or β when it increases) and vice versa. In the present study, the sensitivity analysis is performed for the adopted SP collapse models.

Structural reliability methods have been extensively applied for code calibration. For an implicit safety level of SP, a consistent reliability level is required that can be assessed from partial safety factors. The partial safety factors in the design formats, such as the load and resistance partial safety factor design (LRFD) method, are expressed as:

$$R^c/\gamma_R \geq L^c \cdot \gamma_L \qquad (6)$$

where γ_L and γ_R are the partial safety factors for load (L) and resistance (R) characteristic values (*variable^c*), respectively.

From the probabilistic models for the load and resistance terms, the partial safety factors can be calculated through the following equation:

$$\gamma_L = L^*/L^c \text{ and } \gamma_R = R^c/R^* \qquad (7)$$

where *variable** are the design points obtained by FORM and *variable^c* are the specified characteristic values. The characteristic values are fractiles of the corresponding probability distribution functions. In addition, by multiplication of the two partial safety factors, the design safety factor "k" can be obtained:

$$k = \gamma_L \cdot \gamma_R \qquad (8)$$

It is obvious that the reliability of pipes may be described as a function of the design variable. Unlike single-walled pipes, SPs have a greater number of significant design variables. So, the safety of SPs may become case specific; thus, parametric studies are conducted by varying the random variables of the basic case to different design and operational scenarios. The objective is to investigate the effect of variation of basic design variables on the reliability indices estimated with different strength models.

3. Results and Discussion

3.1. Experimental Validation of Strength Models

Some of the models are purely analytical, while others are empirical and based on numerical simulations. Though these models are validated with small-scale tests, it could be debated the applicability of these models' predictions over a wider range. This section investigates the validity of these models (Appendix A) through a comparison with collapse tests reported in the literature [5,6,9,33].

A set of 24 collapse tests with geometric and material properties is presented in Appendix B, which are conducted on steel and aluminium prototypes. Models' prediction with the experimental test comparison is presented in Figure 3. To assess the overall accuracy of models' prediction, Equation (1) is used; moreover, the scatter can be estimated with Equation (2). The results are displayed in terms of mean and their variation in Figure 3. As T1(FB) and Yg(UB) exhibit very high deviation, they are not included in the figure. By means of minimum error T3(FB) followed by HT(15) are the best models, while with minimum variations T3(OB) is the most consistent one. It is important to note that the minimum error is still very high for all the models in all categories. The present statistical results have revealed the complexity of the problem that it is almost inviable to achieve an accurate prediction at all configurations of an SP. In other words, the use of these models may be better under a limited parametric range to keep conservatism or overestimation to a minimum. This would require a thorough parametric analysis performed in the next section.

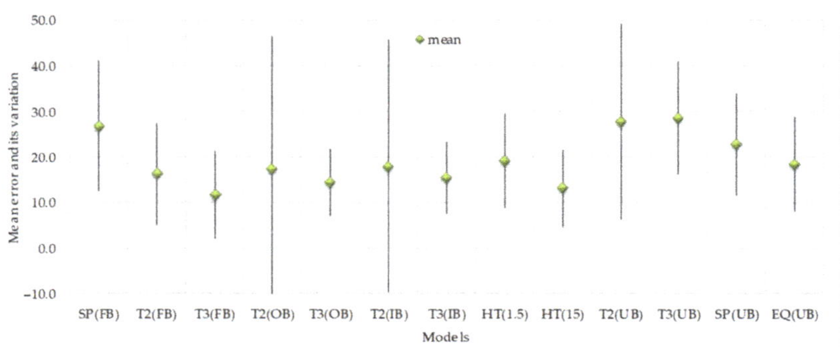

Figure 3. Mean error and their variation for collapse strength models against test data.

3.2. Deterministic Assessment of SP Collapse Strength Models

This section aims to compare the performance of the adopted SP models' collapse strength prediction parametrically. Despite the fact that some parametric studies have been presented by the models' developers, it is imperative to witness such empirical equations' (models) predictions under a common set of geometric and material parameters. The range of parameters used in present deterministic assessment (the parameters assume constant values) are shown in Table 1. The results of such analyses are particularly important to measure the applicability and shortcoming of these models for a broader range of design parameters. Appendix B presents the range of parameters for deterministic assessment (the parameters assume constant values). It is to be noted that the studied range is similar to that used in practical engineering design cases. The Poisson ratio of pipe and core materials are assumed to be constant as it remains constant for steel pipe, while Arjomandi and Taheri [26] have advocated that variation in the Poisson ratio of the core material has an insignificant influence on collapse pressure. In the present study, outer and inner pipes are made up of the same material (steel). However, for the core material, an appreciable elasticity range is assumed.

Table 1. Range of parameters used in deterministic assessment.

$\sigma_1 = \sigma_2$	v_c	v_p	E_c/E_p	t_1/r_1	t_2/r_2	r_2/r_1	t_c/r_c
205 MPa	0.4	0.3	0.001–0.1	0.02–0.08	0.026–0.097	0.6–0.8	0.1–0.4

3.2.1. Effect of Outer Pipe Thickness

Undoubtedly, the thickness of the outer pipe is an important design parameter, so it is varied, and corresponding collapse strengths are evaluated from the adopted models. The results are presented in terms of collapse pressure (P_c) of SPs as a function of t_1/r in Figure 4. The other parameters are assumed as $E_c/E_p = 0.01$, $t_2/r_2 = 0.026$, $t_c/r_c = 0.23$.

In the fully bonded core category (Figure 4a), T1(FB) becomes overestimated after a small increase in outer pipe thickness ($t_1/r_1 > 0.03$); T3(FB) remains almost constant in collapse strength with increasing thickness. T2(FB) and SP(FB) characterise a nearly linear increasing trend in collapse pressure with outer pipe thickness.

For partial bonded core models, three sets of patterns are observed, as presented in Figure 4b. T3(IB) and T3(OB) present close predictions that show a small increase in local collapse pressure. The greatest increase in P_c vs t_1/r_1 with a similar prediction is shown by T2(IB) and T2 (OB). The collapse capacity predictions by HT models fall in between the other two trends.

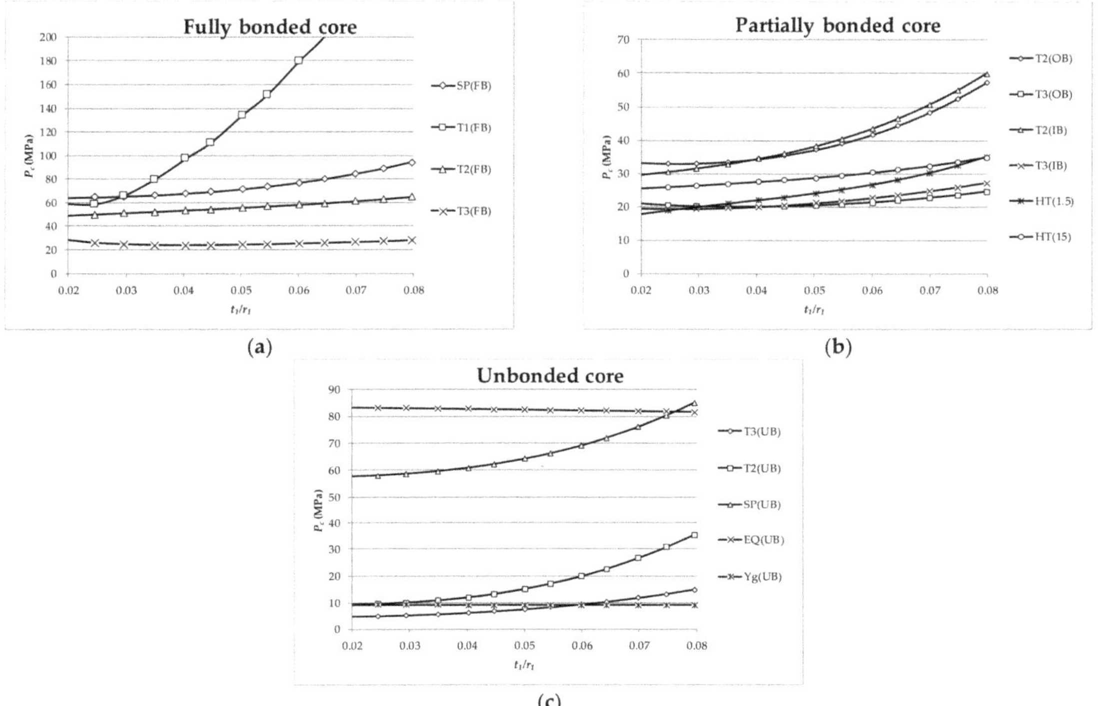

Figure 4. Variation of P_c with t_1/r_1 for (**a**) Fully bonded, (**b**) Partially bonded, and (**c**) Unbonded core.

Lastly, for unbonded core, Yg(UB and EQ(UB)) predict constant collapse pressure at all t_1 while T3(UB) shows little rise in collapse strength. However, EQ(UB) is highly overestimative in this category, followed by SP(UB). T2(UB) seems to provide a better correlation between P_c with an increase in t_1/r_1. In the present scheme of analyses, core thickness has been considered constant; however, if the inner pipe thickness is kept constant instead, analogous trends of P_c variation with outer pipe thickness would have been obtained.

3.2.2. Effect of Core Thickness

In this subsection, the influence of core thickness on the collapse strength of SPs is assessed and plotted in Figure 5 in the three categories of inner-layer adhesion. For this assessment other parameters are kept in the following ratio as $E_c/E_p = 0.01$, $t_1/r_1 = 0.04$, $r_2/r_1 = 0.75$. Figure 5a presents the effect of core thickness on the collapse capacity of SP for a fully bonded core. SP(FB) and T1(FB) are the models independent of expression for core thickness and thus provide a constant reading. Prediction by T2(FB) shows increasing linear trends while T3(FB) shows a near quadratic rise with an increase in core thickness.

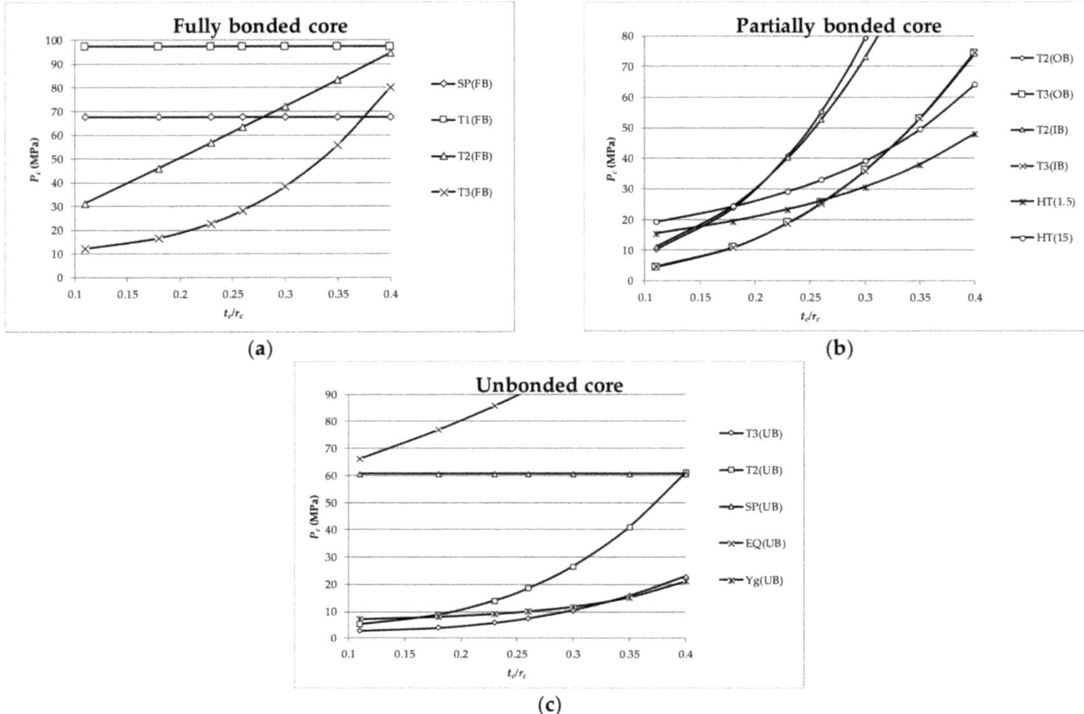

Figure 5. Variation of P_c with t_c/r_c (**a**) Fully bonded, (**b**) Partially bonded, and (**c**) Unbonded core

As shown in Figure 5b, for the case of partially bonded models, three sets of increasing trends are observed (similar to the previous section). T2(OB) and T2(IB) prediction trends are fairly close and increase rapidly with core thickness. T3(OB) and T3(IB) prediction trends overlap and also show a steeper rise at higher core thickness. For both HT models, trends are similar (with minor differences), but their prediction is higher at a lower core thickness ratio ($t_c/r_c < 0.15$), but at higher thickness, lower predictions are observed ($t_c/r_c > 0.3$).

Under the category of unbonded core models, T3(UB) and Yg(UB) show little increase, while T2(UB) shows relatively higher collapse strength estimates as the core thickness increases. From the overall assessment of the results presented in this section, t_c/r_c has a significant effect approximately when $t_c/r_c > 0.25$.

3.2.3. Effect of Inner Pipe Thickness

For the collapse of SP, inner pipe geometry is not as significant as the outer pipe. In fact, it is a crucial factor when estimating the burst strength of the inner pipe and subsequently of SP. Yet, it is interesting to note the extent of influence of inner pipe thickness on P_c, as presented in Figure 6. The ratios, $E_c/E_p = 0.01$, $t_1/r_1 = 0.03$, $r_2/r_1 = 0.75$ are set for the present analysis.

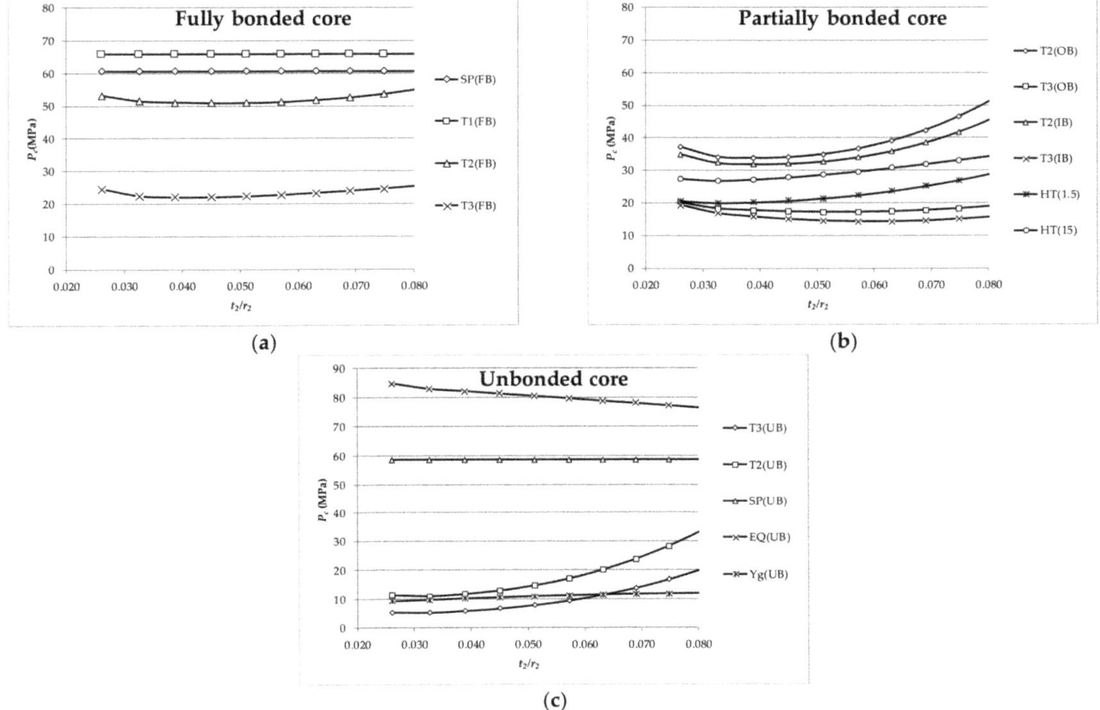

Figure 6. Variation of Pc with t_2/r_2 (**a**) Fully bonded, (**b**) Partially bonded, and (**c**) Unbonded core.

Figure 6a presents the collapse pressure trends for fully bonded models. The effect of t_2 on collapse capacity is nearly nil, as exhibited by all the models in this category. For partially bonded core, a little fall followed by a little increase in P_c is observed for most of the models. The little decrement in collapse strength is due to the compromise in core thickness while increasing the inner pipe thickness. SP(UB) and Yg(UB) models do not acknowledge any effect of t_2/r_2 on the P_c in their model formulation. Surprisingly, EQ(UB) show the negative impact of increasing inner pipe thickness. T2(UB) and T3(UB) perceive a slight rise in the collapse strength at a higher thickness of the inner pipe. This section has indicated that the inner pipe has very little effect on the P_c.

3.2.4. Effect of Core Material Elasticity

One of the typical features of SP is the core material being a structural element of the entire system. In this context, the importance of its structural property on the overall strength of SP is significant. This section reveals that different models' predictions are influenced by core material elasticity. To trace only elasticity effect on predictions, other parameters are kept constant as $t_1/r_1 = 0.04$, $t_2/r_2 = 0.27$, $r_2/r_1 = 0.74$. $t_c/r_c = 0.23$.

For the sake of comparison, two scenarios of SPs are assumed. The first case is when the core is non-structural, implying $E_c = 0$, the corresponding collapse strength of SP will be equal to that of a single walled pipe. The collapse pressure of a single-walled pipe can be calculated by the following equation from DNV [2]:

$$(P_{co} - P_{el})\left(P_{co}^2 - P_p^2\right) = P_{co}P_{el}P_p f_o \frac{D}{t} \tag{9}$$

where P_{el} = elastic pressure; P_p = plastic pressure; D = maximum diameter of pipe; t = maximum thickness of pipe and f_o = ovality of pipe (≥ 0.005).

$$P_{el} = \frac{2E \left(\frac{t}{D}\right)^3}{1 - v_p^2}$$

where E = Young's modulus of elasticity of the pipe material and v = Poisson ratio.

$$P_p = \sigma \, \frac{2t}{D}$$

where σ = yield strength of the pipe.

The collapse pressure for the first case (E_c = 0) with $t = t_1$, $D = D_1$ and $E = E_p$ in Equation (9) is calculated 3.569 MPa. Similarly, an upper bound can be established (E_c/E_p = 1). For this case, the collapse strength of SP can be calculated by Equation (9) with $t = t_1 + t_c + t_2$, $D = D_1$ and $E = E_p$. The collapse strength calculated for this case is 119.27 MPa using Equation (9).

As seen in Figure 7, all models in all categories exhibit a nearly linear trend for change in P_c with E_c/E_p. Another important observation is that many models predict near zero strength when E_c reaches zero, which should be equal to collapse strength in the first case. Surprisingly, many models predict far greater strength than the upper bound of collapse strength way before the point E_c/E_p = 1.

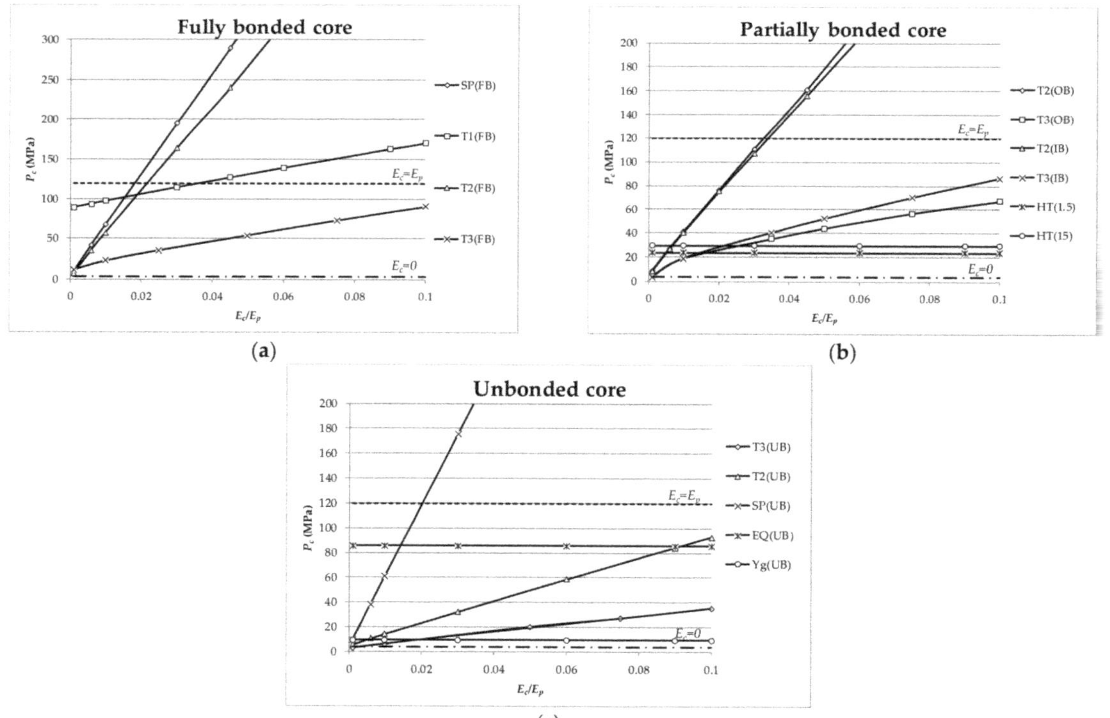

Figure 7. Variation of P_c with E_c/E_p (**a**) Fully bonded, (**b**) Partially bonded, and (**c**) Unbonded core.

As per Figure 7a in fully bonded core models, the SP(FB) and T2(FB) predictions have a large bearing on E_c as they become highly over estimative for $E_c/E_p > 0.02$, which may not be admissible. The model T1(FB) also shows a higher increase in P_c after this level.

In the category of partial bonded condition (Figure 7b), T2(IB) and T2(OB) have a very close prediction that provides high estimates of collapse capacity. T3(IB) and T3(OB) again show similar trends that seem more reasonable in this category. Accounting similar findings in Figure 7c, T2(UB) and T3(UB) also show reasonable maximum strength trends.

HT(1.5), HT(15), EQ(UB) and Yg(UB) models were developed with specific core material, so their predictions are unbiased with changes in E_c.

The inference of this section is that the SP models are vastly dependent on the core material structural property, and their validity lies in a specific range of E_c only. For example, fully bonded models should not be applied when $E_c/E_p > 0.01$. Other models are also to be used under $E_c/E_p < 0.1$ since a linear extrapolation will lead to erroneous results.

3.3. Probabilistic Analysis

The above sections have explored the performance of the models and their distinct behaviour with observed data and have identified some extreme behaviour of models' predictions at different configurations. An SP case study is developed based on a test case from literature with slight modifications for better utilisation [4]. The parameters' uncertainties are provided from the literature, e.g., some variabilities (for diameter, thickness, and strength) are adopted following the probabilistic distributions adopted by Teixeira et al. [15]. Similarly, the stochastic model for pipe and core elasticity follows from Ref. [34]. Poisson ratios for pipe and core material have been provided minimum COV as that of diameter and assumed normally distributed. All the basic random variables and their probabilistic models are shown in Table 2. Log-normal (LN) parameters are also presented for respective distribution.

Table 2. Probabilistic models for basic random variables of SP case study.

	$\sigma_1 = \sigma_2$ (MPa)		v_c	v_p	E_c (GPa)	E_p (GPa)	D_1 (mm)	t_1 (mm)	D_2 (mm)	t_2 (mm)
Mean	413	{19.836}	0.4	0.3	2	200	202.8	2	152.4	1.8
COV	8	{0.0799}	0.1	0.1	5	5	0.1	1	0.1	1
Distribution	LN		N	N	N	N	N	N	N	N
Reference	[15]		Assumed	Assumed	[34]	[34]	[15]	[15]	[15]	[15]

N—Normal, LN—Log-normal, {} lognormal parameters.

It is obvious that an SP is superior in structural strength as compared to an equivalent single-walled pipe. In other words, the collapse strength of an SP must be greater than that of an equivalent single-walled pipe. An equivalent pipe can be conceptualised from SP using an equal amount of steel used and has the same functional requirement (equal inner diameter of inner pipe of an SP). The collapse capability for single-walled equivalent (P_{eq}) pipe is calculated from Equation (9).

3.3.1. Uncertainty Propagation

In order to assess the uncertainty of the SP models, the Monte Carlo simulation method is used, which generates the number of output samples using the probabilistic properties of input variables. In this study, 10,000 simulations are performed using the probabilistic framework outlined in Table 2. The uncertainty in collapse strength in terms of COV is estimated for each model, as presented in Figure 8. Additionally, the figure also illustrates the average model prediction (P_c) normalised with P_{eq}. Monte Carlo simulation is an approximate method, and therefore their predictions are always associated with some error. However, by increasing the number of simulations, the error can be minimised. Figure 9 provides the prediction error (PE in %) of the normalised mean collapse pressure calculated from the 95% confidence interval of the mean. Interestingly, in the present simulations, the error stays less than 0.1% for all the models.

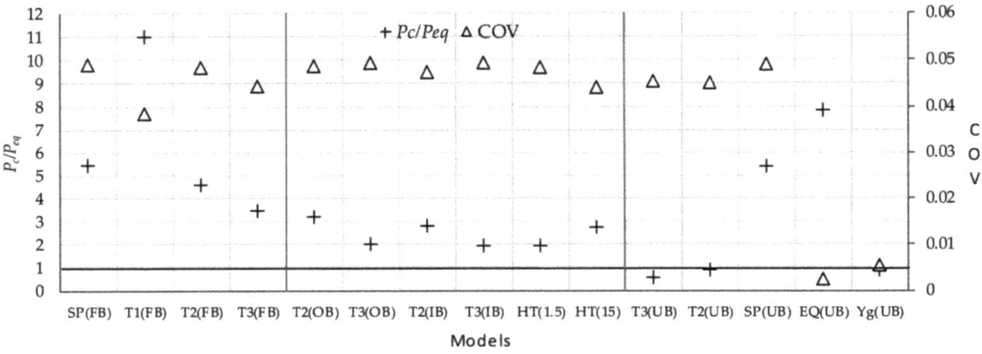

Figure 8. COV and P_c/P_{eq} for SP collapse strength models.

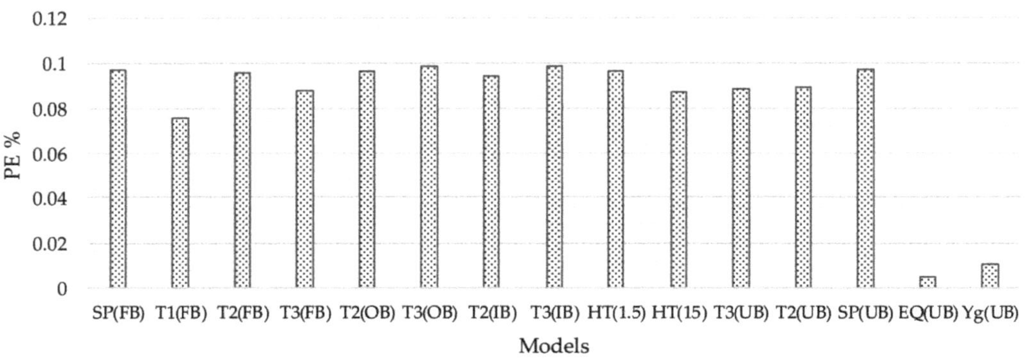

Figure 9. Prediction error (PE in %) of the normalised mean collapse pressure calculated from ±5% confidence interval of the mean.

Interestingly, the collapse strength predicted by some unbonded core models—Yg(UB) and T2(UB)—is quite close to that of the equivalent pipe. SP(UB) and EQ(UB) models are exceptional at predicting the collapse pressure in this category. T3(UB) is the only model that under-predicts the collapse strength. Subsequently, the partially bonded models are capable of larger collapse capability (at least twice) of that of an equivalent pipe. In the fully bonded core category, the models' predictions are notably greater (at least 3.5 times) than of an equivalent pipe. In this category, T1(FB) is over-predictive as the same behaviour seen in previous sections. These findings corroborate the fact that the SPs are always operationally superior and beneficial compared to a single-walled pipe (with the same basic functional requirement).

Minimum COV is associated with EQ(UB) and Yg(UB) models, however, this is due to their relatively lesser dependence on the basic design parameters. For most of the models in all categories, the COV lies between 0.045 to 0.05, except for T1(FB).

3.3.2. Estimation of Reliability Indices

The collapse strength (resistance for external over pressure) of SP has been the focus so far in this study. Furthermore, for the structural reliability analysis of SP, the limit state function $g(\mathbf{X})$ is formed from loading and resistance. In SP, the loading is the hydrostatic pressure acting externally on the outer surface of SP. The external pressure is considered acting alone on the SP while the other loads are neglected.

For the present study, the external pressure on the sandwich pipe is assumed to be the maximum capacity of an equivalent single-walled pipe. The characteristic value of the external pressure (P_{ext}^c) can be calculated by DNV [2] requirement for an intact single-walled pipe under external pressure.

$$P_{ext}^c - P_{min} \leq \frac{P_{eq}(t)}{\gamma_m \cdot \gamma_{SC}} \tag{12}$$

where P_{min} is the minimum internal pressure assumed nil, γ_m and γ_{SC} are the factors accounting for material resistance and safety class, respectively, adopted from DNV [2].

From the analysis conducted so far, some significant conclusions can be drawn for selecting the most appropriate strength model. Clearly, T3 models are the most stable and reasonable in each category and thus adopted for design and reliability analysis.

It is evident from the above section that the partially and fully bonded core condition offers higher collapse pressure than unbonded core. More specifically, in the previous section, it was shown that sandwich pipes with a fully bonded and partially bonded core show at least thrice and twice the collapse capacity with respect to the equivalent pipe and should be realised under higher loading for optimisation.

From Figure 8, T3(FB) has 3.5 times higher strength than a similar equivalent pipe; in the partial bonded category, T3(IB) and T3(OB) provide 2 and 2.1 times the strength. T3(UB) is the only model that predicts inferior strength (around 0.6 times) to the equivalent pipe. Considering this fact and accounting for the additional strength in reliability results, the present study adopts that the loading on three categories of sandwich pipelines are to be multiplied by an over-strength factor. This results in four forms of limit state functions in each category given as:

$$g(X) = \begin{cases} P_{cT3(UB)} - 0.6P_{ext} & \text{for unbonded core models} \\ P_{cT3(IB)} - 2P_{ext} & \text{for partially bonded core models} \\ P_{cT3(OB)} - 2.1P_{ext} & \text{for partially bonded core models} \\ P_{cT3(FB)} - 3.5P_{ext} & \text{for fully bonded core models} \end{cases} \tag{13}$$

The statistical properties of basic design variables are taken from Table 2 with their characteristic values. The characteristic values of diameters, thickness, Poisson ratio and Young's modulus assume their 50 percentiles, while for the strength parameters, it is assumed to be the 5 percentiles. The probabilistic model for P_{ext} is derived similarly to that by Teixeira et al. [15] and is given below in Table 3.

Table 3. Probabilistic model for external pressure.

Random Variable	Distribution	Mean	Standard Deviation	COV(%)	Characteristic Value(X^c), Pr ($x < X^c$)
P_{ext} (MPa)	Gumbel	8.19 {8.09}	0.246 {0.192}	3	8.649 (95% percentile)

Gumbel parameters are in {}.

In probabilistic analysis, it is seldom necessary to introduce model deviations in terms of model uncertainty factors. However, with the limited data set and very high deviations for some models, as clearly observed in Figure 3, model uncertainty factors are not derived. Moreover, the previous study [20] has clearly indicated their inadequacy for the present case.

The reliability index (β) of the outer pipe collapse failure mode due to external pressure is estimated by the SORM method (described in Section 3) and shown in Table 4. The SORM method is deliberately used here to capture the effect of nonlinearity of the limit state function.

Table 4. Reliability indices for SPs with different core adhesion models.

Fully Bonded Core	Partially Bonded Core		Unbonded Core
T3(FB)	T3(OB)	T3(IB)	T3(UB)
4.65	4.31	4.28	4.50

The reliability results are fairly in concordance for the three categories of SPs as comparable reliability indices are obtained. In particular, T3(FB) prescribes the highest reliability, while T3(IB) is associated with the minimum reliability. The aim here is to understand the peculiarities associated with the structural reliability of SP. The results presented here do not claim to be the actual reliability of an operational SP as the external pressure at the installation must be used in limit state function. However, in the present study, external pressure is proportional to the collapse capacity of an equivalent single-walled pipe. Thus, the information displayed in Table 4 signifies the high reliabilities of SP with respect to an equivalent pipe.

3.3.3. Sensitivity Analysis

SP involves several design parameters, and it is always interesting to note which parameters have a greater influence on reliability. In the present section, sensitivity analysis is conducted (using Equation (5)) to quantify the relative importance of each random variable on limit state functions by different models. The importance of each variable is the uncertainty of the limit state functions with corresponding models is assessed using FORM and presented in Table 5 with respect to each model. The highest sensitivity factor is found for P_{ext} for almost all the models in all categories. All models reflect the second highest importance for E_p, which is an important material property. A negative sign in sensitivity factors for a variable implies that an increase in the variable reduces the reliability. Surprisingly, inner pipe geometrical parameters (D_2 and t_2 in some models) contribute negatively. σ_1 and σ_2 have negligible impact on limit state functions.

Table 5. Sensitivity factors (α_i) for the basic variables for the SP models.

Variables	Fully Bonded Core	Partially Bonded Core		Unbonded Core
	T3(FB)	T3(OB)	T3(IB)	T3(UB)
σ_1	0.079	0.012	0.012	0.003
σ_2	0.066	0.011	0.012	0.266
E_p	0.414	0.505	0.560	0.318
D_1	0.092	0.151	0.105	0.097
t_1	−0.076	−0.187	−0.125	-0.105
D_2	−0.079	−0.117	−0.082	-0.079
t_2	−0.055	−0.034	−0.101	-0.078
P_{ext}	−0.891	−0.819	−0.780	-0.892

3.3.4. Design Case for Target Reliability

As for single-walled pipelines, the required safety level of SP can be obtained from the DNV code [2]. SPs are specialised pipes used for extreme conditions, and hence designers should consider high safety class.

A target reliability index of 4.265 (probability of failure = 10^{-5}) is adopted as the design target to satisfy the safety class—high and against the ultimate limit state of SP. The analysis starts by identifying an adequate operational condition for SPs with different core adhesion conditions in the case study for the target safety level. Table 6 presents the safe depths for SP with different core bonding states. Fully bonded core SP having the highest collapse strength can be operational at around 2900 m; for the same dimensions, partially bonded core SPs can work safely at around 1600 m, while unbonded core SPs are suitable for 489 m.

Table 6. Installation depth of various SPs in regard to target reliability.

	Fully Bonded Core	Partially Bonded Core		Unbonded Core
	T3(FB)	T3(OB)	T3(IB)	T3(UB)
Water depth (m)	2912.3	1674.2	1601.7	489.6

3.3.5. Partial Safety Factors

The partial safety factors are interpreted as a factor that should be applied to the characteristic values of the variables to obtain a uniform safety level. This section aims to derive partial safety factors based on the target reliability level. As a continuation of the analysis from the previous section, this section explores the partial safety factors for the load and resistance for maintaining the target reliability. In using Equations (6) and (7), partial safety factors for load and resistance are calculated. In addition, the overall safety factor (k) is also calculated using Equation (8). Table 7 summarises the results of the partial safety factors estimated for design variables for three categories of SPs.

Table 7. Partial safety factors for SP for different inter-layer adhesion condition.

Variables	Fully Bonded Core	Partially Bonded Core		Unbonded Core
	T3(FB)	T3(OB)	T3(IB)	T3(UB)
γ_R	1.148	1.179	1.175	1.139
γ_L	1.109	1.11	1.11	1.119
k	1.273	1.308	1.304	1.275

It is obvious that the models are limited and associated with very high prediction uncertainty. Still, the present approach has indicated means to assess adequate reliability levels and safety factors. Moreover, with more data and research advancement, the present approach can be better utilised.

3.4. Parametric Analysis

In this section, several parametric analyses are performed by varying the design variables, and operating parameters of the limit state functions and corresponding reliabilities are estimated for models in three categories of SPs. Different design and operational scenarios are thus constituted, and the performance of models is observed accordingly. It is to be noted that the characteristic collapse pressure trends for SP are far different from conventional single-walled pipes due to the existence of higher uncertainties (as discussed above in Section 3.2).

3.4.1. Reliability Variation with Outer Pipe Thickness

As for SP, the outer pipe thickness is the most significant design variable, some scenarios around the base case study are depicted by varying t_1. Figure 10 shows the β calculated for three categories of SPs as a function of t_1/r_1. It is observed that the reliability decreases and then increases with the increase in pipe thickness, which can be easily explained by the fact that r_1 also increases with t_1. The t_1 seems to make more impact in unbounded core models as the trends take a higher curve after $t_1/r_1 = 0.02$. In the category of partially bonded core models, the reliability tends to change slightly after $t_1/r_1 = 0.02$. It is clear from Section 3.2.3 that the inner pipe thickness has barely any effect on the collapse capacity of the pipe; consequently, it would have a negligible effect on reliability indices, and thus it is not accounted for.

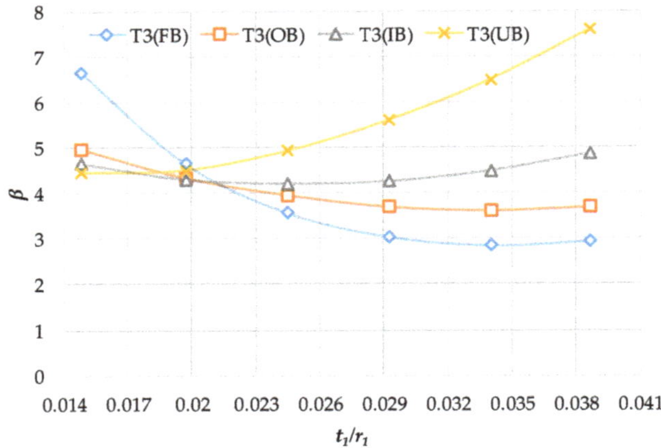

Figure 10. β as a function of outer pipe thickness (t_1/r_1).

3.4.2. Reliability Variation with Core Thickness

As the structural core is a distinct feature of SPs, it is believed necessary to observe the effect of its thickness on the reliability of SPs. Figure 11 presents the trends for β estimated from three categories with increasing core thickness at the compensation of inner pipe thickness. Interestingly, increasing linear trends can be perceived for three categories (4 models). For partially bonded conditions, after $t_c/r_c = 0.32$, no improvement in reliability can be obtained with a further increase in core thickness. Core thickness is the predominant variable as a 3 mm increase in core thickness improves reliability indices by 4.63 for T3(FB), 4.52 for T3(OB), 4.68 for T3(IB), and 5.51 for T3(UB).

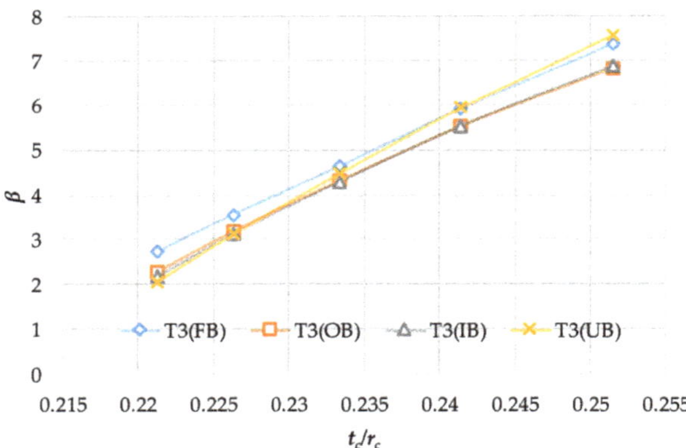

Figure 11. β as a function of core thickness ratio (t_c/r_c).

3.4.3. Reliability Variation with External Pressure

Finally, Figure 12 illustrates β as a function of external pressure. As expected, increasing the external pressure on SP would significantly decrease their reliability. The results trends are rather linear in the three sets of SP design configurations, influencing the reliability to almost the same extent. It should be acknowledged that the reliability indices are very contextual to the design case under consideration. However, the analysis of the trends is important as the models' behaviour is deemed to remain the same. A 40% increase in external pressure reduces the reliability of SP by 74% for T3(FB), 73% for T3(OB), 74% for T3(IB), and 76% for T3(UB).

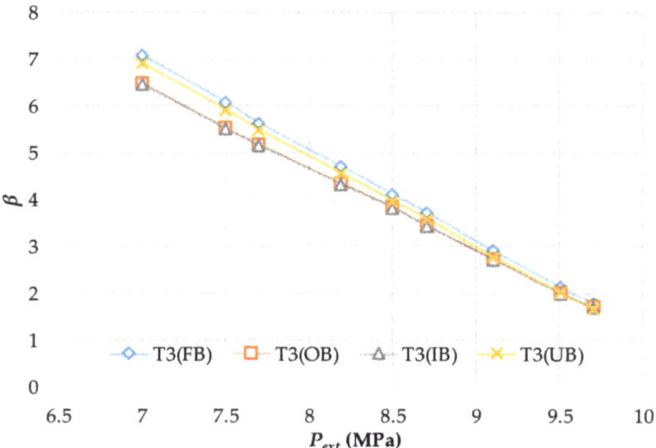

Figure 12. β as a function of external pressure.

The geometric imperfections are not considered in the present study because neither they are design parameters nor are significant with respect to analytical importance.

4. Conclusions

This paper has contributed with a comprehensive probabilistic assessment for collapse failure of sandwich pipelines with available strength models. First, the adopted methodology is illustrated, outlining the main steps of a comprehensive assessment. All results are distinguished in fully bonded, partially bonded and unbonded core categories based on SP interlayer adhesion conditions. In order to verify the validity of adopted collapse strength models, they are compared against experimental data. It is found that all the models deviate significantly from test data, and further research is required for accurate predictions of the collapse strength of SPs.

Later, a detailed deterministic-based analysis is carried out to illustrate prediction capability at different SP design configurations. Outer pipe and core thickness have much influence on collapse strength, while inner pipe thickness has negligible impact. Young's modulus of the core is a very important parameter, and most of the models must be utilised under a given range of Young's modulus for correct predictions. Uncertainty among models' predictions is perceived in different SP design cases.

Probabilistic modelling is applied to the basic random variables, and uncertainty in collapse strength is assessed through the Monte Carlo simulation method. For most of the models in all categories, the COV lies between 0.045 to 0.050.

For the safety assessment of sandwich pipelines, four limit state functions are formulated under external pressure. For the base case scenario, reliability is estimated with Second Order Reliability Methods. Fully bonded core models result in a reliability index of 4.65; in partially bonded conditions, reliability is estimated at around 4.3; and a high-reliability index (4.5) is associated with unbonded core models. Through First Order Reliability Method-based sensitivity analysis, external pressure is found to be the most important parameter for safety, followed by Young's modulus of elasticity. In the absence of a standard or code, the present study defines the target safety level, and the design safety factors are derived accordingly. Some parametric studies evaluating reliability are conducted for different design and operational scenarios.

The objective of the present analysis is to demonstrate that even with limited data and models, it is possible to apply explicit safety standards to SPs.

Author Contributions: Conceptualization, U.B., A.P.T. and C.G.S.; methodology, U.B. and A.P.T.; software, U.B.; validation, U.B.; formal analysis, U.B.; investigation, U.B.; resources, A.P.T. and C.G.S.; data curation, U.B. and A.P.T.; writing—original draft preparation, U.B.; writing—review and editing, A.P.T. and C.G.S.; visualization, U.B.; supervision, A.P.T. and C.G.S.; project administration, C.G.S.; funding acquisition, C.G.S. All authors have read and agreed to the published version of the manuscript.

Funding: This work contributes to the project "Cementitious cork composites for improved thermal performance of pipelines for ultradeep waters—SUPBSEAPIPE, with the reference n.° POCI-01-0145-FEDER-031011 funded by European Regional Development Fund (FEDER) through COMPETE2020—Operational Program Competitiveness and Internationalisation (POCI) and with financial support from the Portuguese Foundation for Science and Technology (Fundação para a Ciência e Tecnologia—FCT), through national funds. This study also contributes to the Strategic Research Plan of the Centre for Marine Technology and Ocean Engineering, which is financed by Portuguese Foundation for Science and Technology (Fundação para a Ciência e Tecnologia—FCT), under contract UIDB/UIDP/00134/2020.

Institutional Review Board Statement: Not applicable.

Informed Consent Statement: Not applicable.

Data Availability Statement: Not applicable.

Conflicts of Interest: The authors declare no conflict of interest.

Nomenclature and Symbols

Symbols	Description
α_i	Sensitivity measure or factor
P_c	Collapse pressure
P_{cal}	Calibrated model prediction
P_{co}	Collapse pressure of outer pipe
P_{el}	Elastic pressure
P_{exp}	Experimental collapse pressure
P_p	Plastic pressure
$f_x(x)$	Joint probability density function
f_o	Ovality of pipe
δ_i	Limit state surface curvatures
$\sigma_{1,2}$	Yield strength of outer (1) and inner pipe (2)
a, b, c, d, e, f, g, h, i, j, k, l, m	Coefficients in T3 model
COV	Coefficient of variation
$D_{1,2}$	Outer diameter of outer (1) and inner pipe (2)
$E_{p,c}$	Young's modulus of pipe (p) and core (c)
EQ()	EUREQA model
FB	Fully bonded core
FORM	First Order Reliability Method
h_{1-16}	Coefficients in HT model
HT()	Models by He et al.
k	Design safety factor
L	Load
LN	Log Normal distribution
LRFD	Load and resistance partial safety factor design
MCS	Monte Carlo Simulation
N	Normal distribution
N	Number of experiments
n	Buckling mode number
PB	Partially bonded core
P_{eq}	Collapse capability for single walled equivalent pipe
P_f	Probability of failure

Symbols	Description
PiP	Pipe in Pipe
P_{min}	Minimum internal pressure
R	Resistance
$r_{1,2,c}$	Maximum radius of outer (1), inner pipe (2) and core (c)
s	Standard deviation
SORM	Second Order Reliability Method
SP	Sandwich pipelines
SP()	Models by Sato and Patel
$t_{1,2,c}$	Maximum thickness for outer (1), inner pipe (2) and core (c) layers of SP
$T_{1,2,3}()$	Models by Arjomandi and Taheri
u*	Design point
UB	Unbonded core
$v_{p,c}$	Poisson ratio of pipe (p) and core (c)
Xc	Model uncertainty factor
Yg()	Models by Yang et al.
γ_L, γ_R	Partial safety factors for load (L) and resistance (R)
γ_m and γ_{SC}	Factors accounting for material resistance (m) and safety class (SC)
θ_{0-15}	Coefficients in Yang et al. model
κ, α_{1-2}, γ_{1-3}, ξ_{1-4}	Coefficients in T2 model
μ	Absolute mean error
g(X)	Limit state function
β	Reliability index
PE	Prediction Error

Appendix A

Table A1. Collapse strength models for sandwich pipes.

Model Description	Symbol Ref.
1. Fully bonded	
$P_c = P_{crs} + \frac{1}{n^2-1}K,$ *where* : $K = E_c \frac{2n(v_c-1)-2v_c+1}{4v_c^2+v_c-3}$ $P_{crs} = \left(\frac{t_1}{r_1}\right)^3 \frac{E_p\left(n^2-1\right)}{\left(1-v_p^2\right)\left(\left(\frac{t_1}{r_1}\right)^2+12\right)}$	SP(FB) [17]
$P_c = \xi_1/\xi_2$ $\xi_1 = 192E_c^2 a_1 r_1^3 \left(v_p^2-1\right)^2 + E_p^2 t_1^4 n^2 \Lambda\left(n^2-1\right)(\Lambda+7)^2 + 2E_c E_p r_1 t_1\left(v_p^2-1\right)(\Lambda+7)$ $\left\{t_1^2 n^2[n(\Lambda-7)-\Lambda-1] - 6r_1 t_1\left[(n+1)^2 + (n-1)^2\Lambda\right] - 12r_1^2[n(\Lambda-1)-\Lambda-1]\right\}$ $\xi_2 = r_1\left(v_p^2-1\right)(\Lambda+7)\left\{-12E_c a_1 r_1^2\left(v_p^2-1\right)[n(\Lambda-7)-\Lambda-1] + E_p t_1 n^2 \Lambda\left(t_1^2+12r_1^2\right)(\Lambda+7)\right\}$ $a_1 = r_1 - t_1/2$ $\Lambda = 4v_p - 3$	T1(FB) [25]
$P_c = \kappa P_{cr} + E_p\left(1+\alpha_1 v_c^2\right)\left(\frac{t_1}{r_1}\right)^{\alpha_1}(\psi_1+\psi_1)$ *where* : $P_{cr} = \frac{E_p}{4\left(1+v_p^2\right)}\left(\frac{t_1}{r_1}\right)^3$ $\psi_1 = \gamma_1\left(\frac{E_c}{E_p}\right)^{\gamma_2}\left(1-\frac{r_2}{r_1}\right)^{\gamma_3}$ $\psi_2 = \xi_1\left(\frac{E_c}{E_p}\right)^{\xi_2}\left(1-\frac{r_2}{r_1}\right)^{\xi_3}\left(\frac{t_2}{r_2}\right)^{\xi_4}$	T2(FB) [26]

κ	α_1	α_2	γ_1	γ_2	γ_3	ξ_1	ξ_2	ξ_3	ξ_4
0.9844	−0.5444	0.1	0.474	0.98	1.062	0.43	0.079	−0.1031	2.8

Table A1. *Cont.*

Model Description								Symbol Ref.

$$P_c = E_p(\varphi_1 + \varphi_2 + \varphi_3)$$
where :
$$\varphi_1 = a\left(\frac{t_1}{r_1}\right)^b \left(\frac{t_c}{r_1}\right)^c \left(\frac{\sigma_1}{E_p}\right)^d$$
$$\varphi_2 = e\left(\frac{t_1}{r_1}\right)^f \left(\frac{t_2}{r_2}\right)^g \left(\frac{t_c}{r_1}\right)^h \left(\frac{\sigma_2}{E_p}\right)^i$$
$$\varphi_3 = j\left(\frac{t_1}{r_1}\right)^k \left(\frac{t_2}{r_2}\right)^l \left(\frac{t_c}{r_1}\right)^m$$

T3(FB) [27]

	a	b	c	d	e	f	g
(Ec = 0.1)	2.032	0.909	0.377	1.055	0.759	0.152	0.884
(Ec = 0.01)	0.765	1.068	0.253	0.89	0.422	0.148	0.947
(Ec = 0.001)	1.406	2.102	−0.134	0.747	0.231	−0.016	2.742

	h	i	j	k	l	m
(Ec = 0.1)	−0.224	1.021	7.25×10^{-2}	−0.067	−0.094	3.833
(Ec = 0.01)	0.112	0.777	6.02×10^{-2}	−0.943	−0.689	3.871
(Ec = 0.001)	−0.515	0.317	2.46×10^{-2}	−0.244	0.206	1.133

2. Partially bonded core

a Outer bonded

P_c same as T2(FB), coefficient given below — T2(OB) [26]

κ	α_1	α_2	γ_1	γ_2	γ_3	ξ_1	ξ_2	ξ_3	ξ_4
1.019	0.2461	−0.0904	0.816	0.982	3.146	0.1792	0.0329	−0.1062	2.929

P_c same as T3(FB), coefficient given below — T3(OB) [27]

		b	c	d	e	f	g
(Ec = 0.1)	0.773	1.757	−1.018	1.148	0.182	−0.032	0.648
(Ec = 0.01)	5.21	2.552	−0.083	0.795	3.029	0.026	2.391
(Ec = 0.001)	3.071	2.617	−0.111	0.662	0.102	−0.234	2.695

	h	i	j	k	l	m
(Ec = 0.1)	0.964	0.586	3.40×10^{-2}	−0.341	−0.198	4.444
(Ec = 0.01)	−0.039	0.741	2.42×10^{-3}	−0.153	−0.033	2.622
(Ec = 0.001)	−0.093	0.176	1.59×10^{-3}	−0.043	0.033	3.272

b Inner bonded

P_c same as T2(FB), coefficient given below — T2(IB) [26]

κ	α_1	α_2	γ_1	γ_2	γ_3	ξ_1	ξ_2	ξ_3	ξ_4
0.9814	1.3922	0.083	0.712	0.962	2.827	0.202	0.041	−0.188	2.913

P_c same as T3(FB), coefficient given below — T3(IB) [27]

	a	b	c	d	e	f	g
(Ec = 0.1)	1.26	0.63	1.54	0.81	0.01	0.52	0.48
(Ec = 0.01)	1.45	2.14	0.14	0.66	4.85	−0.01	2.71
(Ec = 0.001)	1.29	2.33	0.03	0.6	0.01	−0.47	0.69

	h	i	j	k	l	m
(Ec = 0.1)	−1.25	0.75	4.02×10^{-2}	−0.06	−0.29	3.95
(Ec = 0.01)	−0.17	0.77	1.74×10^{-3}	−0.12	−0.21	2.8
(Ec= 0.001)	2.27	0.45	3.66×10^{-2}	−0.25	3.11	−0.46

Table A1. *Cont.*

Model Description	Symbol Ref.
c Inter−layer adhesion	
$P_c = E_p(\phi_1 + \phi_2 + \phi_3)$ *where :* $\phi_1 = h_1\left(\frac{t_1}{r_1}\right)^{h_2}\left(\frac{r_2}{r_1}\right)^{h_3}\left(\frac{t_2}{r_2}\right)^{h_4}\left(\frac{\sigma_1}{E_p}\right)^{h_5}$ $\phi_2 = h_6\left(\frac{t_1}{r_1}\right)^{h_7}\left(\frac{r_2}{r_1}\right)^{h_8}\left(\frac{t_2}{r_2}\right)^{h_9}\left(\frac{\sigma_1}{E_p}\right)^{h_{10}}\Delta_0^{h_{11}}$ $\phi_3 = h_{12}\left(\frac{t_1}{r_1}\right)^{h_{13}}\left(\frac{r_2}{r_1}\right)^{h_{14}}\left(\frac{t_2}{r_2}\right)^{h_{15}}\Delta_0^{h_{16}}$	HT (1.5) [28]

h_1	h_2	h_3	h_4	h_5	h_6	h_7	h_8
28.538	4.249	1.191	−0.028	0.341	0.744	−0.002	0.085
h_9	h_{10}	h_{11}	h_{12}	h_{13}	h_{14}	h_{15}	h_{16}
2.472	0.585	−0.176	9.54×10^{-6}	0.286	−3.071	−0.094	−0.359

P_c same as above, coefficient given below	HT (15) [28]

h_1	h_2	h_3	h_4	h_5	h_6	h_7	h_8
31.791	3.04	0.701	−0.126	0.915	0.059	0.163	−0.221
h_9	h_{10}	h_{11}	h_{12}	h_{13}	h_{14}	h_{15}	h_{16}
1.138	0.53	−0.136	7.29×10^{-6}	0.038	−3.993	−0.319	−0.077

3.   Core unbounded

Model Description	Symbol Ref.
$P_c = P_{crs} + \dfrac{E_c}{[2n(1-v_c)+2v_c-1]\{1+v_c\}}$	SP(UB) [17]

P_c same as T2(FB), coefficient given below   T2(UB) [26]

κ	α_1	α_2	γ_1	γ_2	γ_3	ξ_1	ξ_2	ξ_3	ξ_4
0.9833	1.106	−0.0945	0.336	0.966	3.631	0.1589	0.0184	−0.0837	3.01

P_c same as T3(FB), coefficient given below   T3(UB) [27]

	a	b	c	d	e	f	g
(Ec = 0.1)	3.42	3.26	−0.5	0.59	0.06	−0.15	−0.12
(Ec = 0.01)	0.77	2.63	−0.21	0.47	0.02	−0.04	−0.12
(Ec = 0.001)	2.14	2.7	−0.04	0.54	0.19	−0.06	0.27

	h	i	j	k	l	m
(Ec = 0.1)	3.12	0.3	0.00697	0.07	2.87	−1.53
(Ec = 0.01)	3.21	0.4	0.0766	−0.03	2.91	−0.27
(Ec = 0.001)	4.48	0.6	0.0675	−0.2	3.03	−0.21

Model Description	Symbol Ref.
$P_c = \theta_0 + \left[\theta_1\left(\frac{t_1}{r_1}\right)^{\left(\theta_5+\theta_6\frac{t_c}{r_c}\right)}(1-imp_1)^{\theta_7} + \theta_8\left(\frac{t_2}{r_2}\right)^{\left(\theta_9+\theta_{10}\frac{t_c}{r_c}\right)}(1-imp_2)^{\theta_{11}}\right]$ $\times (\sigma_p)^{\left(\theta_2\frac{t_1}{r_1}+\theta_3\frac{t_2}{r_2}+\theta_4\frac{t_c}{r_c}\right)} + \theta_{12}\left(\frac{t_c}{r_c}\right)^{\left(\theta_{13}+\theta_{14}\frac{t_c}{r_c}\right)}(1-imp_c)^{\theta_{15}}$	Yg(UB) [4]

θ_0	θ_1	θ_2	θ_3	θ_4	θ_5	θ_6	θ_7
−5.8433	7.65762	−0.0226	−0.0877	0.42328	3.240446	−0.6458	0.56457

θ_8	θ_9	θ_{10}	θ_{11}	θ_{12}	θ_{13}	θ_{14}	θ_{15}
14.115	0.72154	0.8829	0.36618	23.4687	4.2532	−0.7719	0.29961

Model Description	Symbol Ref.
$P_c = \left(\frac{r_2}{r_1}\right)\left(\frac{t_2}{r_2}\right)^{\left[3.66+\left(\frac{r_2}{r_1}\right)\right]} + 7.09\left(\frac{t_1}{r_1}\right)^{\left[1+\left(\frac{r_2}{r_1}\right)\right]}\max\left[\left(\frac{t_1}{r_1}\right),\left(\frac{r_2}{r_1}\right)\right]$ $+ \dfrac{705+59.5\left(\frac{t_2}{r_2}\right)+59.5\left(\frac{r_2}{r_1}\right)\left(\frac{\sigma_1}{E_p}\right)}{\left(\frac{t_1}{r_1}\right)+imp_2+6.53\left(\frac{r_2}{r_1}\right)+\left(\frac{t_1}{r_1}\right)imp_1} - 59.5$	EQ(UB) [4]

Appendix B

Table A2. Collapse test data of sandwich pipes [4,6,9,33].

SP	D_1	t_1	t_c	D_2	t_2	P_{exp} (MPa)
	203.2	2	23.4	152.4	2	37.68
	203.2	2	23.4	152.4	2	35.96
	203.2	2	23.4	152.4	2	35.23
	203.2	2	23.4	152.4	2	37.18
Steel pipe, Cementitious core	203.2	2	23.4	152.4	2	36.92
	219.08	2.77	22.63	168.28	2.77	38.71
	219.08	2.77	22.63	168.28	2.77	34.29
	219.08	2.77	22.63	168.28	2.77	37.57
	219.08	2.77	22.63	168.28	2.77	39.05
	74.97	1.62	11.29	49.15	1.62	43.35
	75.92	1.65	11.13	50.36	1.63	34.09
	62.16	1.47	4.23	50.76	1.68	10.98
Aluminium pipe, Polypropylene core	62.25	1.47	4.3	50.71	1.67	12.11
	75.4	1.62	11.26	49.64	1.68	37.64
	75.18	1.61	11.1	49.76	1.62	31.14
	62.1	1.46	4.62	49.94	1.7	20.31
	62.39	1.49	4.69	50.03	1.69	17.13
	69.44	3.34	5.5	50.96	3.19	4.9
Aluminium pipe, Polypropylene core	70	3	5.33	50.96	3.08	11.52
	70	3	5.33	50.96	4.08	13.1
	70	3	5.33	50.96	5.08	11.06
	202.8	2	23.2	152.4	1.8	30.5
Steel pipe, Cementitious core	203	2	23.2	152.6	1.8	30.6
	203	2	23.2	152.6	1.8	29.7

References

1. Bhardwaj, U.; Teixeira, A.P.; Guedes Soares, C. Reliability Assessment of a Subsea Pipe-in-Pipe System for Major Failure Modes. *Int. J. Press. Vessel. Pip.* **2020**, *188*, 104177. [CrossRef]
2. *Standard-ST-F101*; Submarine Pipeline Systems. DNV: Byrum, Norway, 2021; Det Norske Veritas Elendom AS.
3. Netto, T.A.; Santos, J.M.C.; Estefen, S.F. Sandwich Pipes for Ultra-Deep Waters. In Proceedings of the 2002 4th International Pipeline Conference, Calgary, AB, Canada, 29 September–3 October 2002; pp. 2093–2101. [CrossRef]
4. Yang, J.; Estefen, S.F.; Fu, G.; Paz, C.M.; Lourenço, M.I. Collapse Pressure of Sandwich Pipes with Strain-Hardening Cementitious Composite-Part 2: A Suitable Prediction Equation. *Thin-Walled Struct.* **2020**, *148*, 106606. [CrossRef]
5. Yang, J.; Paz, C.M.; Estefen, S.F.; Fu, G.; Lourenço, M.I. Collapse Pressure of Sandwich Pipes with Strain-Hardening Cementitious Composite-Part 1: Experiments and Parametric Study. *Thin-Walled Struct.* **2020**, *148*, 106605. [CrossRef]
6. An, C.; Duan, M.; Toledo Filho, R.D.; Estefen, S.F. Collapse of Sandwich Pipes with PVA Fiber Reinforced Cementitious Composites Core under External Pressure. *Ocean Eng.* **2014**, *82*, 1–13. [CrossRef]
7. Castello, X.; Estefen, S.F.; Leon, H.R.; Fritz, M. Collapse of Sandwich Pipes with Different Annular Materials. In Proceedings of the IBP-Rio Pipeline Conference & Exposition, Rio de Janeiro, Brazil, 2–4 October 2007. [CrossRef]
8. Alrsai, M.; Karampour, H.; Albermani, F. Numerical Study and Parametric Analysis of the Propagation Buckling Behaviour of Subsea Pipe-in-Pipe Systems. *Thin-Walled Struct.* **2018**, *125*, 119–128. [CrossRef]
9. Estefen, S.F.; Netto, T.A.; Pasqualino, I.P. Strength Analyses of Sandwich Pipes for Ultra Deepwaters. *ASME J. Appl. Mech.* **2005**, *72*, 599–608. [CrossRef]
10. Xu, Q.; Gong, S.; Hu, Q. Collapse Analyses of Sandwich Pipes under External Pressure Considering Inter-Layer Adhesion Behaviour. *Mar. Struct.* **2016**, *50*, 72–94. [CrossRef]
11. Castello, X.; Estefen, S.F. Adhesion Effect on the Ultimate Strength of Sandwich Pipes. In Proceedings of the 25th International Conference on Offshore Mechanics and Arctic Engineering, Hamburg, Germany, 4–9 June 2006. ASME Paper No. OMAE2006-92481.
12. Teixeira, A.P.; Zayed, A.; Guedes Soares, C. Reliability of Pipelines with Non-Uniform Corrosion. *J. Ocean Sh. Technol.* **2010**, *1*, 12–30.
13. Teixeira, A.P.; Guedes Soares, C.; Netto, T.A.; Estefen, S.F. Reliability of Pipelines with Corrosion Defects. *Int. J. Press. Vessel. Pip.* **2008**, *85*, 228–237. [CrossRef]

14. Bhardwaj, U.; Teixeira, A.P.; Guedes Soares, C.; Azad, M.S.; Punurai, W.; Asavadorndeja, P. Reliability Assessment of Thick High Strength Pipelines with Corrosion Defects. *Int. J. Press. Vessel. Pip.* **2019**, *177*, 103982. [CrossRef]
15. Teixeira, A.P.; Palencia, O.G.; Guedes Soares, C. Reliability Analysis of Pipelines with Local Corrosion Defects under External Pressure. *ASME J. Offshore Mech. Arct. Eng.* **2018**, *141*, 051601. [CrossRef]
16. Arjomandi, K.; Taheri, F. Elastic Buckling Capacity of Bonded and Unbonded Sandwich Pipes under External Hydrostatic Pressure. *J. Mech. Mater. Struct.* **2010**, *5*, 391–408. [CrossRef]
17. Sato, M.; Patel, M.H. Exact and Simplified Estimations for Elastic Buckling Pressures of Structural Pipe-in-Pipe Cross Sections under External Hydrostatic Pressure. *J. Mar. Sci. Technol.* **2007**, *12*, 251–262. [CrossRef]
18. Hashemian, R.; Mohareb, M. Buckling Finite Element Formulation for Sandwich Pipes under External Pressure. *Int. J. Press. Vessel. Pip.* **2016**, *147*, 41–54. [CrossRef]
19. Hashemian, R.; Mohareb, M. Finite Difference Model for the Buckling Analysis of Sandwich Pipes under External Pressure. *Ocean Eng.* **2016**, *122*, 172–185. [CrossRef]
20. Bhardwaj, U.; Teixeira, A.; Guedes Soares, C. Uncertainty in Collapse Strength Prediction of Sandwich Pipelines. *J. Offshore Mech. Arct. Eng.* **2022**, *144*, 041702. [CrossRef]
21. Pasqualino, I.P.; Pinheiro, B.C.; Estefen, S.F. Comparative Structural Analyses between Sandwich and Steel Pipelines for Ultra-Deep Water. In Proceedings of the ASME 2002 21st International Conference on Offshore Mechanics and Arctic Engineering, Oslo, Norway, 23–28 June 2002. ASME Paper No. OMAE2002-28455.
22. Castello, X.; Estefen, S.F.; Leon, H.R.; Chad, L.C.; Souza, J. Design Aspects and Benefits of Sandwich Pipes for Ultra Deepwaters. In Proceedings of the ASME 2009 28th International Conference on Ocean, Offshore and Arctic Engineering, Honolulu, HI, USA, 31 May–5 June 2009; Volume 3, pp. 453–459. [CrossRef]
23. Castello, X.; Estefen, S.F. Limit Strength and Reeling Effects of Sandwich Pipes with Bonded Layers. *Int. J. Mech. Sci.* **2007**, *49*, 577–588. [CrossRef]
24. Castello, X.; Estefen, S.F. Sandwich Pipes for Ultra Deepwater Applications. In Proceedings of the Offshore Technology Conference, Houston, TX, USA, 5–8 May 2008. [CrossRef]
25. Arjomandi, K.; Taheri, F. A New Look at the External Pressure Capacity of Sandwich Pipes. *Mar. Struct.* **2011**, *24*, 23–42. [CrossRef]
26. Arjomandi, K.; Taheri, F. Stability and Post-Buckling Response of Sandwich Pipes under Hydrostatic External Pressure. *Int. J. Press. Vessel. Pip.* **2011**, *88*, 138–148. [CrossRef]
27. Arjomandi, K.; Taheri, F. The Influence of Intra-Layer Adhesion Configuration on the Pressure Capacity and Optimized Configuration of Sandwich Pipes. *Ocean Eng.* **2011**, *38*, 1869–1882. [CrossRef]
28. He, T.; Duan, M.; Wang, J.; Lv, S.; An, C. On the External Pressure Capacity of Deepwater Sandwich Pipes with Inter-Layer Adhesion Conditions. *Appl. Ocean Res.* **2015**, *52*, 115–124. [CrossRef]
29. SMLH Eureqa Version 0.98 Beta. Available online: https://www.nutonian.com (accessed on 25 May 2022).
30. Melchers, R.E.; Beck, A.T. *Structural Reliability—Analysis and Prediction*; Melchers, R.E., Beck, A.T., Eds.; John Wiley & Sons Ltd.: Chichester, UK, 2018; ISBN 9781119266105.
31. Ditlevsen, O.; Madsen, H.O. *Structural Reliability Methods*; John Wiley & Sons Ltd.: Chichester, UK, 2005.
32. Ang, A.H.S.; Tang, W.H. *Probability Concepts in Engineering: Emphasis on Applications to Civil and Environmental Engineering*; John Wiley & Sons Inc.: Hoboken, NJ, USA, 2006.
33. Gong, S.; Wang, X.; Zhang, T.; Liu, C. Buckle Propagation of Sandwich Pipes under External Pressure. *Eng. Struct.* **2018**, *175*, 339–354. [CrossRef]
34. Blake, J.I.R.; Shenoi, R.A.; Das, P.K.; Yang, N. The Application of Reliability Methods in the Design of Stiffened FRP Composite Panels for Marine Vessels. *Ships Offshore Struct.* **2009**, *4*, 287–297. [CrossRef]

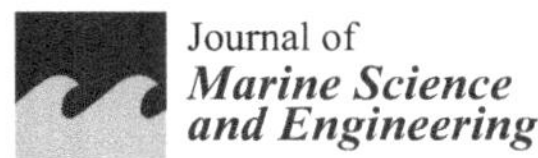

Journal of
Marine Science and Engineering

Article

Uncertainty in the Estimation of Partial Safety Factors for Different Steel-Grade Corroded Pipelines

Utkarsh Bhardwaj, Angelo Palos Teixeira * and C. Guedes Soares

Centre for Marine Technology and Ocean Engineering (CENTEC), Instituto Superior Técnico, Universidade de Lisboa, 1049-001 Lisbon, Portugal
* Correspondence: teixeira@centec.tecnico.ulisboa.pt

Abstract: This paper assesses the uncertainty of the partial safety factors for the design of corroded pipelines against burst failure due to the variability associated with the strength model selection. First, 10 calibrated burst pressure prediction models for corroded pipelines are adopted and duly categorised under low-, medium- and high-grade steel classes. The probabilistic characteristics of the pipe burst strength are studied using Monte Carlo simulation for the selected models. Model uncertainty factors are used to correct the burst pressure predictions by the design equations. Model strength factors are derived for models in each category that will provide coherent reliability. The first-order reliability method is employed to estimate the partial safety factors and their uncertainty as a function of operational time. Finally, the influence of the corrosion growth model on the evaluation of partial safety factors is assessed. The results obtained in this study can provide vital guidance regarding the design and maintenance of different steel-grade pipelines.

Keywords: partial safety factors; uncertainty analysis; first-order reliability method; burst pressure models; corroded pipelines; different steel grades

1. Introduction

The partial safety factor method is a design approach by which a target safety level is achieved and ensured by applying load and resistance factors to characteristic values of the governing variables. These governing variables consist of load effects and the resistance of the structure by virtue of its strength. The characteristic values of loads and resistance variables are selected as specific quantiles of their probability distributions. The provision of partial safety factors reflects the uncertainty in load effects and resistance and the target safety class. Structural reliability analysis methods are used in probabilistic design to evaluate or calibrate partial safety factors for a given design case.

The set of partial safety factors is an important element for the design and safer operation of pipelines for a prescribed time. Thus, the safety factor estimation of oil and gas corroded pipelines needs high-fidelity models and appropriate design methods. During operations, pipelines might face several uncertainties relative to degradation, which influence the calculation of safety factors. This implies that an appropriate partial safety factor is a key measure to ensure sufficient strength reserve from early design to the operational scenario. The partial safety factors account for pipe defects and other uncertainties; moreover, they also relate the maximum allowable pressure to the hydrostatic test pressure. The design standards-based assessment procedures are easy to apply, yet they may introduce large safety margins.

The design codes and standards lack a dedicated partial safety factor estimation method, although some supplementary explanations are provided [1,2]. On the other hand, investment costs and the safety of the pipeline's infrastructure are greatly influenced by partial safety factors. For example, design and maintenance requirements are inflated with high values of safety factors. On the other hand, low safety factors could underrate the

Citation: Bhardwaj, U.; Teixeira, A.P.; Guedes Soares, C. Uncertainty in the Estimation of Partial Safety Factors for Different Steel-Grade Corroded Pipelines. *J. Mar. Sci. Eng.* **2023**, *11*, 177. https://doi.org/10.3390/jmse11010177

Academic Editors: Erkan Oterkus and Cristiano Fragassa

Received: 25 November 2022
Revised: 20 December 2022
Accepted: 6 January 2023
Published: 10 January 2023

actual risks of pipeline failure. Therefore, a successful pipeline project requires meticulous estimation of partial safety factors. A sophisticated reliability analysis is admissible not only to determine the most practical partial safety factor but also to reflect the intrinsic uncertainties of the design and operational variables.

The design, construction, operation and maintenance of pipelines is regulated through pipeline integrity management approaches [3–6]. Through well-established management systems and operational procedures, such approaches vastly cover the issues related to pipeline defect identification, assessments of defect growth based on inline inspection and theoretical methods, and developing risk-based maintenance and repair strategies [1,2].

Time-dependent degradation processes such as corrosion are generally neglected in the probabilistic design of new structures. Conversely, corrosion degradation has a detrimental effect on pipeline integrity. Corrosion is an intricate phenomenon that has random characteristics and, consequently, it has attracted significant attention. Some significant studies in this context have been presented by Teixeira et al. [7–9].

The pipelines are susceptible to burst failure when undergoing corrosion and are internally pressurised. For the safety of corroded pipelines, accurate burst pressure prediction is the key factor and thereupon, strength prediction models have been proposed by using numerical, theoretical and experimental techniques [10]. In the past decade, several burst strength models have been developed for new generations of pipelines; however, they have not been specifically assessed for different steel grades [11].

Efficient pipeline systems with modified geometric and material properties have been used to provide a solution to the challenges posed by the oil and gas industries. As per the API 5L definition, Grade B, X42, X46, X52, X56, X60, X65, X70, X80, and X100 have been developed for oil and gas transmission at onshore, offshore and sour services [11,12].

Recent developments have intensified the fact that a corroded pipe's burst behaviour changes with the steel grade [12–17]. Previous studies [11,18] have implemented a comparative study on burst strength models and have provided recommendations for model selection. Burst pressure prediction models must provide an adequate safety margin for corroded pipelines in all steel grades [19]. Although the DNV code is the most extensively used recommended practice for corroded pipelines, the quality of other burst strength models cannot be neglected. More importantly, the adoption of a model concerning the pipe material grade is important and has been assessed in earlier studies. A recent study has provided an approach for structural reliability assessment involving calibrated strength models [20]. This approach has illustrated a framework to reduce the uncertainties associated with the strength model's selection. However, uncertainties in estimating partial safety factors have not been addressed.

Previous reliability studies [7,18,21–26] have incorporated the topic of pipe safety with the derivation of the reliability index (probability of failure), but the partial safety factor estimation for defective pipelines has not been paid much attention [27]. The DNV recommended practice [2] specifies safety factors to accomplish yearly reliability levels that are attuned to capture uncertainties associated with pipe design variables. However, the impact of pipe corrosion defects and uncertainties in calculating partial safety factors are not explicitly described.

Partial safety factors for structures undergoing corrosion are not frequently estimated and the literature is limited in this regard [28,29]. One recent study has proposed a safety factor estimation scheme based on risk factors for corroded pipelines and their range [29]. A standard known as HPI FFS has been developed by the high-pressure institute of Japan that includes the effect of metal loss in the safety factor estimation of pressure vessels [30]. Machida et al. [31] have evaluated partial safety factors for circumferential flawed pipes used in fitness-for-service assessments.

The probability of failure is a function of defects and other conditions; thus, it is necessary to estimate partial safety factors to clarify the relation between the probability of failure and defect impact. The adopted strength model is an important aspect that impacts the estimation of safety factors.

Following from the above discussion, the present study aims to calculate partial safety factors and assess prediction uncertainty for pressurised corroded pipelines using probabilistic concepts. A combination of Monte Carlo simulation and the First Order Reliability Method (FORM) is used in this paper. The primary objective of this paper is to assess the variability (or prediction uncertainty) in partial safety factors due to models adopted in each material category.

The remaining paper is arranged in the following way. The methodology is explained stepwise in Section 2. Using an appropriate case study, Section 3 discusses the obtained results. In particular, important findings in terms of partial safety factors are described in Section 3. A brief discussion is then provided. Finally, the conclusions of the conducted study are provided. The ultimate objective of the present study is to evaluate the variability in partial safety factors in the most practical sense within the context of time-varying corrosion in pipelines. The results may help to achieve suitable safety levels of corroded pipelines of different steel grades and consequently, improve design formats and optimise maintenance policies.

2. Methodology

The proposed methodology for the uncertainty assessment of the partial safety factors of corroded pipelines for burst failure is presented in this section. The burst capacity is the maximum pressure sustained by the pipeline before the plastic collapse. Burst capacity depends on the dimension, material of the pipe and corrosion defect parameters. The burst pressure prediction models adopted in this paper are presented in Appendix A.

Several studies have endorsed the idea of the classification of burst models with pipe material (grades). For example, Cosham et al. [32] and Amaya-Gómez et al. [33] have advocated for three material classes as low, medium and high grades. A similar three-material classification has been adopted in some recent studies pertaining to pipe reliability [34] and the uncertainty quantification of burst strength prediction [11]. This study also assumes the above-mentioned classification scheme.

The top 10 models are suggested in each category and model uncertainty factors are derived and characterised in order to acknowledge mathematical and methodological uncertainty [20]. Table 1 presents the top 10 models and statistical moments for the model uncertainty factors adopted from the previous study [20]. For any random variable, specifying the probability distribution bears a large importance. It has been revealed by statistical testing that the Frechet probability distribution provides the best fitting to the data [20]. Table 1 also shows the parameters of the Frechet distributions used in the analysis hereafter.

Table 1. Stochastic description of model uncertainty factor (X_m).

Rank	Low Steel Grade					Medium Steel Grade					High Steel Grade				
	Model	μ	SD	η	ρ	Model	μ	SD	η	ρ	Model	μ	SD	η	ρ
1	CUP	1.16	0.17	1.07	7.76	DNV F101	1.18	0.23	1.07	6.66	Ma et al. [13].	1	0.07	0.97	19.8
2	PCORRC	1.1	0.18	1.00	7.37	NETTO	1	0.22	0.91	6.65	SHELL-92	1.30	0.13	1.23	13.5
3	CHOI	1.24	0.21	1.13	7.31	CHLNG	1	0.22	0.91	6.65	ORYNYAK	1.07	0.07	1.03	20.0
4	NETTO	1	0.17	0.93	8.40	Z & L	1	0.22	0.91	6.65	PCORRC	1.04	0.08	1.00	17.0
5	Z & L	1	0.17	0.93	8.40	PCORRC	1.21	0.25	1.09	6.23	W & Z	1.10	0.10	1.05	14.5
6	Petrobras	1	0.17	0.93	8.40	Petrobras	1	0.22	0.91	6.65	NG-18	1.06	0.09	1.02	16.0
7	CPS	1.13	0.22	1.03	6.71	ORYNYAK	1.3	0.24	1.18	6.76	SIMS	1.07	0.07	1.04	20
8	CHLNG	1	0.17	0.93	8.40	CUP	1.18	0.22	1.07	6.81	Fitnet FFS	1.07	0.07	1.03	20
9	Ma et al. [13]	1	0.17	0.93	8.40	CSA	1.24	0.23	1.13	6.66	Z & L	1.00	0.07	0.97	19.8
10	DNV F101	1.1	0.2	1.00	6.83	Fitnet FFS	1.27	0.24	1.15	6.67	CHOI	1.12	0.08	1.08	18

μ—mean; SD—standard deviation; η and ρ—scale and shape parameters of the Frechet probability distribution.

First, the groups of random samples from basic variables related to burst strength are generated using stochastic models. From this generated set of samples, an output array of burst strength can be obtained. Further, the distribution type and corresponding parameters of burst strength can be obtained using appropriate statistical techniques.

Each model prediction (P_b) can be calibrated by multiplying a model uncertainty factor (X_m) with the basic model prediction, as shown below

$$P_b = X_m \cdot P \tag{1}$$

where P is the model prediction using the mathematical form given in Appendix A.

A limit state function is required to assess the structural reliability of a corroded pipeline. The limit state function defined in this paper is based on the bursting of the pipe, which happens when the internal operating pressure exceeds the pipe capacity. A generalised form of the limit state function $g(\mathbf{X})$ can be formed using the corrected model prediction with the operating pressure as

$$g(\mathbf{X}) = X_m \cdot P - P_o \tag{2}$$

where $\mathbf{X}$ is the vector of the basic random variables of the limit state function associated with the uncertainties. If $g(\mathbf{X})$ is less than or equal to zero, then the pipeline is assumed to have failed. Conversely, the pipe is safe in a region where $g(\mathbf{X})$ is greater than 0. Now, the failure probability can be described using a joint probability density function $f_\mathbf{X}(=$ of vector $\mathbf{X}$ as

$$P_f = \int_{g(\mathbf{X}) \leq 0} f_\mathbf{X}(\mathbf{x}) d\mathbf{x} \tag{3}$$

Evaluating P_f using Equation (3) is computationally arduous, as a multidimensional integration is required. Thus, approximation or simulation methods are employed to solve such problems. This paper utilises the First-Order Reliability Method (FORM) that assumes a linear approximation of the limit state function. This technique can efficiently estimate reliability with non-normally distributed random basic variables. FORM first transforms the original space of the variable into a normal Gaussian space and calculates the shortest distance from its origin point to the failure surface. The corresponding point in the failure surface is known as the design point ($\mathbf{u}^*$) and the shortest distance is the reliability index.

$$\mathbf{u}^* = \min\{\| \mathbf{u} \| \mid g(\mathbf{u}) = 0\} \text{ and } \beta = \| \mathbf{u}^* \| \tag{4}$$

The pipe failure probability (P_f) can be estimated from the reliability index (β) [35] by

$$P_f = \Phi(-\beta) \tag{5}$$

where Φ is the standard normal distribution function.

The design method uses safety factors as a margin of safety to compensate for any uncertainty in the load bearing capacity of the structure. The pipe design follows the load and safety factor design (LRFD) method, which involves partial safety factors as seen below:

$$R^c / \gamma_R \geq L^c \cdot \gamma_L \tag{6}$$

where γ_L and γ_R are the partial safety factors for the load ($L = P_o$) and resistance ($R = P_b = X_m \cdot P$), respectively. Generally, the load and resistance factors are relative to the type of load and resistance. A higher uncertainty in the load variables corresponds to higher load factors; conversely, a higher uncertainty in the resistance implies a lower resistance strength factor. In the present context, Equation (6) can be expressed for pipe failure criteria as:

$$X_m^c \cdot P^c / \gamma_R \geq P_o^c \cdot \gamma_L \tag{7}$$

Pipe resistance ($X_m \cdot P$) must be characterised for the estimation of partial safety factors. The stochastic modelling of X_m has already been discussed and presented in Table 1 [20]. Uncertainties can never be absolutely eliminated when estimating structural resistance and loading. The uncertainties in the prediction of resistance and load effects are accounted for via partial safety factors. The direct applicability of reliability algorithms in estimating the safety factors is dubious, given the number of models and several defect parameters. Thus, Monte Carlo simulations are used to characterise the burst strength of the pipeline.

When the probabilistic models for the load and resistance terms are known, it is easy to calculate partial safety factors using the following equation:

$$\gamma_L = P_o^* / P_o^c \quad \text{and} \quad \gamma_R = X_m^c \cdot P^c / (X_m^* \cdot P^*) \tag{8}$$

where *variable** is the value at the design point calculated using the FORM approach and *variable^c* is the specified characteristic values. The characteristic value can be obtained from the specified fractiles of the corresponding probability distribution function.

The integrity of operational offshore pipelines is assured by identifying hazards and adopting adequate inspection methods. For example, DNV-RP-F116 suggests some inspection methods for marine pipelines [36]. A reliability analysis is particularly vital when inspection results confirm the corrosion defect. Design codes specify the year until no defects are evidenced, known as year 0. However, this year is not the time from the first year of installation of the pipeline. Thus, T_o is the time of the latest inspection when a corrosion defect is reported.

This paper assumes linear corrosion growth with time and identifies partial safety factors as a function of time. Let d_r and l_r be the corrosion rate and T is the given time instance; the steady-state corrosion depth can be defined as

$$d(T) = d_o + d_r(T - T_o) \tag{9}$$

where d_o is the corrosion depth at T_o. Similarly, the steady-state corrosion length can be defined as

$$l(T) = l_o + l_r(T - T_o) \tag{10}$$

where l_o is the corrosion length at T_o. Using Equations (9) and (10) in the models' expression (Appendix A), the partial safety factors of a corroded pipe can be predicted for a future time.

3. Results

This section is dedicated to the implementation of the above-mentioned methodology to access uncertainty in the safety factors of corroded pipes for a given time. The detailed results are shown stepwise in the following discussion. Case studies of low-, medium- and high-grade pipelines are adopted. To perform a reliability analysis, it is necessary to account for the variability in random variables by providing probabilistic models.

Table 2 presents the pipe geometry and material as well as the defect parameters for three grades of pipes. The values for corrosion growth rates (d_r and l_r) are based on the previous study [24,37] with minor modifications. Notably, the real corrosion rate depends on various factors and is highly case-specific. Factors such as transporting fluid, electrolyte, environment, pipe material composition and operating conditions are the primary drivers for corrosion growth. Furthermore, the probabilistic models for the above-stated parameters are also shown in Table 2. A normal distribution is assumed for geometric variables, while the coefficient of variation (*COV*) is adopted from Teixeira et al. [7].

For intact pipelines, design codes prescribe a characteristic value of maximum operating pressure to be 72% of the burst capacity. The same assumption is applied to corroded pipelines. Material properties (yield and tensile strength) are assumed to be log-normally distributed with parameters estimated from their characteristic values [38]. For the loading parameter (operating pressure), the Gumbel distribution and COV are assumed from Teixeira et al. [7]. For the corrosion depth and length at T_o, Weibull and a log-normal distribution are adopted following Teixeira et al. [9]. The corrosion growth rates (d_r and l_r) are adopted to be normally distributed. The above-stated assumptions are considered in this paper to demonstrate the case under study.

Table 2. Stochastic models of random variables of the design equation.

Parameters	Low Steel Grade		Medium Steel Grade		High Steel Grade			Distribution
	μ	SD	μ	SD	μ	SD	COV	
D—Diameter (mm)	324	3.24	508	5.08	1320	13.2	0.01	Normal
t—Thickness (mm)	6.45	0.064	6.35	0.063	15.88	0.16	0.01	Normal
l_0—length of corrosion (mm)	100	57	100	57	300	171	0.57	Log-Normal
w—width of corrosion (mm)	50	25	50	25	100	50	0.50	Log-Normal
Y—Yield strength (MPa)	328.8	26.3	429.6	34.4	782.5	62.6	0.08	Log-Normal
T—Tensile strength (MPa)	456.6	36.53	672.5	53.8	803.3	64.3	0.08	Log-Normal
P_o—operating pressure (MPa)	9.1	0.64	7.44	0.52	13.06	0.91	0.07	Gumbel
d_0—depth of corrosion (mm)	1	0.17	1	0.17	3.6	0.61	0.17	Weibull
d_r—rate of radial corrosion (mm/yr)	0.2	0.02	0.2	0.02	0.2	0.02	0.10	Normal
l_r—rate of axial corrosion (mm/yr)	0.2	0.02	0.2	0.02	0.2	0.02	0.10	Normal
X_m—model uncertainty factor			Model dependent—(as shown in Table 1)					Frechet

μ—mean; *SD*—standard deviation; *COV*—coefficient of variation.

3.1. Modelling of Resistance

By using Monte Carlo simulation and substituting generated samples (10,000) of random variables for the models, the corresponding series of burst pressures are obtained for them. The Chi-squared testing approach is adopted here, with the results suggesting that a normal distribution is well suited for defining the distribution of burst pressure. A sample result is presented in Figure 1 with the fitting of distribution and Chi-squared test results [39]. For the sake of brevity, the statistical properties of the burst pressures for different models and different corrosion levels are not listed here. It should be noted that for the entire operational life and all models, the normal distribution is found to be suitable for describing the probabilistic distribution of *P*.

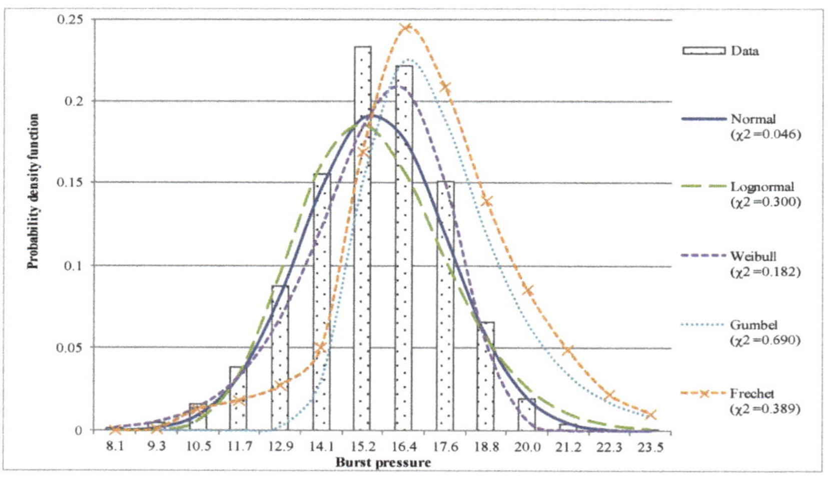

Figure 1. Distribution fitting of and Chi-squared test results for a sample resistance (burst strength) for the top model low-grade steel.

Further, this study investigates the effect of changing the *COV* of the random variables that constitute pipe burst resistance. As presented in Figure 2, the influence of uncertainty associated with geometric parameters such as diameter (a), thickness (b) and material parameters such as strength(c) on uncertainty in burst strength is considered separately. When changing the *COV* of diameter from 1 to 4, the maximum change in *P*'s uncertainty remains 1.79, 0.98 and 0.92 for low, medium and high grades. Changes of 1.83 (for LS), 0.94 (for MS) and 1.21 (for HS) in the *COV* of *P* are observed while changing the *COV* of thickness from 1 to 4. Strength (yield or tensile) is a key parameter that significantly impacts the resistance, and variations in its *COV* (6 to 10) exhibit a linearly increasing trend in the *COV* of burst strength for low, medium and high steel grades with 3.75, 3.59 and 3.82 variation in the *COV*, respectively.

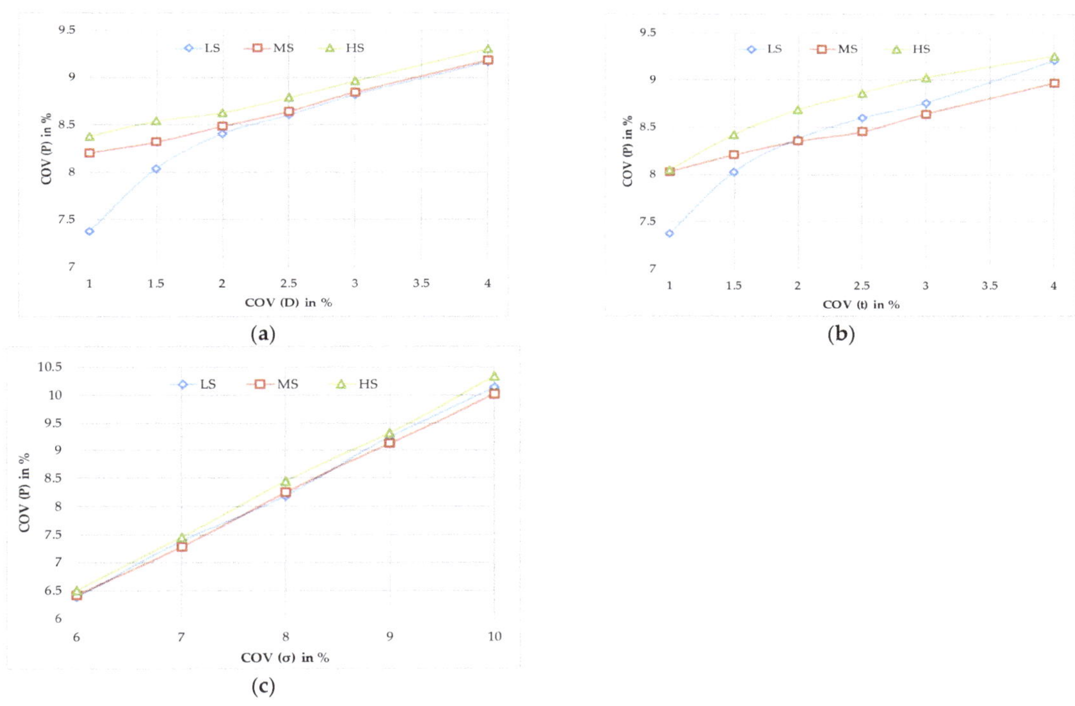

Figure 2. Effect of *COV* of (**a**)diameter, (**b**) thickness and (**c**) strength variables on the model burst strength (*P*) for the top model in each category.

3.2. Development of Model Correction Factors for the Model to Attain Target Reliability

The models' predictions and inherent uncertainty vary at every instance of time. Using FORM, the reliability indices are calculated for the 10 models in three categories at T_0 (intact pipe). As evident in Table 3, the reliability indices are subjective with respect to the models in each category and show appreciable scatter.

Table 3. Reliability indices (β) calculated using 10 top strength models in three steel categories at T_0.

Low-Grade Steel									
CUP	PCORRC	CHOI	NETTO	Z & L	Petrobras	CPS	CHLNG	Ma et al. [13].	DNV F101
2.51	2.29	2.23	2.19	2.20	2.14	2.20	2.37	2.16	2.05
Medium-Grade Steel									
DNV F101	NETTO	CHLNG	Z & L	PCORRC	Petrobras	ORYNYAK	CUP	CSA	Fitnet FFS
2.22	1.93	2.00	2.00	2.23	1.97	2.47	2.33	2.29	2.29
High-Grade Steel									
Ma et al. [13].	SHELL-92	ORYNYAK	PCORRC	W & Z	NG-18	SIMS	Fitnet FFS	Z & L	CHOI
2.25	2.26	2.39	2.29	2.41	2.19	2.68	2.03	2.57	2.42

This study assumes a target reliability level to be achieved by each model at this instance of time. To obtain a pragmatic value of the target reliability index, the β sample presented in Table 3 is gathered irrespective of material class. The normal distribution is fitted to the β dataset, and the 95-percentile value is selected (β_{target} = 2.47) as the target safety level.

To correct each model in such a way that each model predicts the same target reliability level, an additional strength factor is introduced. The model strength factor (X_r) for the reliability adjustment is introduced in the LSF as

$$g(\mathbf{X}) = X_r \cdot X_m \cdot P - P_o \tag{11}$$

Several iterations of reliability analysis are conducted with varying X_r values in the above equation. Table 4 presents the values of the factors (X_r) for the 10 models in three categories at T_o that provide the target reliability of 2.47 (=β_{target}). The values are slightly higher than 1 for most cases, with the highest correction factor being required for DNV F101 in low-grade, Petrobras in medium-grade, and NG-18 in high-grade steel. Only a few models have shown X_r less than 1.

Table 4. Strength factors (X_r) for the 10 top models in three categories at T_o.

Low-Grade Steel									
CUP	PCORRC	CHOI	NETTO	Z & L	Petrobras	CPS	CHLNG	Ma et al. [13].	DNV F101
0.95	1.05	1.16	1.07	1.06	1.11	1.07	1.02	1.05	1.22
Medium-Grade Steel									
DNV F101	NETTO	CHLNG	Z & L	PCORRC	Petrobras	ORYNYAK	CUP	CSA	Fitnet FFS
1.17	1.27	1.29	1.23	1.14	1.33	1	1.07	1.13	1.14
High-Grade Steel									
Ma et al. [13].	SHELL-92	ORYNYAK	PCORRC	W & Z	NG-18	SIMS	Fitnet FFS	Z & L	CHOI
1.05	1.05	1.04	1.05	1.01	1.06	0.97	1.10	0.99	1.01

This study further investigates X_r overtime by similarly targeting new reliability indices as identified above. However, the results are very close to those obtained in Table 4.

3.3. Development of Partial Safety Factors

Structural reliability methods, particularly FORM, have been extensively used for code calibration, i.e., for the evaluation of partial safety factors (PSFs) to achieve a uniform structural safety level. The calculation of PSFs is very much dependent on the prediction models specified in industrial practices. The PSFs are to be used with characteristic values to account for uncertainty and maintain a specific level of safety as described by the code. This paper focuses on reducing the subjectivity of design models from the perspective of PSFs and thus, the formal calibration of a PSF for a given target reliability is out of scope. The focus is rather on their uncertainty with respect to different steel grades of corroded pipes.

First, this study calculates PSFs for intact pipes of low, medium and high steel grades with the results shown in Table 5. FORM is employed for the design equation expressed in Equation (6), and the burst pressure adopts a normal distribution, as explained in Section 3.1, for all the models. The design points evaluated by FORM, together with the design variables' characteristic values, allow assessing the PSFs using Equation (8). Generally, a lower fractile is used for resistance and a higher fractile for the load term. In the present study, a 5% fractile for resistance and 95% fractile for load are used as characteristic values.

Table 5. PSFs of intact pipes of low, medium and high steel grades.

PSF	Low Grade Steel									
	CUP	PCORRC	CHOI	NETTO	Z & L	Petrobras	CPS	CHLNG	Ma et al. [13].	DNV F101
γ_L	0.910	0.909	0.908	0.909	0.909	0.909	0.907	0.909	0.909	0.908
γ_R	1.491	1.363	1.422	1.273	1.309	1.246	1.291	1.195	1.339	1.224
	Medium Grade Steel									
	DNV F101	NETTO	CHLNG	Z & L	PCORRC	Petrobras	ORYNYAK	CUP	CSA	Fitnet FFS
γ_L	0.891	0.891	0.891	0.890	0.890	0.891	0.891	0.891	0.891	0.891
γ_R	1.437	1.204	1.204	1.297	1.543	1.248	1.545	1.579	1.496	1.538
	High Grade Steel									
	Ma et al. [13].	SHELL-92	ORYNYAK	PCORRC	W & Z	NG-18	SIMS	Fitnet FFS	Z & L	CHOI
γ_L	0.929	0.920	0.932	0.928	0.922	0.924	0.930	0.932	0.929	0.930
γ_R	1.183	1.210	1.149	1.114	1.242	1.211	1.257	1.110	1.227	1.183

The factors γ_L and γ_R are calculated and shown in Table 5, using 10 models in low-, medium- and high-grade steel. The variation in γ_L values is negligible and remains around 0.909, whereas γ_R appreciably varies with the models. The most suitable γ_R may still be reckoned from the mean of 1.315. Similarly, for medium-grade steel (Table 5), the γ_L should be taken as 0.891, while 1.409 is the optimum value for γ_R. The models, however, offer higher uncertainties in the prediction of γ_R. The highest load safety factors are observed for high-strength pipes with a value of 0.928 (Table 5). With minimum uncertainty among the model's prediction of γ_R, a value of 1.189 can be prescribed for the high-grade pipe. In addition, the mean and uncertainty (COV) are given in Table 6. The uncertainty in γ_L is negligible for all three categories. The uncertainty in γ_R for medium-grade steel is the highest.

Table 6. Mean and COV of PSFs of intact pipes of low, medium and high steel grades.

PSF	Low Steel Grade		Medium Steel Grade		High Steel Grade	
	Mean	*COV*	Mean	*COV*	Mean	*COV*
γ_L	0.909	0.001	0.891	0.0004	0.928	0.004
γ_R	1.315	0.066	1.409	0.104	1.189	0.041

This study assesses the effect of corrosion degradation on partial safety factors. The same approach as described above in this section (intact pipes) is adopted to estimate partial safety factors for time-dependent corrosion degradation. The partial safety factors are derived in each steel grade for the top 10 models. The results also provide limits of partial safety factors corresponding to the elapsed life of the pipeline after the first detection of corrosion.

3.3.1. Low-Grade Steel Pipes

As described above, the γ_R shows greater variation with models, so it is only presented first in the present set of analyses. Comparative plots are constructed in Figure 3a to study the influence of corrosion evolution on the γ_R for the low-grade category. Commonly, a higher corrosion level corresponds to a lower partial safety factor of resistance. CUP and NETTO models are responsible for the upper bound of safety factors, while CHLNG and DNV F101 models dictate the lower bound. Z & L, Petrobras and Ma et al. [13]. provide a closer prediction of γ_R with elapsed time.

The results are further presented in three sets: mean, upper and lower limits in Figure 3b with uncertainty in PSF predictions. The mean γ_R varies from 1.24 to 0.646 over 16 years. The most conservative and non-conservative estimates of safety factors are also shown. The uncertainties (COV) become appreciable after the 8th year.

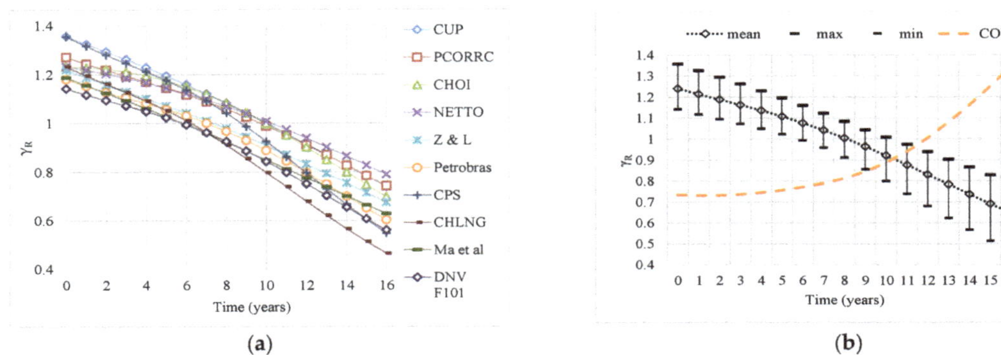

(a)   **(b)**

Figure 3. (**a**) PSF (γ_R) for low-grade steel as a function of elapsed life and (**b**) mean, range and uncertainty of PSF.

3.3.2. Medium-Grade Steel Pipes

Here the analyses are conducted using medium-grade steel models and the results (γ_R vs. time) are plotted in Figure 4. In particular, the models' evaluation of γ_R is presented in Figure 4a and their cumulative results, such as mean and *COV*, are shown in Figure 4b. In this material class, PCORRC followed by CUP always realises the non-conservative calculation of safety factors while CHLNG shows the most discrete behaviour, calculating the minimum γ_R over the operational life. ORYNYAK, CSA and Fitnet FFS have similar estimations as characterised by NETTO, Z & L and Petrobras with the same pattern of variations. Figure 4b shows a similar behaviour as described above for low-strength models in Figure 3b. This figure is focused on the uncertainty caused due to different models; the uncertainty observes a rising trend while passing year 9. It is also revealed that the γ_R varies from 1.37 to 0.71 over 16 years.

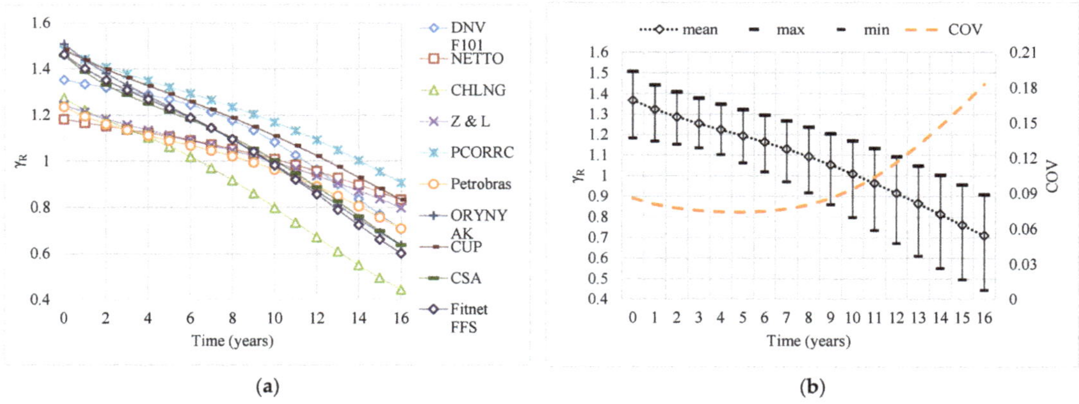

(a)   **(b)**

Figure 4. (**a**) PSF (γ_R) for medium steel grade as a function of elapsed life and (**b**)mean, range and uncertainty of PSF.

3.3.3. High-Grade Steel Pipes

Figure 5 shows the partial safety factors calculated for high-grade steel models. As revealed in Figure 5a, the trends of all the models can be recognised as linear variations. The Z & L followed by the W & Z models specify higher values of γ_R for the elapsed life of corroded pipes. The γ_R obtained by the traditional model, NG-18, represents the most conservative results. The dispersion among the models' predictions is significant and as time elapses, tends to disperse slightly more. The same is better reflected and depicted in Figure 5b. The prediction uncertainties are high and tend to increase with time, whereas the mean γ_R varies from 1.02 to 0.83.

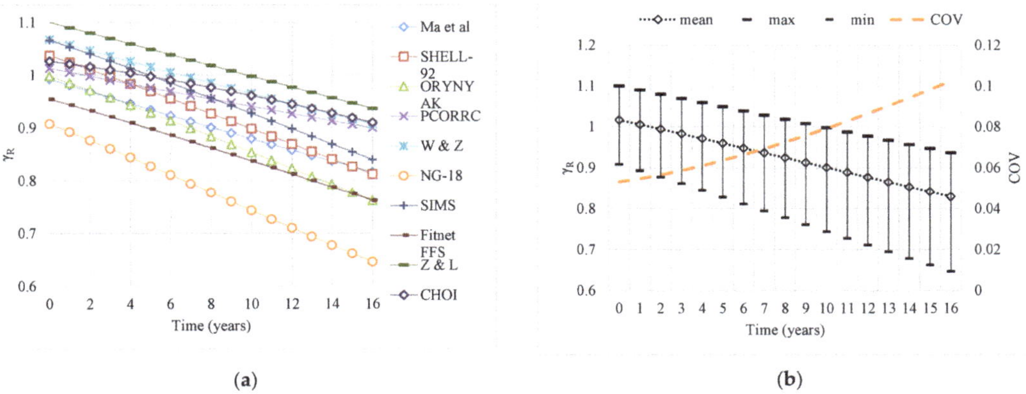

Figure 5. (**a**) PSF (γ_R) for high-grade steel as a function of elapsed life and (**b**) mean, range and uncertainty of PSF.

This study further estimates the load safety factors (γ_L) using the top 10 models in each category in a similar manner as conducted above. It is found that the uncertainty in the prediction of γ_L is negligible and remains around 0.001 (COV) in all three categories. Therefore, the γ_L values predicted by different models are not shown here. Figure 6 presents the γ_L as a function of time in three material categories. Despite γ_L being a vital variable, it can be understood from the figure that it is nearly constant with time for low- and medium-grade models. However, for high-grade pipe models, a little drop in γ_L is evident.

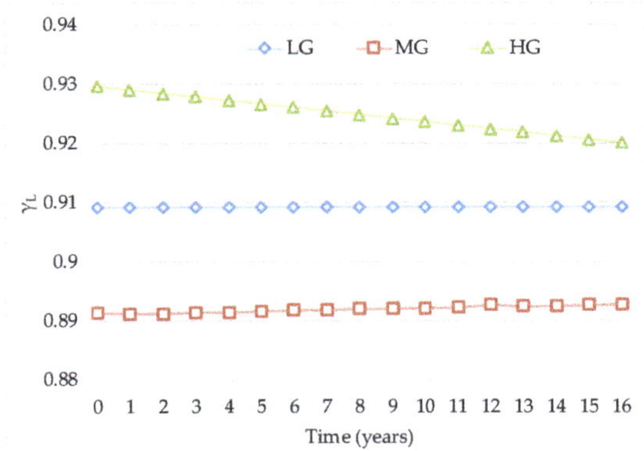

Figure 6. PSF (γ_L) for the load as a function of time.

3.4. Effect of Corrosion Growth Models

Corrosion is the primary threat to oil and gas pipelines and understanding its behaviour over time is vital for their design and adequate maintenance. Earlier in the study, the corrosion is assumed to grow linearly as presented mathematically in Equations (9) and (10). However, there are other models [40] and approaches [22] that suggest non-linear corrosion growth in pipes. The linear corrosion model is regarded as CM1 in this section; two more corrosion models are adopted to understand their effect on the estimation of partial safety factors. A non-linear trend has been proposed by Guedes Soares and Garbatov [41] that characterises corrosion as growing exponentially with time, given as:

$$
\begin{aligned}
d(T) &= d_\infty \left(1 - e^{-\left(\frac{T - T_0 - \tau_c}{\tau_l} \right)} \right) & T &> \tau_c \\
d(T) &= 0 & T &\leq \tau_c \\
l(T) &= l_\infty \left(1 - e^{-\left(\frac{T - T_0 - \tau_c}{\tau_l} \right)} \right) & T &> \tau_c \\
l(T) &= 0 & T &\leq \tau_c
\end{aligned}
\tag{12}
$$

where τ_c is the total time for which anti-corrosion protection lasts, τ_l is the transition time of the corrosion phenomenon, which is expressed as a time from the beginning up to 63% of the maximum thickness (d_∞). This study refers to this model as CM2. The length of corrosion is also followed to assume the same growth model.

In actual situations, the corrosion growth rate may vary significantly. For instance, corrosion growth rates are very high compared to common corrosion and, thus, are responsible for especially localised deep holes in pipe surfaces. Romanoff [42] has provided a model that depicts the corrosion process following a power law with time to consider pitting corrosion growth; the same is referred to as CM3 in the present study given by

$$
\begin{aligned}
d(T) &= d_o + k(T - T_o)^m \\
l(T) &= l_o + k(T - T_o)^m
\end{aligned}
\tag{13}
$$

where k is the pitting proportionality constant and m is the exponent factor; both are material constants.

To compare CM1, CM2 and CM3, some assumptions are required. The parameters are adjusted in the corrosion models so that they deliver an equal reduction in thickness over 16 years. Table 7 presents the values of these parameters. However, some additional assumptions are made, such as no corrosion thickness allowance and zero initial depth at T_o.

Table 7. Parameters of the different corrosion models.

Corrosion Growth Models	Equation	Parameter		Units
CM2	12	d_∞	4.6	mm
		l_∞	4.6	mm
		τ_l	13.45	years
CM3	13	k	0.374	
		m	0.774	

Figure 7 presents the adopted corrosion growth models CM1, CM2 and CM3, depicting equal reduction over 16 years for a low-grade corroded pipe sample case. The trends by the three models differ significantly until 16 years. Other cases also show the same level of reduction. Moreover, the length of corrosion for all the cases shows similar corrosion growth.

Next, partial safety factors are estimated using the above-stated corrosion models with the parameter values given in Table 7. The best model from each category is used and the influence of corrosion over time is shown in Figure 8 (years). The safety factors of resistance show a distinct pattern corresponding to CM1, CM2 and CM3. In all the categories CM1 always estimates higher values of safety factors, followed by CM3 and CM2.

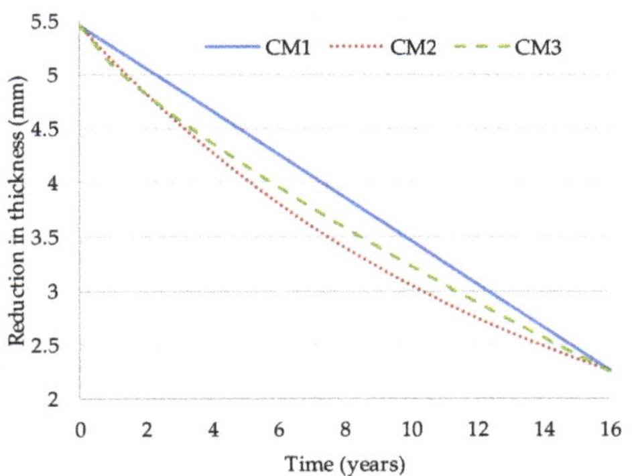

Figure 7. Corrosion-induced reduction in thickness provided by the three models as a function of time (years).

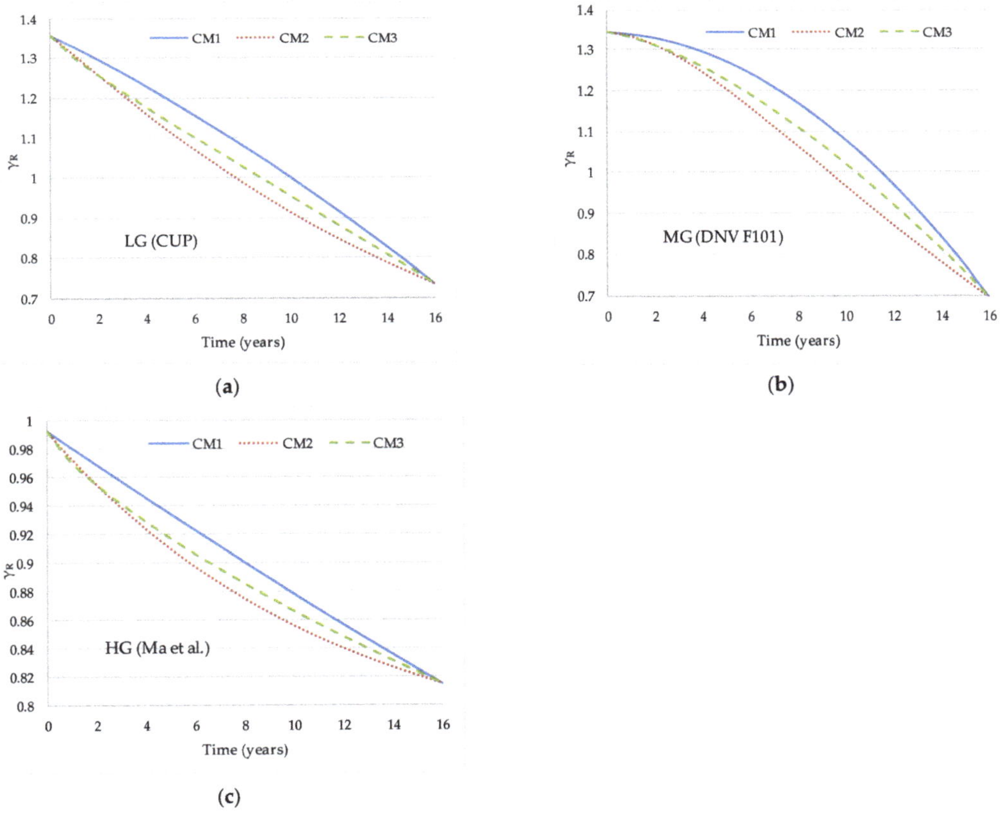

(a)

(b)

(c)

Figure 8. Influence of corrosion growth models on partial safety factors (years) for (**a**) low-grade, (**b**) medium-grade, and (**c**) high-grade steel categories.

4. Discussion

Pipeline integrity management relies on proper design and code formulations for an adequate decision-making process. This requires adequate strength models that are deterministically calibrated with test data and code formats and probabilistically calibrated to account for uncertainty in the design variables in order to achieve uniform levels of structural safety. The safety factor calculations are probabilistic in nature, subjective to the choice of strength models, and thus associated with uncertainties. The pipelines, once installed, remain in service for many decades and are exposed to natural degradation processes such as corrosion. Therefore, this paper seeks the assessment of uncertainty in safety factor estimation for such pipelines using an appropriate case study. It is demonstrated how strength models and corrosion growth influence the calculation of partial safety factors and their uncertainty. This study adopts a uniform corrosion growth model (CM1) for depth and length. However, the overall corrosion growth will not be linear. Similarly, the overall corrosion growth for other models (CM2 and CM3) may not follow the same idealisation of growth as adopted for a single parameter.

The partial safety factors are estimated in three classes of steel grades; however, they are associated with the case study adopted. Therefore, the obtained results are a function of geometrical parameters, defect parameters and model uncertainty factors. A systematic approach covering a wide range of design cases is still necessary to properly assess partial safety factors regarding the burst strength design of corroded pipelines of different steel grades.

5. Conclusions

The accurate estimation of partial safety factors is of the utmost importance in mitigating the risk of pipeline burst failure and making any pipeline project successful. The present method is mainly useful to increase the confidence in safety and decision regarding the maintenance investment of different steel-grade pipes.

First, a stepwise approach is proposed to properly account for the uncertainties involved in the partial safety factor prediction of corroded pipelines in three steel grades. From the literature, the top 10 calibrated burst models are adopted in low, medium and high steel grades of pipes. Moreover, model uncertainty factors for burst models are adopted with their statistical moments. Limit state function and load and resistance factor design equations are formulated using model-uncertainty-factor-calibrated burst strength models and operating pressures. Case studies for corroded pipelines of three steel grades are developed with stochastic properties based on the literature.

The probabilistic characteristics of the burst strength of all the models are described using Monte Carlo simulation and the normal distribution is selected based on Chi-squared test results. Corrosion is assumed to grow linearly with time and its effect on the estimation of partial safety factors is assessed. The First-Order Reliability Method is employed to estimate the partial safety factors for the intact and corroded pipes using the top 10 models in each steel grade. Model strength factors are derived at a time instance for all the selected models to attain the same target reliability.

Moreover, the approach is refined by showing their mean, range and uncertainty. Nevertheless, the grade of intact pipe partial safety factors of a load assumes identical values with respect to the model, and 0.909, 0.891 and 0.928 for low, medium and high grades, respectively. Conversely, the resistance partial safety factors constitute high uncertainty regarding the models used, with means of 1.315, 1.409 and 1.189 for low, medium and high steel grades, respectively. For corroded pipelines, resistance safety factors are presented as a function of the time since the inspection. The mean and coefficient of variation of resistance factors in three categories are shown. The mean varies from 1.24 to 0.65 for low, 1.37 to 0.7 for medium and 1.02 to 0.83 for high grades for a 16-year operational period.

The effect of linear, exponential and power law corrosion growth models on the variation of partial safety factors is assessed with time. Some obvious differences in partial safety factors are evidenced corresponding to the corrosion growth model. Particularly, higher values of safety factors are obtained using the linear corrosion growth model.

Author Contributions: Conceptualisation, U.B., A.P.T. and C.G.S.; methodology, U.B. and A.P.T.; software, U.B.; validation, U.B.; formal analysis, U.B.; investigation, U.B.; resources, A.P.T. and C.G.S.; data curation, U.B. and A.P.T.; writing—original draft preparation, U.B.; writing—review and editing, A.P.T. and C.G.S.; visualisation, U.B.; supervision, A.P.T. and C.G.S.; project administration, C.G.S.; funding acquisition, C.G.S. All authors have read and agreed to the published version of the manuscript.

Funding: This work contributes to the project "Cementitious cork composites for improved thermal performance of pipelines for ultradeep waters—SUPBSEAPIPE, with the reference no. POCI-01-0145-FEDER-031011 funded by European Regional Development Fund (FEDER) through COMPETE2020—Operational Program Competitiveness and Internationalization (POCI) and with financial support from the Portuguese Foundation for Science and Technology (Fundação para a Ciência e Tecnologia—FCT), through national funds. This study also contributes to the Strategic Research Plan of the Centre for Marine Technology and Ocean Engineering, which is financed by Portuguese Foundation for Science and Technology (Fundação para a Ciência e Tecnologia—FCT), under contract UIDB/UIDP/00134/2020.

Institutional Review Board Statement: Not applicable.

Informed Consent Statement: Not applicable.

Data Availability Statement: Not applicable.

Conflicts of Interest: The authors declare no conflict of interest.

Appendix A Burst Strength Models of Corroded Pipes

Table A1. Burst Strength Models of Corroded Pipes.

1	NG-18 [43]	$P = 2.2\left(\sigma_y + 68.95\right)\dfrac{t}{D}\left[\dfrac{1 - \dfrac{d}{t}}{1 - \dfrac{d}{tM}}\right]$ $\quad M = \sqrt{1 + 2.51\left(\dfrac{L}{2\sqrt{Dt}}\right)^2 - 0.54\left(\dfrac{L}{2\sqrt{Dt}}\right)^4}$
2	SIMS [44]	$P = 2.22\sigma_y\dfrac{t}{D}\left[\dfrac{1 - \dfrac{d}{t}}{1 - \dfrac{d}{tM}}\right]$ $\begin{cases} for\ w > 6d + 0.1D, & M = \sqrt{1 + 2.5\dfrac{L^2}{Dt}} \\ for\ w \le 6d + 0.1D, & M = \sqrt{1 + 0.8\dfrac{L^2}{Dt}} \end{cases}$
3	CSA [45]	$P = 1.8\sigma_t\dfrac{t}{D}\left[\dfrac{1 - \dfrac{d}{t}}{1 - \dfrac{d}{tM}}\right]$ $\begin{cases} M = \sqrt{1 + 0.6275\dfrac{L^2}{Dt} - 0.003375\left(\dfrac{L^2}{Dt}\right)^2} & for\ \dfrac{L^2}{Dt} \le 50 \\ M = 3.3 + 0.032\dfrac{L^2}{Dt} & for\ \dfrac{L^2}{Dt} > 50 \end{cases}$
4	SHELL-92 [46]	$P = 1.8\sigma_t\dfrac{t}{D}\left[\dfrac{1 - \dfrac{d}{t}}{1 - \dfrac{d}{tM}}\right]$ $\quad M = \sqrt{1 + 0.805\dfrac{L^2}{Dt}}$
5	DNV F101 [2]	$P = 1.8\sigma_t\dfrac{t}{(D - t)}\left[\dfrac{1 - \dfrac{d}{t}}{1 - \dfrac{d}{tM}}\right]$ $\quad M = \sqrt{1 + 0.31\dfrac{L^2}{Dt}}$
6	PCORRC [47]	$P = 2\sigma_t\dfrac{t}{D}\left[1 - \dfrac{d}{t}\left\{1 - \exp\left(-0.157\dfrac{L}{\sqrt{D(t - d)/2}}\right)\right\}\right]$
7	Petrobras [48]	$P = 2\sigma_t\dfrac{t}{(D - t)}\left[\dfrac{1 - \dfrac{d}{t}}{1 - \dfrac{d}{tM}}\right]$ $\quad M = \sqrt{1 + 0.217\dfrac{L^2}{Dt} + \dfrac{1}{1.15 \times 10^6}\left(\dfrac{L^2}{Dt}\right)^2}$

Table A1. *Cont.*

#		Equation
8	CPS [49]	$$P = P_{LG} + g(P_{PP} - P_{LG})$$ $$P_{PP} = 0.9\left(\frac{E\sigma_y^{m-1}}{\sqrt{3}\alpha m}\right)^{1/m}\frac{4}{\sqrt{3}(D-t)}\frac{t}{\left[exp\left(\frac{1}{2m}\right)\right]^2}$$ $$P_{LG} = \frac{4\sigma_t}{\sqrt{3}(D-t)}(t-d)\exp\left(-\sqrt{\frac{3}{4}}\varepsilon_{crit}\right)$$ $$g = \frac{4tan^{-1}\left[exp\left(-\frac{L}{2\sqrt{D(t-d)}}\right)\right]}{\pi}$$
9	CHOI [14]	$$P = \begin{cases} 1.8\sigma_t\frac{t}{D}\left[C_2\left(\frac{L}{\sqrt{Dt/2}}\right)^2 + C_1\left(\frac{L}{\sqrt{Dt/2}}\right) + C_0\right] & for\ \frac{L}{\sqrt{Dt/2}} < 6 \\ 2\sigma_t\frac{t}{D}\left[C_4\left(\frac{L}{\sqrt{Dt/2}}\right) + C_3\right] & for\ \frac{L}{\sqrt{Dt/2}} \geq 6 \end{cases}$$ $$C_0 = 0.06\left(\frac{d}{t}\right)^2 - 0.1035\left(\frac{d}{t}\right) + 1$$ $$C_1 = -0.6913\left(\frac{d}{t}\right)^2 + 0.4548\left(\frac{d}{t}\right) - 0.1447$$ $$C_2 = 0.1163\left(\frac{d}{t}\right)^2 - 0.1053\left(\frac{d}{t}\right) + 0.0292$$ $$C_3 = -0.9847\left(\frac{d}{t}\right) + 1.1101$$ $$C_4 = 0.0071\left(\frac{d}{t}\right) - 0.0126$$
10	Z&L [50]	$$P = \frac{4\sigma_t}{(\sqrt{3})^{n+1}}\frac{t}{D}\left[1 - \frac{d}{t}\left\{1 - \exp\left(-0.157\frac{L}{\sqrt{D(t-d)/2}}\right)\right\}\right]$$
11	NETTO [51]	$$\frac{P}{P_{in}} = \left[1 - 0.9435\left(\frac{d}{t}\right)^{1.6}\left(\frac{l}{D}\right)^{0.4}\right]$$
12	Fitnet FFS [52]	$$P = 2\sigma_t\frac{t}{D-t}\left(\frac{1}{2}\right)^{\frac{65}{\sigma_y}}\left[\frac{1-\frac{d}{t}}{1-\frac{d}{tM}}\right] \quad M = \sqrt{1 + 0.8\frac{L^2}{Dt}}$$
13	ORYNYAK [53]	$$P = 2\sigma_t\frac{t}{D}\left[\frac{1 + \frac{L^2}{Dt}\left(1 - \frac{d}{t}\right)\frac{d}{t}}{1 + \frac{L^2}{Dt}\left(\frac{d}{t}\right)}\right]$$
14	Ma et al. [13]	$$P = \frac{4\sigma_t}{(\sqrt{3})^{\frac{m+1}{m}}}\frac{t}{D}\left[1 - \frac{d}{t}\left\{1 - 0.7501\exp\left(-0.4174\frac{L}{\sqrt{Dt}}\right)\left(1 - \frac{d}{t}\right)^{-0.1151}\right\}\right]$$
15	W&Z [54]	$$\frac{P}{P_{in}} = \begin{cases} 1 - 0.886\left(\frac{d}{t}\right)^1\left(\frac{l}{D}\right)^{0.3} & if\ D < 610\ mm \\ 1 - 1.12\left(\frac{d}{t}\right)^{1.15}\left(\frac{l}{D}\right)^{0.3} & if\ D \geq 610\ mm \end{cases}$$

Table A1. *Cont.*

16	CHLNG [15]	$$P = 2\sigma_t \frac{t}{(D-t)}\left[C_0 + C_1\left(\frac{L}{\sqrt{Dt}}\right) + C_2\left(\frac{L}{\sqrt{Dt}}\right)^2\right]\left[G_0 + G_1\left(\frac{2w}{\pi D}\right) + G_2\left(\frac{2w}{\pi D}\right)^2\right]$$ $$\text{for } L < \sqrt{20Dt}\begin{cases}C_0 = 0.8816 + 0.7942\left(\frac{d}{t}\right) - 0.05329\left(\frac{d}{t}\right)^2\\ C_1 = 0.03982 - 0.3946\left(\frac{d}{t}\right) - 0.1901\left(\frac{d}{t}\right)^2\\ C_2 = -0.0044248 + 0.02983\left(\frac{d}{t}\right) + 0.03091\left(\frac{d}{t}\right)^2\\ G_0 = 1.065 - 0.2992\left(\frac{d}{t}\right) - 0.248\left(\frac{d}{t}\right)^2\\ G_1 = 0.06604 + 0.7039\left(\frac{d}{t}\right) - 2.027\left(\frac{d}{t}\right)^2\\ G_2 = -0.000185 - 1.211\left(\frac{d}{t}\right) + 2.356\left(\frac{d}{t}\right)^2\end{cases}$$ $$\text{for } L \geq \sqrt{20Dt}\begin{cases}C_0 = 1.061 - 0.4754\left(\frac{d}{t}\right) - 0.5692\left(\frac{d}{t}\right)^2\\ C_1 = 0.03102 - 0.1621\left(\frac{d}{t}\right) + 0.1343\left(\frac{d}{t}\right)^2\\ C_0 = -0.002118 + 0.009434\left(\frac{d}{t}\right) - 0.006719\left(\frac{d}{t}\right)^2\\ G_0 = G_1 = G_2 = 0\end{cases}$$
17	CUP [23]	$$P = 2\sigma_t \frac{t}{D}\left[1 - \frac{d}{t}\left\{1 - \left\langle 0.1075\left(1 - \left(\frac{w}{\pi D}\right)^2\right)^6 + 0.8925\exp\left(-0.4103\frac{L}{\sqrt{Dt}}\right)\right\rangle\left(1 - \frac{d}{t}\right)^{0.2504}\right\}\right]$$

D—maximum diameter of pipe; t—maximum pipe thickness; d—maximum depth of corrosion; L—length of corrosion; w—width of corrosion; M—Follias factors; σ_t—ultimate tensile strength; σ_y—yield strength; $SMYS$—specified minimum yield strength; $SMTS$—specified minimum tensile strength; E—Young's modulus; ε—critical strain; n—strain hardening coefficient (based on stress–strain power law); α & m—Ramberg–Osgood material constant.

References

1. ASME B31G. *Manual for Determining the Remaining Strength of Corroded Pipelines. A Supplement to ANSI/ASME B31 Code for Pressing Piping*; Revision of ASME B31G-2009; American Society of Mechanical Engineers (ASME): New York, NY, USA, 2012.
2. DNV. *Corroded Pipelines—Recommended Practice RP-F101*; Det Norske Veritas: Bærum, Norway, 2019; ISBN 978-0-08-044566-3.
3. Cosham, A.; Hopkins, P. The Assessment of Corrosion in Pipelines–Guidance in the Pipeline Defect Assessment Manual (PDAM). In Proceedings of the 4th International Pipeline Technology Conference, Oostende, Belgium, 9–13 May 2004.
4. Da Cunha Bisaggio, H.; Netto, T.A. Predictive Analyses of the Integrity of Corroded Pipelines Based on Concepts of Structural Reliability and Bayesian Inference. *Mar. Struct.* **2015**, *41*, 180–199. [CrossRef]
5. Pinheiro, B.; Guedes Soares, C.; Pasqualino, I. Generalized Expressions for Stress Concentration Factors of Pipeline Plain Dents under Cyclic Internal Pressure. *Int. J. Press. Vessel. Pip.* **2019**, *170*, 82–91. [CrossRef]
6. Vosooghi, N.; Sriramula, S.; Ivanović, A. Response Surface Based Reliability Analysis of Critical Lateral Buckling Force of Subsea Pipelines. *Mar. Struct.* **2022**, *84*, 103246. [CrossRef]
7. Teixeira, A.P.; Guedes Soares, C.; Netto, T.A.; Estefen, S.F. Reliability of Pipelines with Corrosion Defects. *Int. J. Press. Vessel. Pip.* **2008**, *85*, 228–237. [CrossRef]
8. Teixeira, A.P.; Guedes Soares, C.; Wang, G. Probabilistic Modelling of the Ultimate Strength of Ship Plates with Non-Uniform Corrosion. *J. Mar. Sci. Technol.* **2013**, *18*, 115–132. [CrossRef]
9. Teixeira, A.P.; Palencia, O.G.; Guedes Soares, C. Reliability Analysis of Pipelines with Local Corrosion Defects under External Pressure. *ASME J. Offshore Mech. Arct. Eng.* **2018**, *141*, 051601. [CrossRef]
10. Zhu, X.K.; Leis, B.N. Evaluation of Burst Pressure Prediction Models for Line Pipes. *Int. J. Press. Vessel. Pip.* **2012**, *89*, 85–97. [CrossRef]
11. Bhardwaj, U.; Teixeira, A.P.; Guedes Soares, C. Quantification of the Uncertainty of Burst Pressure Models of Corroded Pipelines. *Int. J. Press. Vessel. Pip.* **2020**, *188*, 104208. [CrossRef]
12. Chauhan, V.; Crossley, J. *Corrosion Assessment Guidance for High Strength Steels*; R9017; GL Industrial Services: Loughborough, UK, 2009.
13. Ma, B.; Shuai, J.; Liu, D.; Xu, K. Assessment on Failure Pressure of High Strength Pipeline with Corrosion Defects. *Eng. Fail. Anal.* **2013**, *32*, 209–219. [CrossRef]
14. Choi, J.B.; Goo, B.K.; Kim, J.C.; Kim, Y.J.; Kim, W.S. Development of Limit Load Solutions for Corroded Gas Pipelines. *Int. J. Press. Vessel. Pip.* **2003**, *80*, 121–128. [CrossRef]

15. Su, C.-L.; Li, X.; Zhou, J. Failure Pressure Analysis of Corroded Moderate-to-High Strength Pipelines. *China Ocean Eng.* **2016**, *30*, 69–82. [CrossRef]

16. Chen, Z.; Yan, S.; Ye, H.; Shen, X.; Jin, Z. Effect of the Y/T on the Burst Pressure for Corroded Pipelines with High Strength. *J. Pet. Sci. Eng.* **2017**, *157*, 760–766. [CrossRef]

17. Kuanhai, D.; Yang, P.; Bing, L.; Yuanhua, L.; Jiandong, W. Through-Wall Yield Ductile Burst Pressure of High-Grade Steel Tube and Casing with and without Corroded Defect. *Mar. Struct.* **2021**, *76*, 102902. [CrossRef]

18. Bhardwaj, U.; Teixeira, A.P.; Guedes Soares, C. Reliability Assessment of Corroded Pipelines with Different Burst Strength Models. In *Developments in Maritime Technology and Engineering*; Guedes Soares, C., Santos, T.A., Eds.; Taylor and Francis: London, UK, 2021; pp. 687–696.

19. Zhou, W.; Huang, G.X. Model Error Assessments of Burst Capacity Models for Corroded Pipelines. *Int. J. Press. Vessels Pip.* **2012**, *99–100*, 1–8. [CrossRef]

20. Bhardwaj, U.; Teixeira, A.P.; Guedes Soares, C. Probabilistic Safety Assessment of the Burst Strength of Corroded Pipelines of Different Steel Grades with Calibrated Strength Models. *Mar. Struct.* **2022**, *86*, 103310. [CrossRef]

21. Kawsar, M.R.U.; Youssef, S.A.; Faisal, M.; Kumar, A.; Seo, J.K.; Paik, J.K. Assessment of Dropped Object Risk on Corroded Subsea Pipeline. *Ocean Eng.* **2015**, *106*, 329–340. [CrossRef]

22. Teixeira, A.P.; Zayed, A.; Guedes Soares, C. Reliability of Pipelines with Non-Uniform Corrosion. *J. Ocean Sh. Technol.* **2010**, *1*, 12–30.

23. Shuai, Y.; Shuai, J.; Xu, K. Probabilistic Analysis of Corroded Pipelines Based on a New Failure Pressure Model. *Eng. Fail. Anal.* **2017**, *81*, 216–233. [CrossRef]

24. Li, S.-X.; Zeng, H.-L.; Yu, S.-R.; Zhai, X.; Chen, S.-P.; Liang, R.; Yu, L. A Method of Probabilistic Analysis for Steel Pipeline with Correlated Corrosion Defects. *Corros. Sci.* **2009**, *51*, 3050–3056. [CrossRef]

25. Witek, M. Gas Transmission Pipeline Failure Probability Estimation and Defect Repairs Activities Based on In-Line Inspection Data. *Eng. Fail. Anal.* **2016**, *70*, 255–272. [CrossRef]

26. Leira, B.J.; Næss, A.; Brandrud Næss, O.E. Reliability Analysis of Corroding Pipelines by Enhanced Monte Carlo Simulation. *Int. J. Press. Vessel. Pip.* **2016**, *144*, 11–17. [CrossRef]

27. Bai, Y.; Song, R. Fracture Assessment of Dented Pipes with Cracks and Reliability-Based Calibration of Safety Factor. *Int. J. Press. Vessel. Pip.* **1997**, *74*, 221–229. [CrossRef]

28. Blomfors, M.; Larsson Ivanov, O.; Honfí, D.; Engen, M. Partial Safety Factors for the Anchorage Capacity of Corroded Reinforcement Bars in Concrete. *Eng. Struct.* **2019**, *181*, 579–588. [CrossRef]

29. Zhang, H.; Dong, S.; Ling, J.; Zhang, L.; Cheang, B. A Modified Method for the Safety Factor Parameter: The Use of Big Data to Improve Petroleum Pipeline Reliability Assessment. *Reliab. Eng. Syst. Saf.* **2020**, *198*, 106892. [CrossRef]

30. Kaida, T.; Sakai, S. Application of Partial Safety Factors for Fitness-for-Service Assessment of Pressure Equipment with Local Metal Loss. In Proceedings of the Proceedings of the ASME 2017 Pressure Vessels and Piping Conference, Waikoloa, HI, USA, 16–20 July 2017.

31. Machida, H.; Chitose, H.; Arakawa, M. Partial Safety Factors Assessment of Pipes with a Circumferential Surface Flaw. In Proceedings of the ASME 2010 Pressure Vessels and Piping Division/K-PVP Conference, Bellevue, WA, USA, 18–22 July 2010; pp. 513–520.

32. Cosham, A.; Hopkins, P.; Macdonald, K.A. Best Practice for the Assessment of Defects in Pipelines—Corrosion. *Eng. Fail. Anal.* **2007**, *14*, 1245–1265. [CrossRef]

33. Amaya-Gómez, R.; Sánchez-Silva, M.; Bastidas-Arteaga, E.; Schoefs, F.; Muñoz, F. Reliability Assessments of Corroded Pipelines Based on Internal Pressure—A Review. *Eng. Fail. Anal.* **2019**, *98*, 190–214. [CrossRef]

34. El Amine Ben Seghier, M.; Keshtegar, B.; Elahmoune, B. Reliability Analysis of Low, Mid and High-Grade Strength Corroded Pipes Based on Plastic Flow Theory Using Adaptive Nonlinear Conjugate Map. *Eng. Fail. Anal.* **2018**, *90*, 245–261. [CrossRef]

35. Melchers, R.E.; Beck, A.T. (Eds.) *Structural Reliability—Analysis and Prediction*; John Wiley & Sons Ltd.: Chichester, UK, 2018; ISBN 9781119266105.

36. DNV. *Recommended Practice DNV-RP-F116—Integrity Management of Submarine Pipeline Systems*; DNV: Bærum, Norway, 2021.

37. Keshtegar, B.; Miri, M. Reliability Analysis of Corroded Pipes Using Conjugate HL-RF Algorithm Based on Average Shear Stress Yield Criterion. *Eng. Fail. Anal.* **2014**, *46*, 104–117. [CrossRef]

38. Bhardwaj, U.; Teixeira, A.P.; Guedes Soares, C.; Azad, M.S.; Punurai, W.; Asavadorndeja, P. Reliability Assessment of Thick High Strength Pipelines with Corrosion Defects. *Int. J. Press. Vessel. Pip.* **2019**, *177*, 103982. [CrossRef]

39. Ross, S.M. *Introduction to Probability and Statistics for Engineers and Scientists*, 4th ed.; Academic Press: Cambridge, MA, USA, 2009; ISBN 978-0-12-370483-2.

40. Bhardwaj, U.; Teixeira, A.P.; Guedes Soares, C. Uncertainty in Reliability of Thick High Strength Pipelines with Corrosion Defects Subjected to Internal Pressure. *Int. J. Press. Vessel. Pip.* **2020**, *188*, 104170. [CrossRef]

41. Guedes Soares, C.; Garbatov, Y. Reliability of Maintained, Corrosion Protected Plates Subjected to Non-Linear Corrosion and Compressive Loads. *Mar. Struct.* **1999**, *12*, 425–445. [CrossRef]

42. Romanoff, M. *Underground Corrosion*; NACE International: Washington, DC, USA, 1957.

43. Kiefner, J.; Maxey, W.; Eiber, R.; Duffy, A. The failure stress levels of flaws in pressurised cylinders. Progress in flaw growth and fracture toughness testing. In Proceedings of the 1972 National Symposium on Fracture Mechanics, STP 536, Philadelphia, PA, USA, 28–30 August 1972; American Society for Testing and Material: West Conshohocken, PA, USA, 1972; pp. 461–481.

44. Hantz, B.F.; Sims, J.R.; Kenyon, C.T.; Turbak, T.A. *Fitness for Service: Groove like Local Thin Areas on Pressure Vessels and Storage Tanks*; Sinnappan, J., Ed.; American Society of Mechanical Engineers: New York, NY, USA, 1993.

45. *CSA Z662-07*; Oil and Gas Pipeline Systems. Canadian Standards Association: Toronto, ON, Canada, 2007; p. 2771.

46. Ritchie, D.; Last, S. Shell 92—Burst Criteria of Corroded Pipelines—Defect Acceptance Criteria. In Proceedings of the EPRG PRC Biennial Joint Technical Meeting on Line Pipe Research, Cambridge, UK, 18–21 April 1995; p. 32.

47. Stephens, D.R.; Leis, B.N. Development of an Alternative Criterion for Residual Strength of Corrosion Defects in Moderate-to High-Toughness Pipe. In Proceedings of the 3rd International Pipeline Conference, Calgary, AB, Canada, 1–5 October 2000; ASME: Calgary, AB, Canada, 2000; Volume 2, pp. 781–792.

48. Benjamin, A.C.; Vieira, R.D.; Freire, J.L.F.; De Castro, J.T.P. Modified Equation for the Assessment of Long Corrosion Defects. In Proceedings of the 20th International Conference on Offshore Mechanics and Arctic Engineering, Rio de Janeiro, Brazil, 3–8 June 2001; Volume 4, p. OMAE01/PIPE-4111.

49. Cronin, D.S.; Pick, R.J. Prediction of the Failure Pressure for Complex Corrosion Defects. *Int. J. Press. Vessel. Pip.* **2002**, *79*, 279–287. [CrossRef]

50. Zhu, X.K.; Leis, B.N. Influence of Yield-to-Tensile Strength Ratio on Failure Assessment of Corroded Pipelines. *J. Press. Vessel Technol.* **2005**, *127*, 436–442. [CrossRef]

51. Netto, T.A.; Ferraz, U.S.; Estefen, S.F. The Effect of Corrosion Defects on the Burst Pressure of Pipelines. *J. Constr. Steel Res.* **2005**, *61*, 1185–1204. [CrossRef]

52. Koçak, M. *FITNET European Fitness for Service Network. Final Technical Report*; GTC 2001 43049; FITNET: Geesthacht, Germany, 2008.

53. Orynyak, I.V. Leak and Break Models of Ductile Fracture of Pressurized Pipe with Axial Defects. In Proceedings of the 6th International Pipeline Conference, Calgary, AB, Canada, 29 September–3 October 2008; ASME International: Calgary, AB, Canada, 2008; pp. 41–55.

54. Wang, N.; Zarghamee, M.S. Evaluating Fitness-for-Service of Corroded Metal Pipelines: Structural Reliability Bases. *J. Pipeline Syst. Eng. Pract.* **2014**, *5*, 04013012. [CrossRef]

Journal of
*Marine Science
and Engineering*

Article

Prediction Method and Validation Study of Tensile Performance of Reinforced Armor Layer in Marine Flexible Pipe/Cables

Hualin Wang [1], Zhixun Yang [1], Jun Yan [2], Gang Wang [3], Dongyan Shi [1], Baoshun Zhou [4] and Yanchun Li [5],*

[1] Institute of Marine Machinery, College of Mechanical and Electrical Engineering, Harbin Engineering University, Harbin 150001, China; wanghualin@hrbeu.edu.cn (H.W.); yangzhixun@hrbeu.edu.cn (Z.Y.); shidongyan@hrbeu.edu.cn (D.S.)

[2] Department of Engineering Mechanics, Faculty of Vehicle Engineering and Mechanics, Dalian University of Technology, Dalian 116024, China; yanjun@dlut.edu.cn

[3] School of Civil Engineering, Dalian Jiaotong University, Dalian 116028, China; wanggang_1977@163.com

[4] Department of Mechanical Engineering, Technical University of Denmark, 2800 Copenhagen, Denmark; baozh@mek.dtu.dk

[5] Institute of Advanced Technology, Heilongjiang Academy of Science, Harbin 150001, China

* Correspondence: liyanchun0451@163.com; Tel.: +86-13946157815

Abstract: Marine flexible pipe/cables, such as umbilicals, flexible pipes and cryogenic hoses, are widely adopted in ocean engineering. The reinforcing armor layer in these pipe/cables is the main component for bearing loads, which is a typical multi-layer helically wound slender structure with different winding angles for different devices. There has been no general theoretical methodology to describe the tensile performance of these flexible pipe/cables. This paper first introduces a theory to solve the tensile mechanical behavior of a helically wound structure. Based on the curved beam theory, a solution of the tensile stress of a helically wound slender is derived. Then, the deformation mechanism of the marine flexible pipe/cables structure with different winding angles is studied. Through comparing theoretical and numerical results, the deformation characteristic of the helically wound slender structure is further explained. It is found that a sectional torsional deformation generates when the structure with a larger winding angle is under tension condition, while the sectional deformation of the structure with a smaller winding angle is mainly tension. Finally, a couple types of marine flexible pipe/cables under the tension condition are provided to analyze the mechanical performance and compare the difference between different theoretical models. The research conclusion from this paper provides a useful reference for the structural analysis and design of marine flexible pipe/cables.

Keywords: marine flexible pipe/cables; helically wound structures; tensile performance

Citation: Wang, H.; Yang, Z.; Yan, J.; Wang, G.; Shi, D.; Zhou, B.; Li, Y. Prediction Method and Validation Study of Tensile Performance of Reinforced Armor Layer in Marine Flexible Pipe/Cables. *J. Mar. Sci. Eng.* **2022**, *10*, 642. https://doi.org/10.3390/jmse10050642

Academic Editors: Fuping Gao and Bruno Brunone

Received: 13 April 2022
Accepted: 4 May 2022
Published: 8 May 2022

1. Introduction

Marine flexible pipe/cables are important devices in the ocean resource exploitation [1]. For example, umbilicals connect the floater and the manifold, which can be used to transmit the power and control signal, and flexible pipes can transport oil or gas. Cryogenic hoses link FLNG and LNG to deliver the liquefied natural gas. In order to improve the structural performance during operation, a single-layer or double-layer steel armor, the so-called "reinforced armor layer" is assembled inside the outer sheath, which is a usually helically wound structure. Helically wound structures can bear both axial and bending loads, which have been widely used in various industrial equipment [2]. For example, wire rope with a high tensile and fatigue strength is a key load-bearing component in special equipment, such as lifts and cranes [3]. Considering the complex loading condition in the sea environment, helically wound structures, such as umbilicals, flexible pipes and cryogenic hoses, have been introduced to ocean engineering [4]. Helically wound structures almost bear the overall axial load, and the accurate analysis of the structural tensile behavior is meaningful to the analysis and design of flexible pipe/cables [5].

In past research, the tensile behavior of helically wound structures has been a popular topic. Based on Hruska's [6,7] assumption that the axial deformation of the helical wire is small and neglects the cross section shape, the helical wire can be taken as a whole component to derivate the tensile stiffness. Cardou [8] summarized multiple theoretical models of a single-strand and double-strand helical wire rope to evaluate the structural tensile property. In these models, the structural winding angles are usually small ($0°\sim20°$) so that the structure can bear the larger axial stretching. Knapp [9–11] proposed a theoretical model to calculate the axial strain and winding angle of the deformed structure, and in this model, the core deformation is not considered. The helically wound slender structure can be regarded as a spring model when the winding angle reaches $70°\sim90°$, and the mechanical property of this type of structural model can be solved easily. In 2006, Feyrer [12] found that the local deformation of a single steel wire is affected by the fatigue and damage behavior of steel wire rope. For marine flexible pipe/cables, Kirchhoff's nonlinear mechanical model was developed, and the theoretical accuracy was verified experimentally by Liu [13]. Some theoretical models are proposed to calculate the local pressure, radial force and contact force between the cable armor layer [14,15].

Many numerical methods have been gradually proposed to analyze the mechanical behavior of helically wound structures. In 2019, Chang and Chen [16] simulated the mechanical behavior of submarine cables under combined loads of tension, torsion and compression, and the coupling effects of loads generated a great influence on the analysis and design of submarine cables. In 2021, Yang [17] proposed a numerical method to calculate the nonlinear tension–torsion coupling effect. In the method, the stiffness of the structure inside the reinforced armor layer, or the so-called 'core' can be obtained easily. Then, the tension–torsion coupled stiffness can be calculated more accurately than the theoretical solution.

Recently, many scholars have proposed theoretical analysis models to analyze helically wound structures with different helix angles. However, there is a small number of discussions on the applicability of these models, especially when considering the helically wound structure in the cylindrical core. This paper discusses a tensile model of the helically wound structure wrapped around a cylindrical core. Based on the curved beam theory [12], a tensile solution to the helically wound structure with different winding angles is derived. Comparing with numerical results of three winding angle structures, the theoretical model is validated. Through comparing the error and application range of different theoretical analysis methods, the deformation mechanism is further explained.

2. Structural Characteristics of Reinforced Armor Layer in Marine Flexible Pipe/Cables

The different types of reinforced armor layers are required when marine flexible pipe/cables work in various conditions. The reinforced armor layer is typically a helically wound structure with different winding angles, as shown in Figure 1. For cables or umbilicals, the function of a reinforced armor layer is to resist the tensile load. Therefore, the winding angle of this type of structure is small, usually between $15°$ and $25°$ [18]. For flexible pipe, the reinforced armor layer is adopted for bearing the tension and inner pressure loads, and the winding angle is usually between $30°$ and $40°$ [19,20]. For cryogenic flexible hose, the reinforced armor layer mainly aims to bear the inner pressure and outer impacting loads, and the winding angle is usually between $75°$ and $80°$ [21]. The tensile strength of helically wound structures with a small angle is high, while the structure with an angle has a strong radial resistance capacity and is more flexible.

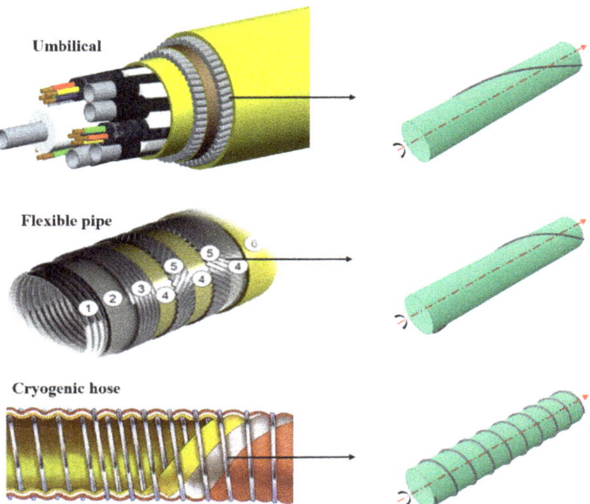

Figure 1. Structural characteristics of the marine flexible pipe/cables armoring layer.

The mechanical model of helically wound structures can be generally established as shown n Figure 2. A helically wound structure with an undeformable cylindrical core is proposed. Considering the complexity of the reinforced armor layer, the following assumptions are given:

1. Ignoring the interaction between layers, only one of the armoring layers is taken to conduct the property analysis. The core is an undeformable cylinder.
2. The diameter of the helically wound structure is much smaller than that of the core.
3. Under the axial load, the section of the deformed structure remains flat.
4. The material property of the helically wound structure is isotropic.
5. The helically wound structure cannot be affected by the external bending moment per unit length.

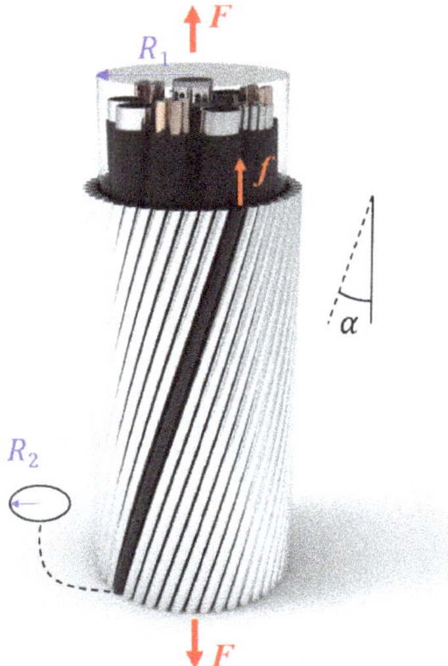

Figure 2. Schematic diagram of the helically wound structure subjected to tensile load.

Given that the armored layer is subjected to the axial tension F, and the average force on each helically wound wire is f. The winding angle is α, the thread pitch before the deformation is $h = 2\pi r \tan\alpha$ and r is the distance between the central line of the core and the central line of the wounded wire. R_2 is the radius of the helically wound wire, R_1 is the radius of the cylindrical core and $r = R_1 + R_2$.

3. Theoretical Model of Tensile Behavior of the Helically Wound Structure

3.1. Mechanical Model of Helically Wound Structures

In order to explain the mechanical model clearly, a helically wound steel wire was taken, and the tensile loading condition is shown in Figure 3.

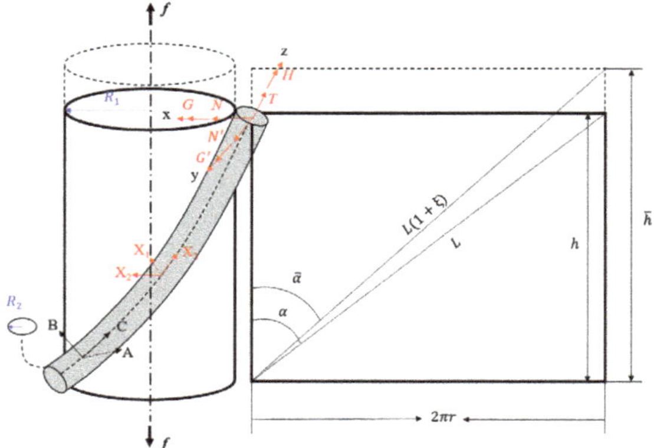

Figure 3. Schematic diagram of the deformation of the helically wound structure.

Assuming that the core is undeformable, the winding angle after stretching is $\bar{\alpha}$, the changed value of the winding angle because of stretching is $\Delta\alpha = \bar{\alpha} - \alpha$, the thread pitch after stretching is $\bar{h}$ and the amplitude of stretching deformation is $\Delta h = \bar{h} - h$.

The axial strain ε can be expressed as:

$$\varepsilon = \frac{\bar{h} - h}{h} \tag{1}$$

A micro-segment of the helically wound structure can be regarded as a 3D curve. Combining the curvature and torsion of the spatial helical curves, the subcomponents of the angular velocity vector are projected to the A, B, and C axes. κ, κ' and τ indicate deformation curvature and torsion in per unit length. The curvature and torsion before deformation are expressed as:

$$\kappa = 0, \ \kappa' = \frac{\sin^2\alpha}{r}, \ \tau = \frac{\sin\alpha\cos\alpha}{r} \tag{2}$$

The curvature and torsion after deformation are expressed as:

$$\bar{\kappa} = 0, \ \bar{\kappa'} = \frac{\sin^2\bar{\alpha}}{r}, \ \bar{\tau} = \frac{\sin\bar{\alpha}\cos\bar{\alpha}}{r} \tag{3}$$

When the helically wound structure is only subjected to the tensile load, the force in each section along the winding path after canceling the boundary is the same. Three subcomponents of forces, two bending moments and one torque can be generated in the cross section.

N and N' are the shear force components in the A- and B-axis directions on the cross-section of the helical structure, respectively. T is the axial tension of the central axis of the helically wound slender structure. G and G' are the bending moments along the A- and

B axis, respectively. H represents the torsion moment on the central axis of the helically wound slender structure. X_1, X_2, and X_3 represent the linear load components per unit length of helically wound structures.

According to Hypothesis 4, its internal force can be expressed linearly through the curvature change and torsion per unit length as:

$$G = \frac{\pi R_2^4}{4} E(\overline{\kappa} - \kappa), \ G' = \frac{\pi R_2^4}{4} E(\overline{\kappa'} - \kappa'), \ H = \frac{\pi R_2^4}{4} E(\overline{\tau} - \tau) \tag{4}$$

The thin rod theory of Love (1944) [22] gives the force and moment balance equations as Equation (5).

$$\begin{cases} \frac{\mathrm{d}N}{\mathrm{d}S} - N'\tau + T\kappa' + X_1 = 0 \\ \frac{\mathrm{d}N'}{\mathrm{d}S} - T\kappa + N\tau + X_2 = 0 \\ \frac{\mathrm{d}T}{\mathrm{d}S} - N\kappa' + N'\kappa + X_3 = 0 \\ \frac{\mathrm{d}G}{\mathrm{d}S} - G'\tau + H\kappa' - N' = 0 \\ \frac{\mathrm{d}G'}{\mathrm{d}S} - H\kappa + G\tau + N = 0 \\ \frac{\mathrm{d}H}{\mathrm{d}S} - G\kappa' + G'\kappa + \Theta = 0 \end{cases} \tag{5}$$

where Θ represents the external moment per unit length of the helically wound structure. According to Hypothesis 5, the balanced equation is simplified as:

$$\begin{cases} -N'\overline{\tau} + T\overline{\kappa'} + X_1 = 0 \\ X_2 = 0 \\ X_3 = 0 \\ -G'\overline{\tau} + H\overline{\kappa'} - N' = 0 \\ N = 0 \\ \Theta = 0 \end{cases} \tag{6}$$

The axial tension of the helically wound structure is $T = \pi R_2^2 E \zeta$, where E is the elasticity modulus and ζ is the axial strain.

The length of the helically wound structure before stretching is L, and after stretching, the length becomes $L(1 + \zeta)$. Therefore, Equation (1) can be rewritten as:

$$\varepsilon = \frac{\overline{h} - h}{h} = (1 + \zeta)\frac{\cos \overline{\alpha}}{\cos \alpha} - 1 \tag{7}$$

Similarly, the rotation strain β of the helically wound structure can be expressed as:

$$\beta = (1 + \varepsilon)\tan \overline{\alpha} - \tan \alpha \tag{8}$$

and $\cos \overline{\alpha}$ can be expressed as:

$$\cos \overline{\alpha} = \cos(\alpha - \Delta\alpha) = \cos \alpha + \Delta\alpha \sin \alpha \tag{9}$$

ζ is assumed to be small and higher-order terms are ignored, Equations (8) and (9) are rewritten as:

$$\varepsilon = \zeta + \Delta\alpha \tan \alpha \tag{10}$$

$$\beta = \zeta \tan \alpha - \Delta\alpha = 0 \tag{11}$$

Based on Equations (10) and (11), the axial strain ζ and the change value $\Delta\alpha$ of the winding angle can be obtained by:

$$\zeta = \frac{\varepsilon}{1 + \tan^2 \alpha} \tag{12}$$

$$\Delta\alpha = \frac{\varepsilon \tan \alpha}{1 + \tan^2 \alpha} \tag{13}$$

The curvature change and the torsion change per unit length can be linearized as:

$$R_2 \Delta \kappa = -\frac{2 \sin \alpha \cos \alpha}{r/R_2} \frac{\varepsilon \tan \alpha}{1 + \tan^2 \alpha} \tag{14}$$

$$R_2 \Delta \tau = \frac{1 - 2 \cos^2 \alpha}{r/R_2} \frac{\varepsilon \tan \alpha}{1 + \tan^2 \alpha} \tag{15}$$

where $\Delta \kappa$ and $\Delta \tau$ are the curvature change and the torque change. The internal force of the helically wound structure can be calculated by:

$$
\begin{cases}
G' = -\dfrac{E \pi R_2^4 \varepsilon \cos^2 \alpha \tan^2 \alpha}{2r\left(1+\tan^2 \alpha\right)} \\[2mm]
H = \dfrac{E \pi R_2^4 \varepsilon \tan \alpha \sin^2 \alpha}{2r\left(1+\tan^2 \alpha\right)} \\[2mm]
N' = \dfrac{\left[E \pi R_2^4 \varepsilon \sin \alpha \cos \alpha \left(1-2\cos^2 \alpha\right)+2E\pi R_2^4 \varepsilon \cos^3 \alpha \sin \alpha\right] \tan^2 \alpha}{4r^2\left(1+\tan^2 \alpha\right)} \\[2mm]
T = \dfrac{E \pi R_2^2 \varepsilon}{1+\tan^2 \alpha}
\end{cases}
\tag{16}
$$

The stress caused by the axial tension T is:

$$\sigma_T = \frac{T}{\pi R_2^2} = \frac{E\varepsilon}{1 + \tan^2 \alpha} \tag{17}$$

The maximum normal stress caused by the bending moment component G' in the B-axis direction is:

$$\sigma_{G'} = \frac{4G'}{\pi R_2^3} = -\frac{2E R_2 \varepsilon \cos^2 \alpha \tan^2 \alpha}{r\left(1 + \tan^2 \alpha\right)} \tag{18}$$

The maximum shear force caused by the torsion moment H on the central axis is:

$$\sigma_H = \frac{2H}{\pi R_2^3} = \frac{E R_2 \varepsilon \tan \alpha \sin^2 \alpha}{r\left(1 + \tan^2 \alpha\right)} \tag{19}$$

3.2. Theoretical Model of Tensile Mechanical Behavior of Helically Wound Structures with Typical Winding Angles

If the winding angle is between $0°$ and $20°$, the helically wound structure can usually regard as a thin rod. One of typical appliccations of this type of helically wound slender structure is the tensile armor steel wire of dynamic submarine. The winding angle α is small, and $\sin \alpha$ approaches zero. Therefore, some terms related to α can be dropped in the calculation. Equations (12) and (13) can be rewritten as the formula [9], which can be used to solve the axial strain ξ and the changed value of the winding angle $\Delta \alpha$ by

$$\xi = \varepsilon \cos^2 \alpha \tag{20}$$

$$\frac{\cos \overline{\alpha}}{\cos \alpha} = 1 + \varepsilon - \xi \tag{21}$$

The internal force solution of the steel wire can be written as:

$$
\begin{cases}
G' = \frac{\pi}{4} E R_2^3 \left(-\frac{2 \sin \alpha \cos \alpha}{r/R_2}\left\{\cos^{-1}\left[(1+\varepsilon - \varepsilon \cos^2 \alpha)\cos \alpha\right] - \alpha\right\}\right) \\[2mm]
H = \frac{\pi}{4} E R_2^3 \left(\frac{1-2\cos^2 \alpha}{r}\left\{\cos^{-1}\left[(1+\varepsilon - \varepsilon \cos^2 \alpha)\cos \alpha\right] - \alpha\right\}\right) \\[2mm]
N' = \frac{H}{R_2}\frac{\sin^2 \alpha}{r/R_2} - \frac{G'}{R_2}\frac{\sin \alpha \cos \alpha}{r/R_2} \\[2mm]
T = \pi \varepsilon \cos^2 \alpha E R_2^2
\end{cases}
\tag{22}
$$

The stress caused by the axial tension T, the maximum normal stress caused by the bending moment component G' in the B-axis direction and the maximum shear force caused by the torque H can be calculated through the intermediate winding angle theory.

When the winding angle is between 70° and 90°, the helically wound slender structure will evolve into spring, and the mathematical model is shown in Figure 4. Here, α is a large value. Therefore, the terms related to $\cos \alpha$ can be dropped, and the helically wound structures can be regarded as a classical spring model, the mechanical property of which can be solved briefly based on some existing methods. The tensile force f along the central axis of the core is expressed as [23]:

$$f = k\Delta h \tag{23}$$

$$k = \frac{GR_2^4}{4N_c r^3} = \frac{GR_2^4}{4r^3} \tag{24}$$

where the shear modulus is $G = E/2(1+\nu)$, E is the elasticity modulus, ν is the Poisson's ratio, k is the spring constant and N_c is the coil number. The tension force f can be obtained by:

$$f = \frac{\varepsilon E R_2^4 \pi}{4r^2(1+\nu)\tan\alpha} \tag{25}$$

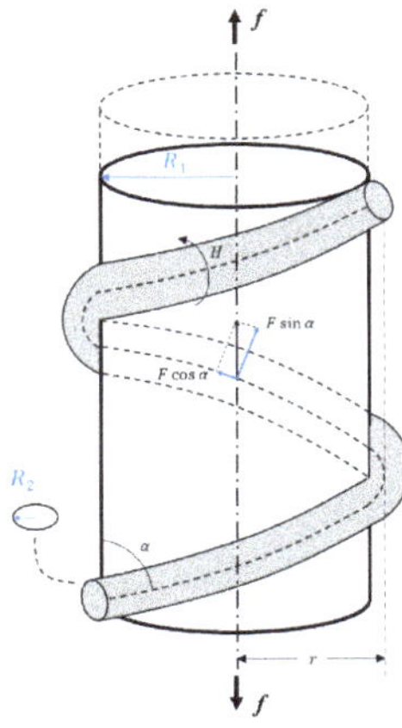

Figure 4. Schematic diagram of spring parameters.

The torsion moment H of the spring is expressed as:

$$H = fr = \frac{\varepsilon E R_2^4 \pi}{4r(1+\nu)\tan\alpha} \tag{26}$$

The tension force f of the core is decomposed into the spring axial tension T as:

$$T = f\cos\alpha = \frac{\varepsilon E R_2^4 \pi \cos\alpha}{4r^2(1+\nu)\tan\alpha} \tag{27}$$

The axial stress and the torsion stress can be expressed as:

$$\sigma_T = \frac{T}{\pi R_2^2} = \frac{\varepsilon E R_2^2 \cos\alpha}{4r^2(1+\nu)\tan\alpha} \tag{28}$$

$$\sigma_H = \frac{2H}{\pi R_2^3} = \frac{\varepsilon E R_2}{2r(1+\nu)\tan\alpha} \tag{29}$$

4. Numerical Simulation Verification Analysis

Currently, Knapp's [8] theory is used in the offshore engineering industry to solve the tensile properties of reinforced armored layers, while the spring theory is adopted for reinforced armor layers with large winding angles. Therefore, a general theoretical model of helically wound slender structures with different winding angles is derived in this paper. Thus, the accuracy of theoretical estimation in practical engineering can be improved. In order to validate the proposed theory models, an accurate 3D finite element model was built as a benchmark.

4.1. Establishing the Numerical Model

Using commercial software ABAQUS [24] to build a beam element of helically wound structures, helically slender structures were wound on a cylindrical shell model at different winding angles, from 10° to 90°, with a difference of 10°. The length of the model was a helical pitch, and the beam section was a circle with a radius of 2.5 mm. A model with a 45° winding angle was taken as an example to explain the proposed method. The material was elastoplastic, and the material parameters are shown in Table 1. [25]

Table 1. Model material parameters.

Elasticity Modulus (MPa)	Poisson's Ratio	Density (kg·m^{-3})
210,000	0.3	7800

When building a numerical model, a B31 beam element and a S4R shell element are taken to define the sectional property of the beam. There are interactions, such as the contact and extrusion, between the helically wound structure and the cylindrical core. In the numerical model, a universal contact between the beam element and the shell element is set up and the friction is not considered. Through a grid convergence analysis, high-efficient and accurate numerical models are given as the model meshed equally by 1000 grids along the axial direction, and the S4R shell model is divided by quadrilateral grids. After a numerical calculation, it was found that this numerical model could describe the contact between the beam and the shell very well. The finite element model is shown in Figure 5.

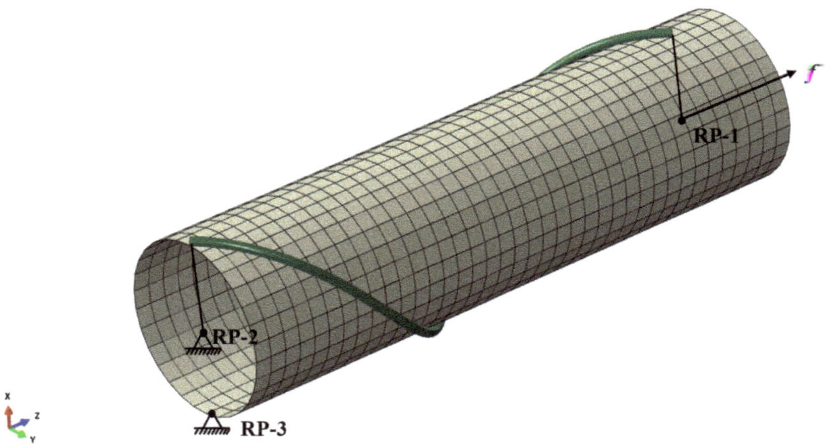

Figure 5. Finite element model and boundary conditions.

4.2. Loads and Boundary Conditions

Considering the axial periodicity of the helically wound structure, a proper way to apply loads on the ends is required for various loading conditions. An improper method may lead to an undesired displacement and a stress concentration. Therefore, we compared the mechanical property of different numerical models. The degrees of freedom of all points on the ending surface were respectively coupled with the ones of the central points as reference points RP-1 and RP-2. Shell elements were defined as rigid elements and were completely fixed with RP-2, and a displacement load with the amplitude of a thousandth the length of the pitch was applied to the RP-1. The loading and constraint conditions are shown in Figure 5. The quasi-static loading condition was applied to smooth the analysis step and avoid sudden stress generation. Because of the axial periodicity of the structure, the internal force and stress changed periodically. Therefore, the model with the length of a pitch was taken to conduct the numerical calculation.

4.3. Tensile Mechanical Behavior Deformation Mechanism

Figure 6 presents the numerical results of the mechanical property of helically wound structures with a winding angle of 45°, involving various loading conditions such as the axial tension, displacement, bending and torsion. Under the axial tension condition, the axial strain and torsional and bending deformation may appear. For this type of helically wound structures, the deformation is mainly the axial translation. The torsion direction at two ends is opposite and the bending strain is small during the middle segment, while the one is larger at ending parts and with an opposite direction. Due to the periodicity of helically wound structures, the mechanical behavior along the axial direction should be the same. Therefore, a further study on the tensile behavior was carried out in the following section.

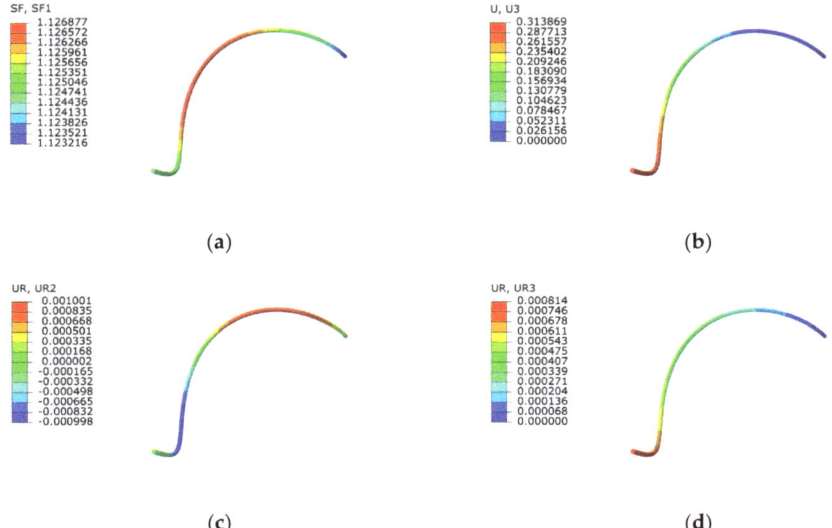

Figure 6. Finite element cloud map. (**a**) Axial tension, (**b**) stretch, (**c**) bending, (**d**) torsion.

5. Validation of Model Analysis

The winding angle of the helically wound slender structure is usually required to be changed for different service conditions. The angle in the steel wire rope or the umbilical is usually small, while the one in the vibration damping device is larger. Figure 7 presents the mechanical property of helically wound structures with different angles, where 10° and 20° are defined as smaller winding angles; 30°, 40°, 45°, 50°, and 60° are defined as intermediate winding angles; and 70° and 80° are defined as larger winding angles.

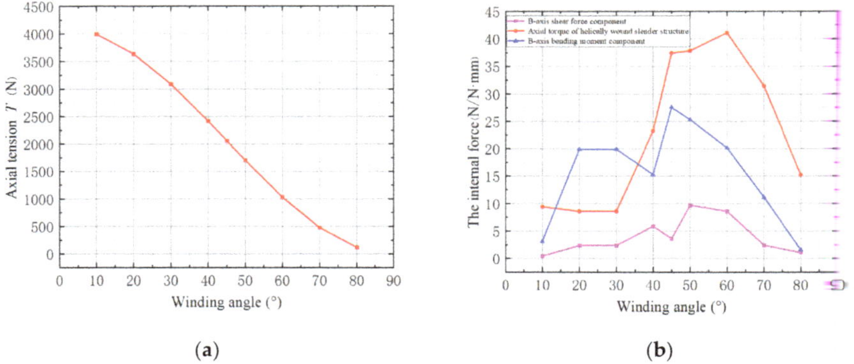

Figure 7. Numerical solutions of internal forces at different winding angles. (**a**) Axial tension, (**b**) other internal forces.

5.1. Mechanical Tensile Performance of Helically Wound Structures

Figure 7 shows the numerical solutions of the helically wound structure with different winding angles under stretching loads. From Figure 7a, the axial tensile force decreased linearly as the winding angle increased. The normal stress σ_T was equal to the ratio of the axial force T to the cross-sectional area of the helically wound structure, so the stress change trend was similar to the tension change trend. From Figure 7b, the changing trend of the shear force along the B-axis direction was the same as that of the bending moment and the axial torque as the winding angle increased, and the internal force tended to the maximum when the winding angle was about 45°. Based on theoretical mechanics, the slender structure was tensile but easy to bend. With the increase of the winding angle, the load on the curved beam gradually changed from tension to shear, resulting in a linear decrease of the axial tension T, while the tensile strength of the structure was weakened. When the winding angle was about 45°, the structure was more comprehensive and the force was uniform, resulting in the internal force tending to the maximum value. From Figure 7b, not normally distributed, because the numerical solution outputted the data, the direction of the discrete nodal force was different, which caused the data to produce periodic fluctuations. Therefore, the limited output precision of the internal force of some angles caused the curve to fluctuate. As the winding angle changes, the theoretical results should also change regularly, which is also confirmed below.

Based on the simulation, the tensile force and stress of the helically wound structure increased linearly with the increasing winding angle. Another interesting behavior is that other internal forces and stresses may reach a peak when the winding angle reaches 45°. Therefore, a micro-element analysis method was applied to conduct a study on the symmetry of the mechanical behavior of the helically wound slender structure. When the winding angle increased, the torsion–stretch ratio changed, as shown in Figure 8. The cloud image was a torsion cloud image. The dotted line shows the undeformed state. As the winding angle increased, the torsional strain gradually increased and the stretching strain gradually decreased. When the winding angle was 45 degrees, the torsion–stretch ratio tended to the middle value. Both the torsional and tensile deformations were large, so the internal force and stress tended to the maximum when the winding angle was 45°.

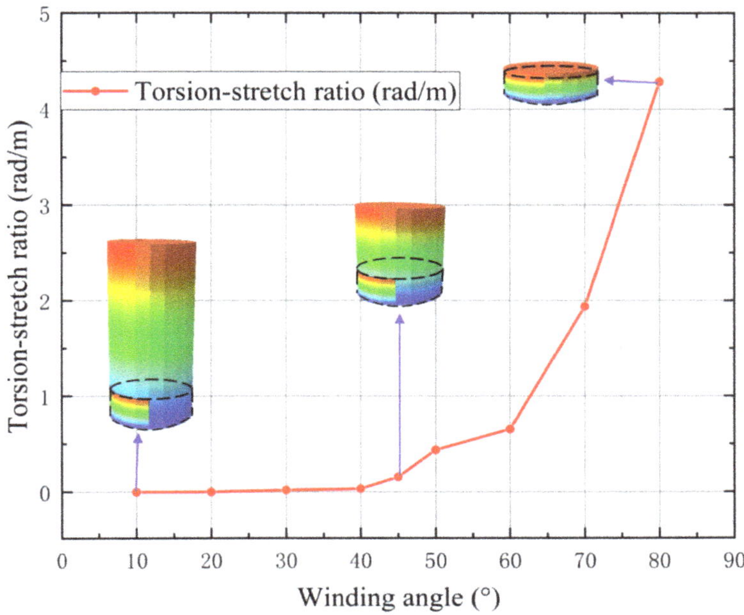

Figure 8. The torsion–stretch ratio diagram with different winding angles.

5.2. Theory Suitability Analysis

In order to discuss the application scope of different theories, the mechanical behavior was further analyzed based on different methods, and various solutions are plotted in Figure 9. It was observed that the intermediate winding angle theory and the small angle theory were basically consistent with the numerical solution in solving the axial tension T, and the solution error was less than 2%, which meets the requirements of engineering practice for the accuracy of the theoretical solution. For other internal forces, the changing trend of these two theoretical solutions are same as the numerical solution, but the accuracy of the intermediate winding angle solution is significantly better than that of the smaller winding angle theory. As the above theoretical models ignore the effect of torsion under the stretching condition, the accuracy of the solution for the smaller winding angle is better than that for the other winding angle. The smaller winding angle has a stronger tensile strength, which leads to a smaller axial strain. The classical spring theory assumes that the inner core can be contracted, and the tensile member is not subject to the reaction force of the inner core, resulting in a large error in the calculation of the axial stress T. The smaller winding angle theory ignores the influence of the changing value of the winding angle, resulting in only T with higher accuracy. The theoretical formulas of N', G' and H have sine and cosine trigonometric functions, so the calculation results of the complementary winding angles are the same, resulting in a normal distribution of the curves. The stresses due to T, the maximum positive stress due to G' and the maximum shear due to the torque H are obtained using the internal forces of the helically wound structure described above. Among them, the maximum normal stress due to N' is small, which is not a priority strength requirement in engineering.

The theoretical formula for a middle winding angle has a higher accuracy and the widest applicability. In the theory for a large winding angle, the core is assumed to be deformable, which is more suitable for the spring than for the theory without considering the radial deformation.

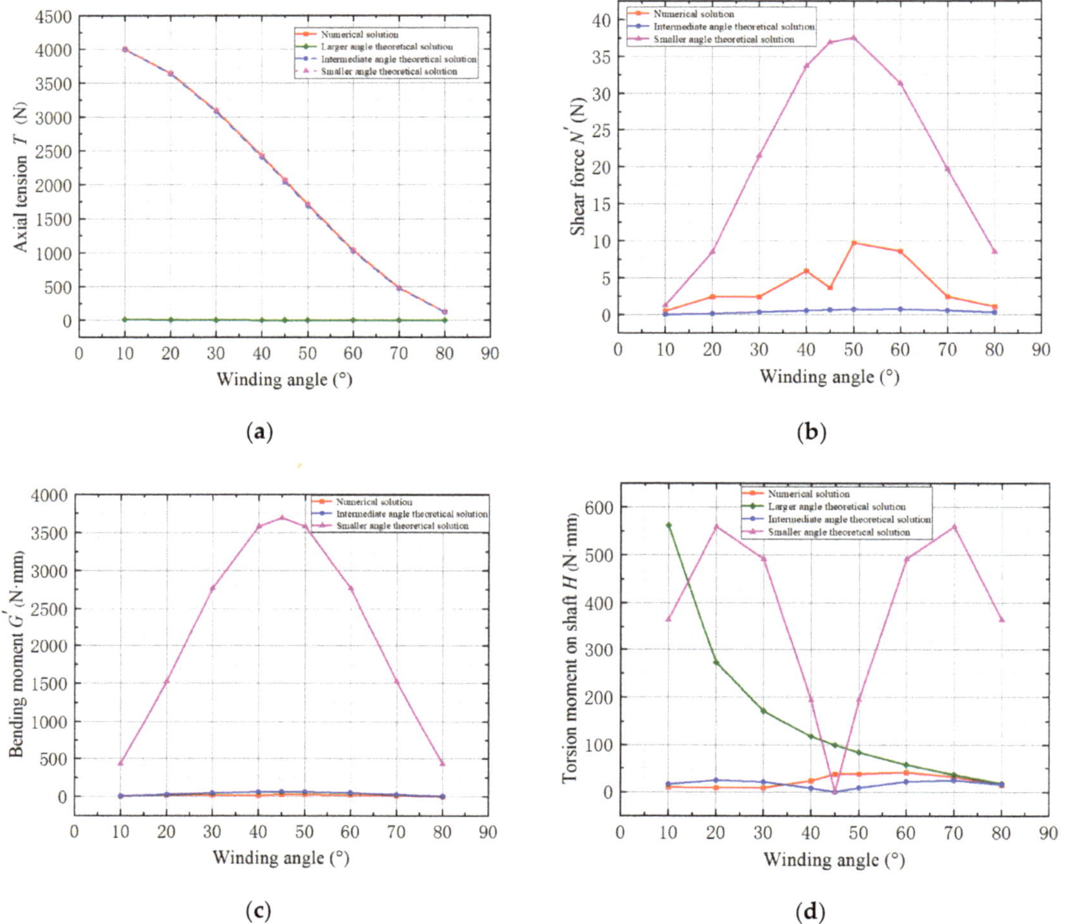

Figure 9. Internal force comparison point line diagram. (**a**) Axial tension, (**b**) shear force in the B-axis direction, (**c**) B-axis direction bending moment, (**d**) torsion moment on the shaft.

5.3. Error Analysis of Tensile Behavior of Marine Flexible Pipe/Cables

The winding angles of helically wound structures in the engineering field are not all small or large, and structures with intermediate winding angles have been applied in some fields. The various types of helically wound armored or braided layers have been used in marine devices such as umbilicals, flexible pipes and cryogenic hoses, as shown in Figure 10. The winding angles of 20°, 30° and 80° are used for different functions.

The marine flexible pipe/cables mainly bear the axial tensile load during operation. The axial tensile bearing capacity is mainly affected by the tensile stiffness and strength. Tensile stiffness $EA = \sigma A/\varepsilon$. σ is the axial normal, ε is the axial strain, and A is the section area. In the numerical calculation, the axial strain ε and section area A were the same, so the tensile stiffness was only related to axial normal stress σ. The tensile strength was the minimum breaking force, which can be calculated by $F = \varepsilon_{max}EA/cos\alpha = \sigma\varepsilon_{max}A/cos\alpha$. Here, ε_{max} is the yield strain of the material, and α is the winding angle of the helically wound slender structure. Therefore, the tensile strength was mainly affected by the normal stress σ.

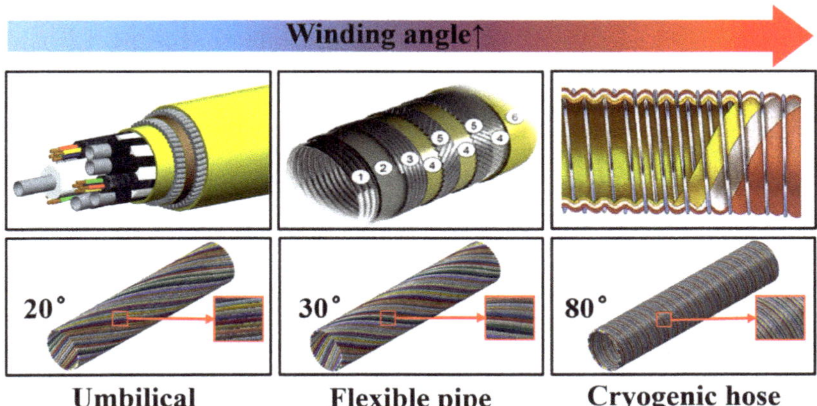

Figure 10. Three pipe/cables winding angles.

In the design stage, the theoretical model of a small winding angle is unable to satisfy the design requirements because of the ignorance of the torsion deformation, which not only increases the manufacturing cost but also easily causes safety accidents. Therefore, this paper analyzed the error of stress estimation of marine flexible pipe/cables. The tensile, bending and torsional deformation may appear when stretching helically wound slender structure. The normal stress $\sigma = \sigma_T + \sigma_{G'}$, as shown in Figure 11, and σ_T were the normal stress due to tensile deformation, and $\sigma_{G'}$ was the normal stress due to bending deformation.

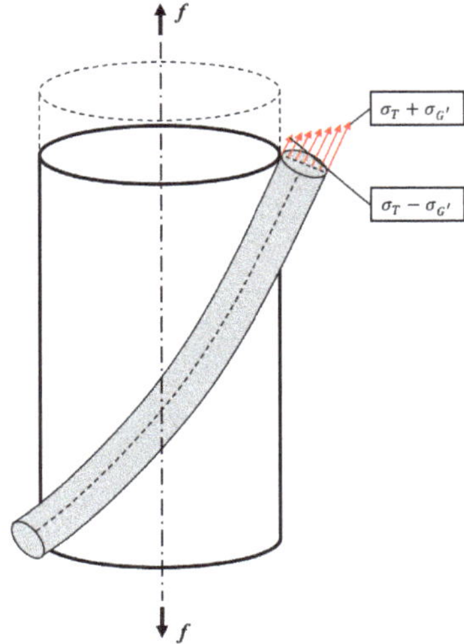

Figure 11. Stress diagram.

Figure 12 presents an error comparison result. From Figure 12a, the error of the intermediate winding angle theory was smaller than that of the smaller angle theory. Because these theories ignore torsion, while the winding angle increased, the error increased, and this phenomenon is more obvious for the smaller winding angle theory. Because the Poisson effect was considered in the simulation, the theoretical normal stress of the intermediate winding angle was slightly larger than that of the numerical solution, and the error was less than 2%. From Figure 12b, when the winding angle was 45°, the shear stress error of the intermediate winding angle was the largest, reaching up to 100%. However, it was much smaller than the normal stress and had little effect on the resultant force. In order to improve the theoretical estimation accuracy of marine flexible pipe/cables, the more general method calculation error of three representative pipe/cables is proposed:

1. The normal stress of the umbilical increases by 0.36%, and the shear stress decreases by 50.51%.
2. The normal stress of the flexible pipe increases by 0.62%, and the shear stress decreases by 67.62%.
3. The normal stress of the cryogenic hose increases by 0.72%, and the shear stress decreases by 8.44%.

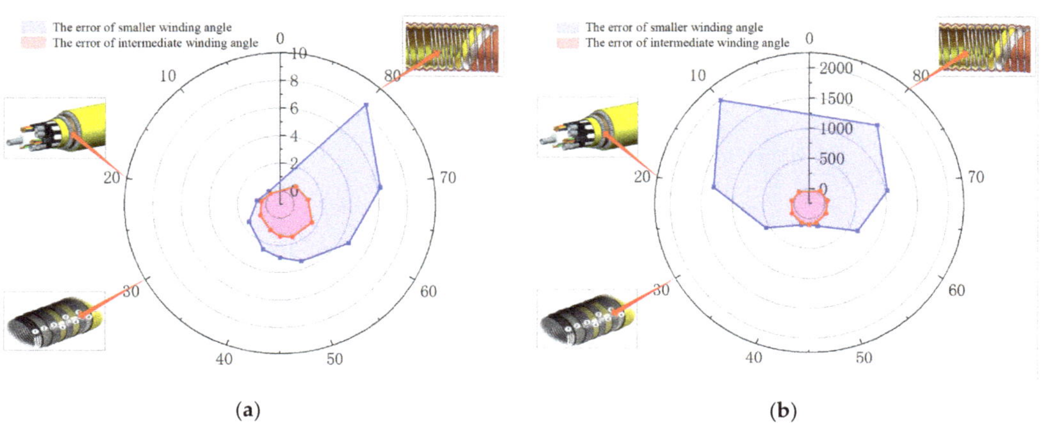

Figure 12. Error comparison. (**a**) Normal stress error, (**b**) shear stress error.

In summary, although the error is within the allowable range, when the tensile stress of helically wound structures is large, the effect cannot be underestimated. Therefore, the influence of the above errors should be considered in the calculation of engineering to improve the calculation accuracy.

6. Conclusions

In this paper, combining theoretical and numerical methods, the mechanical mechanism of the helical winding structure under tension was deeply studied. The conclusions are as follows:

1. A more general method was deduced for different winding angles, which solves the problem of poor applicability of previous theoretical formulas. The theoretical calculation errors of different marine flexible pipe/cables were analyzed, and the theoretical calculation accuracy was improved.
2. Under the premise of the same axial strain, the tensile–torsion ratio of different winding angles was analyzed. It was found that with the increase of winding angle, the torsion of the structure gradually replaced the stretch, leading to increased error, so the effect of torsion should be fully considered.
3. When the increase of winding angle T decreased linearly, the tensile strength decreased, and the theoretical formulas of N', G' and H had sine and cosine trigonometric functions. Therefore, the calculation results of the complementary winding angles were the same, resulting in a normal distribution of the curves.

In summary, the intermediate winding angle theoretical provides positive suggestions for the design and verification of helically wound structures. Future work will consider the mechanical mechanism of helically wound structures under bending and torsion.

Author Contributions: Conceptualization, H.W., Z.Y., J.Y., G.W., D.S., B.Z. and Y.L.; Data curation, H.W.; Formal analysis, H.W. and J.Y.; Funding acquisition, Y.L.; Investigation, H.W. and Z.Y.; Methodology, Z.Y.; Project administration, Y.L.; Resources, Y.L.; Supervision, Z.Y. and J.Y.; Visualization, H.W. and G.W.; Writing—original draft, H.W. and Z.Y.; Writing—review & editing, H.W. and Z.Y. All authors have read and agreed to the published version of the manuscript.

Funding: This research was funded by the National Natural Science Foundation of China, grant number 52001088, the Natural Science Foundation (Key Joint Fund of Shandong Province), grant number U1906233, the Major Science and Technology Innovation Project in Shandong Province, grant number 2019JZZY010801, Natural Science Foundation of Heilongjiang Province, grant number LH2021E050, the State Key Laboratory of Structural Analysis for Industrial Equipment, grant number GZ20105.

Institutional Review Board Statement: Not applicable.

Informed Consent Statement: Not applicable.

Acknowledgments: The authors highly appreciate the financial support provided by the Natural Science Foundation. We would like to thank research center of marine equipment in Dalian University of Technology for all of their support for permitting us to use their facilities and/or resources to make this project possible.

Conflicts of Interest: The authors declare no conflict of interest.

References

1. Xia, R.; Zhang, X.; Zheng, L.J. Reliability analysis of an umbilical under ultimate tensile load based on response surface approach. *Underw. Technol.* **2020**, *37*, 87–93.
2. Meng, X.B.; Gao, K.; Sun, Y.H. Mathematical modeling and geometric analysis for wire rope strands. *Appl. Math. Model.* **2015**, *39*, 1019–1032.
3. Xiang, L.; Wang, H.Y.; Chen, Y. Modeling of multi-strand wire ropes subjected to axial tension and torsion loads. *Int. J. Solids Struct.* **2015**, *58*, 233–246. [CrossRef]
4. Inagaki, K.; Ekg, J.; Zahrai, S. Mechanical analysis of second order helical structure in electrical cable. *Int. J. Solids Struct.* **2007**, *44*, 1657–1679. [CrossRef]
5. Usabiaga, H.; Pagalday, J.M. Analytical procedure for modelling recursively and wire by wire stranded ropes subjected to traction and torsion loads. *Int. J. Solids Struct.* **2008**, *45*, 5503–5520. [CrossRef]
6. Hruska, F.H. Radial forces in wire ropes. *Wire Wire Prod.* **1952**, *27*, 459–463.
7. Hruska, F.H. Tangential forces in wire ropes. *Wire Wire Prod.* **1953**, *28*, 455–460.
8. Cardou, A.; Jolicoeur, C. Mechanical models of helical strands. *Appl. Mech. Rev.* **1997**, *50*, 1–14. [CrossRef]
9. Knapp, R.H. Derivation of a new stiffness matrix for helically armoured cables considering tension and torsion. *IJNME* **1979**, *14*, 515–529. [CrossRef]
10. Knapp, R.H. Structural modeling of undersea cables. *J. Offshore Mech. Arct. Eng.* **1989**, *111*, 323–330. [CrossRef]
11. Knapp, R.H. Design methodology for undersea umbilical cables. In Proceedings of the OCEANS 91 Proceedings, Honolulu, HI, USA, 1–3 October 1991; pp. 1319–1327.
12. Feyrer, K. *Wire Ropes: Tension, Endurance, Reliability*; Springer: Berlin/Heidelberg, Germany, 2007; pp. 1–322.
13. Liu, Y.W.; Chen, J.B.; Liu, J.G. Nonlinear mechanics of flexible cables in space robotic arms. *Orig. Pap.* **2018**, *94*, 649–667.
14. Lanteigne, J. Theoretical estimation of the response of helically armored cables to tension, torsion, and bending. *J. Appl. Mech.* **1985**, *52*, 423–432. [CrossRef]
15. Li, X.; Liang, L.; Wu, S. Analysis of mechanical behaviors of internal helically wound strand wires of stranded wire helical spring. *Proc. Inst. Mech. Eng. C J. Mech. Eng. Sci.* **2018**, *232*, 1009–1019. [CrossRef]
16. Chang, H.C.; Chen, B.F. Mechanical behavior of submarine cable under coupled tension, torsion and compressive loads. *J. Appl. Mech.* **2019**, *189*, 106272. [CrossRef]
17. Yang, Z.X.; Su, Q.; Yan, J. Study on the nonlinear mechanical behaviour of an umbilical under combined loads of tension and torsion. *Ocean Eng.* **2021**, *238*, 109742. [CrossRef]
18. Tang, M.G.; Yan, J.; Wang, Y. Tensile stiffness analysis on ocean dynamic power umbilical. *China Ocean Eng.* **2014**, *28*, 259–270. [CrossRef]
19. Zhu, X.H.; Lei, Q.L.; Meng, Y. Tensile response of a flexible pipe with an incomplete tensile armor layer. *J. Offshore Mech. Arct. Eng.* **2021**, *143*, 1–9. [CrossRef]
20. Yue, Q.J.; Lu, Q.Z.; Yan, J. Tension behavior prediction of flexible pipelines in shallow water. *Ocean Eng.* **2013**, *58*, 201–207. [CrossRef]
21. Yang, Z.X.; Yan, J.; Chen, J.L. Multi-objective shape optimization design for liquefied natural gas cryogenic helical corrugated steel pipe. *J. Offshore Mech. Arct. Eng.* **2017**, *139*, 1–11. [CrossRef]

22.	Love, A.E.H. *A Treatise on the Mathematical Theory of Elasticity*; Cambridge University Press: Cambridge, UK, 1944; p. 305.
23.	Lemaitre, J.; Chaboche, J.L.; Maji, A.K. Mechanics of solid materials. *J. Eng. Mech.* **1992**, *119*, 642–643. [CrossRef]
24.	ABAQUS CAE. *Analysis User's Manual Version 6.12*; Dassault Systemes Simulia, Inc.: Johnston, RI, USA, 2012.
25.	Yin, Y.C.; Lu, Q.Z.; Wu, S.H. Experimental study on the interlayer friction and wear mechanism between armor wires of umbilicals. *Mar. Struct.* **2021**, *80*, 103102. [CrossRef]

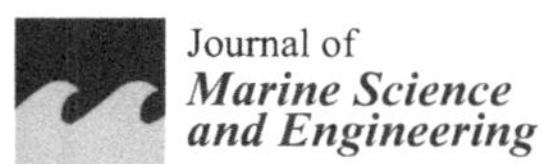

Journal of
Marine Science and Engineering

Article

Lateral Buckling of Subsea Pipelines Triggered by Sleeper with a Nonlinear Pipe–Soil Interaction Model

Zhenkui Wang [1,2] and C. Guedes Soares [2,*]

[1]   Ocean College, Zhejiang University, Zhoushan 316021, China; zhenkui.wang@zju.edu.cn
[2]   Centre for Marine Technology and Ocean Engineering (CENTEC), Instituto Superior Técnico, Universidade de Lisboa, 1049-001 Lisbon, Portugal
*   Correspondence: c.guedes.soares@centec.tecnico.ulisboa.pt

Abstract: Buckle-initiation techniques, such as sleepers, are usually installed to trigger lateral buckling at pre-designated locations to release the axial compressive forces induced by thermal loading. Taking the nonlinear pipe–soil interaction model into account, a mathematical model is proposed to investigate the lateral buckling of subsea pipelines triggered by a sleeper. The numerical solution is validated by comparing the model with solutions in the literature, and the model shows good agreement. The discrepancy between them is analysed by presenting the effect of mobilisation distance during buckling. The influence of the breakout resistance, sleeper height, and sleeper friction coefficient on the buckled configuration, post-buckling behaviour, and minimum critical temperature difference is discussed parametrically. The results show that the deformation of the buckled pipeline shrinks, and both the minimum critical temperature difference and the maximum stress along the buckled pipeline enlarge when the nonlinear pipe–soil interaction model is incorporated. However, the influence of the nonlinear pipe–soil interaction reduces with increasing sleeper height.

Keywords: subsea pipeline; nonlinear pipe–soil interaction model; breakout resistance; sleeper

Citation: Wang, Z.; Guedes Soares, C. Lateral Buckling of Subsea Pipelines Triggered by Sleeper with a Nonlinear Pipe–Soil Interaction Model. *J. Mar. Sci. Eng.* **2022**, *10*, 757. https://doi.org/10.3390/jmse10060757

Academic Editor: Fuping Gao

Received: 22 April 2022
Accepted: 22 May 2022
Published: 30 May 2022

Publisher's Note: MDPI stays neutral with regard to jurisdictional claims in published maps and institutional affiliations.

1. Introduction

Subsea pipelines may buckle laterally due to the excessive axial compressive force due to high-temperature and high-pressure conditions. Lateral buckling occurs when the axial compressive force reaches critical levels. Lateral buckling, if not controlled, can lead to serious accidents involving local buckling, fracture, and fatigue [1]. To control this phenomenon, buckle initiation techniques, such as sleepers, are employed along pipelines to trigger buckles at predesigned locations. A sleeper is a pipe segment that is installed underneath and perpendicular to the pipeline, which typically has a low friction surface to reduce the lateral friction force. Thus, the pipeline is uplifted vertically. A combination of the vertical out-of-straightness and low lateral resistance results in reduced critical buckling force. When the sleeper is used as the buckle initiation facility, part of the pipeline is suspended. The pipeline segment at the end of the suspended section has a larger embedment into the seabed, since a vertical concentrated force exists. This embedment affects the lateral breakout resistance, which is a key design parameter governing the initiation of the lateral buckle. Thus, a nonlinear pipe—soil interaction model is considered in the mathematical model to investigate the influence of breakout resistance on post-buckling behaviour.

The global buckling of subsea pipelines was investigated by numerous researchers. Hobbs' solutions for a straight pipeline were derived by assuming specific buckling mode shapes and constant lateral soil resistance [2]. Based on this, an analytical model was proposed by Taylor and Gan [3] with a consideration of initial imperfection. A simplified analytical model was proposed by Croll [4] for upheaval buckling. The interaction between propagation buckling and global buckling in subsea pipelines was investigated by Karampour et al. [5]. This interaction leads to a significant reduction in buckle design capacity. The lateral buckling of imperfect pipelines was studied by Liu et al. [6], using FEM. The analytical solutions for the high-order lateral buckling of a pipeline with symmetric and anti-symmetric initial imperfection were derived by Hong et al. [7] and Liu et al. [8], respectively. Lateral buckling was investigated by Konuk [9,10] with coupled lateral and axial pipe–soil interactions. Zhang et al. [11,12] derived unified formulas for the critical buckling forces of the upheaval and lateral buckling of subsea pipelines with different types of initial imperfection. The influence of the pipe length on the lateral buckling behaviour of imperfect pipelines was investigated through FEM [13].

More recently, researchers investigated the influence of the nonlinear pipe–soil interaction model on lateral buckling. Zeng and Duan [14] used a quintic polynomial formula to simulate nonlinear pipe–soil interactions. Incorporating the tri-linear pipe–soil interaction model, Chee et al. [15] investigated the effect of imperfections on the buckling response through FEM. Considering both the initial imperfection and the nonlinear lateral soil resistance model, the critical force of the lateral buckling was analysed by assuming that the length of the buckled region equals the wavelength of initial imperfection [16].

To increase the reliability of buckle formation predictions, buckle initiation facilities were incorporated into the mathematical models. A single buoyancy or distributed buoyancy with a specific length installed along the pipeline was considered to derive some simple analytical solutions [17]. The critical load of lateral buckling triggered by a single buoyancy was investigated by Shi and Wang [18]. Moreover, dual distributed buoyancy sections with a gap between them were employed to initiate lateral buckling [19]. A new way to trigger lateral buckling is to introduce a pre-deformed section along the pipeline before installation [20]. Lateral buckling triggered by a sleeper was investigated experimentally by Silva-Junior et al. [21] and de Oliveira Cardoso and Solano [22]. Bai et al. [23] studied the lateral buckling triggered by dual sleepers through FEM. Analytical solutions for antisymmetric buckling modes triggered by a sleeper were obtained by Wang and Tang [24]. They found that the symmetric buckling mode was more likely to occur with lower sleeper friction or smaller sleeper height. Hong and Liu [25] investigated the vertical deflection of a pipeline on a sleeper by FEM.

By assuming constant lateral soil resistance, analytical solutions were derived for the lateral thermal buckling triggered by a sleeper in [26]. In practice, the pipeline always has an initial embedment into the soil and the lateral soil resistance is not constant. However, there are no studies about lateral thermal buckling triggered by sleepers that consider nonlinear lateral soil resistance.

The innovative aspect of this study is the nonlinear pipe–soil interaction model that is incorporated into the governing equations. In previous published studies about lateral buckling triggered by sleepers, the lateral soil resistance, $f(w_2)$, is assumed to be constant. However, in this study, this function is nonlinear and it includes the effect of breakout resistance. This is due to the fact that in practice, pipe–soil interactions are nonlinear.

2. Mathematical Modelling

To avoid rogue buckles along subsea pipelines, buckle-initiation techniques, such as installing sleepers along the pipeline, are usually employed to trigger the pipeline to buckle in a controlled way at the predesignated location. For a pipeline laid on a sleeper and subjected to a temperature difference T_0, the axial compressive force is accumulated. The axial compressive force, P_0, is expressed as

$$P_0 = EA\alpha T_0 \tag{1}$$

where E is the elastic modulus, A is the cross-sectional area of the pipeline, and α is the coefficient of linear thermal expansion.

When P_0 is larger than the critical value, lateral buckling can be triggered at sleeper. The configuration and load distribution of lateral buckling are illustrated in Figure 1. From Figure 1a, it is clear that part of the pipe segment within $-l_1 \leq x \leq l_1$ is uplifted by the sleeper. In $-l_1 \leq x \leq l_1$, the soil resistances are zero. However, there are concentrated contact forces F_s between pipeline and sleeper at the sleeper and F_t between the pipe and seabed at the end of the suspended section, respectively. The vertical configuration of the pipeline laid on a sleeper was solved by Wang et al. [26]. From their derivation, F_s and F_t can be expressed as

$$F_s = \frac{4}{3}W_{pipe}l_1 \text{ and } F_t = \frac{1}{3}W_{pipe}l_1 \tag{2}$$

where W_{pipe} is the submerged weight per unit length and l_1 is the half-length of the free span, solved by

$$l_1 = \sqrt[4]{\frac{72EIv_{om}}{W_{pipe}}} \tag{3}$$

where I is the moment of inertia and v_{om} is the sleeper height. Therefore, the value of l_1 can be obtained by Equation (3) when v_{om} is specified. Furthermore, F_s and F_t can be solved.

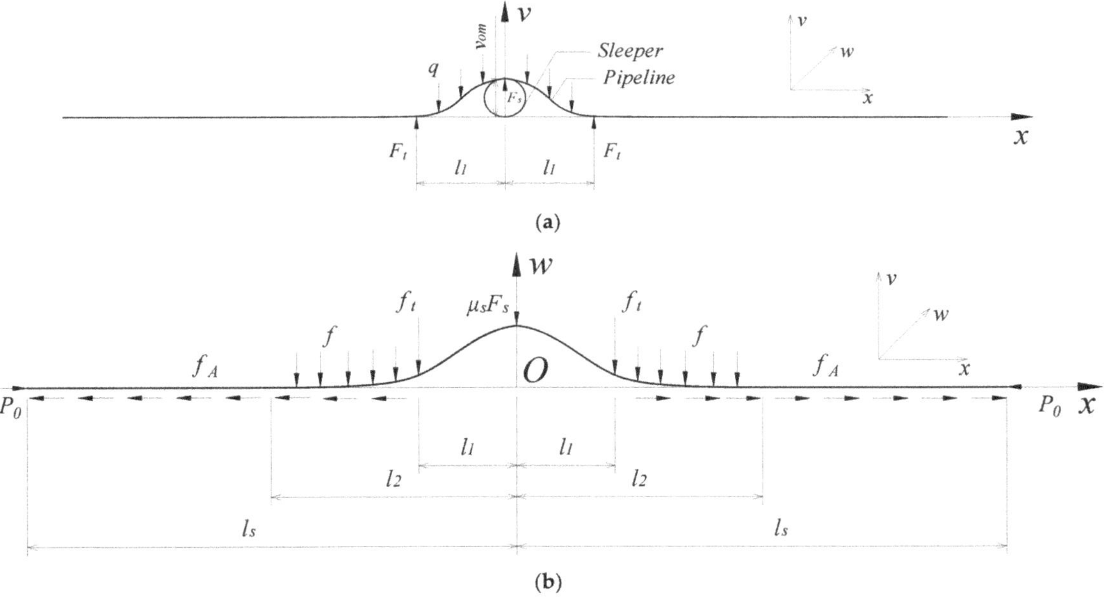

Figure 1. Configuration and load distribution. (**a**) Vertical plane. (**b**) Horizontal plane.

After the pipeline buckles, additional pipe coming from the thermal expansion is fed into the buckled section. Therefore, the axial force reduces partially due to the release of axial strain (see Figure 2). The axial force within the suspended region $-l_1 < x < l_1$, denoted by P, is constant. At $x = \pm l_1$, there is a jump in axial force with an amplitude of f_{At} induced by F_t. Within the region where the pipeline makes contact with the seabed, the axial force increases because of the restraint of axial soil resistance. The axial force will reach P_0 at $x = \pm l_s$. From Figure 2, the axial force distribution $\overline{P}(x)$ is

$$\overline{P}(x) = \begin{cases} P \ (0 \leq x < l_1) \\ P + f_{At} + f_A(x - l_1) \ (l_1 \leq x \leq l_s) \end{cases} \tag{4}$$

where $f_A = \mu_A W_{pipe}$ is the axial soil resistance per unit length and μ_A is the axial friction coefficient. The force $f_{At} = \mu_A F_t$ is induced by F_t.

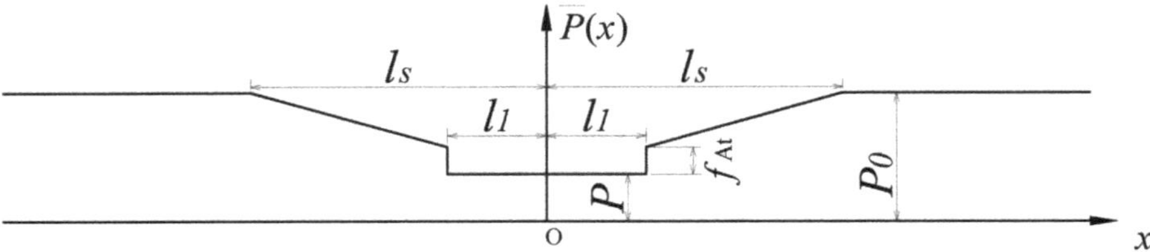

Figure 2. Axial compressive force distribution.

The axial force at $x = l_s$ is

$$\overline{P}(l_s) = P_0 = P + f_{At} + f_A(l_s - l_1) \tag{5}$$

Linear beam theory is used to simulate pipeline buckling. Thus, the equilibrium equations governing lateral deformation are [27]:

$$\begin{cases} EI\dfrac{d^4 w_1}{dx^4} + P\dfrac{d^2 w_1}{dx^2} = 0 & (0 \le x < l_1) \\ EI\dfrac{d^4 w_2}{dx^4} + \overline{P}(x)\dfrac{d^2 w_2}{dx^2} = -f(w_2) & (l_1 \le x \le l_2) \end{cases} \tag{6}$$

where w_1 and w_2 are lateral deflections, EI is bending stiffness, and $f(w_2)$ is the nonlinear lateral soil resistance determined by the nonlinear pipe–soil interaction model, as shown in Figure 3. Here, the variation in the axial force within the buckled region $l_1 \le x \le l_2$ is ignored when solving lateral deformations. This assumption is acceptable [26,28].

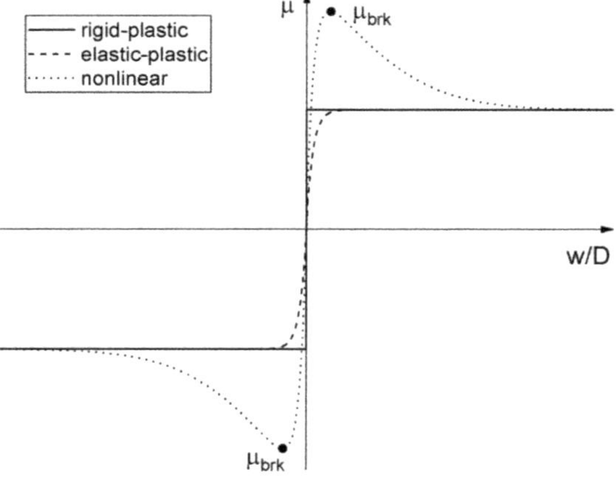

Figure 3. Pipe–soil interaction models.

Here, the nonlinear pipe–soil interaction model proposed by Chatterjee et al. [29] is employed. It can simulate breakout resistance and is given by

$$\mu = \frac{w}{|w|}\left(\mu_{brk}\left(1 - e^{-a_1\left(\frac{|w|}{D}\right)^{a_2}} \right) + (\mu_{res} - \mu_{brk})\left(1 - e^{-a_3\left(\frac{|w|}{D}\right)^{a_4}} \right) \right) \tag{7}$$

where μ, μ_{brk} and μ_{res} are the equivalent friction coefficients, and D is the external diameter of the pipeline. The quantities $f(w) = \mu W_{pipe}$, $F_{brk} = \mu_{brk} W_{pipe}$ and $F_{res} = \mu_{res} W_{pipe}$ are, therefore, respectively, the nonlinear lateral soil resistance, the breakout resistance, and the residual resistance. The value of the coefficient a_3 in [29] is given as

$$a_3 = a_5 \left(\frac{W_{pipe}}{V_{max}} \right) + a_6 \tag{8}$$

where W_{pipe} is the weight of the pipe and V_{max} is the vertical bearing capacity. The values of a_5 and a_6 are calculated by

$$a_5 = 8.2 \frac{v_{init}}{D} - 4.9, \qquad a_6 = -5.8 \frac{v_{init}}{D} + 4.5 \tag{9}$$

where v_{init} is the initial embedment of the pipe into the soil. In [29], $a_1 = 25$ and $a_4 = 1.5$ are employed, but $a_2 = 1$ is used here in order to have a finite linear resistance, which is physically realistic. $V_{max} = 5W_{pipe}$ and $v_{init} = 0.3D$ are adopted so that $a_3 = 2.272$, and set $\mu_{res} = 0.5$.

Due to symmetry, half a pipeline is considered. The slope of the deflection at $x = 0$ is zero, while the shear force $f_{ow} = \mu_s F_s / 2$ at $x = 0$ is induced by the friction force $\mu_s F_s$. Here, μ_s is the friction coefficient between pipeline and sleeper. The displacement, slope, and moment at $x = l_2$ are also zero. The boundary conditions at $x = 0$ and $x = l_2$ are

$$\begin{cases} \frac{dw_1}{dx}(0) = 0 \\ \frac{d^3 w_1}{dx^3}(0) + \frac{f_{ow}}{EI} = 0 \\ w_2(l_2) = 0 \\ \frac{dw_2}{dx}(l_2) = 0 \\ \frac{d^2 w_2}{dx^2}(l_2) = 0 \end{cases} \tag{10}$$

The displacement, slope, and bending moment must be continuous at the touchdown point $x = l_1$, while there is a jump in shear force with an amplitude of $f_t = \mu_{res} F_t$ at $x = l_1$ induced by the force F_t. Thus, additional conditions at $x = l_1$ are

$$\begin{cases} w_1(l_1) = w_2(l_1) \\ \frac{dw_1}{dx}(l_1) = \frac{dw_2}{dx}(l_1) \\ \frac{d^2 w_1}{dx^2}(l_1) = \frac{d^2 w_2}{dx^2}(l_1) \\ \frac{d^3 w_1}{dx^3}(l_1) = \frac{d^3 w_2}{dx^3}(l_1) + \frac{f_t}{EI} \end{cases} \tag{11}$$

With Equations (10) and (11), the nonlinear governing equations are solved numerically by the shooting method [30]. Once the lateral deflections are known, the geometric shortening u_2 is obtained by

$$u_2 = \frac{1}{2} \int_0^{l_1} \left(\frac{dw_1}{dx} \right)^2 dx + \frac{1}{2} \int_{l_1}^{l_2} \left(\frac{dw_2}{dx} \right)^2 dx \tag{12}$$

The following compatibility condition is employed to link the lateral deflection and the thermal loading induced deflection:

$$u_1 = u_2 \tag{13}$$

where u_1 is thermal expansion in $0 < x < l_s$.

We have

$$u_1 = \int_0^{l_s} \frac{\Delta \overline{P}(x)}{EA} dx \tag{14}$$

where $\Delta \overline{P}(x) = P_0 - \overline{P}(x)$.

Thus, this leads to

$$u_1 = \frac{f_A(l_s - l_1)^2}{2EA} + \frac{(P_0 - P)l_1}{EA}$$

(15)

The following formula is obtained by combining Equations (5), (13), and (15).

$$l_s = \sqrt{\frac{1}{3}l_1^2 + \frac{2EAu_2}{f_A}}$$

(16)

With Equations (5) and (16), one finally obtains

$$P_0 = P + f_A\left(\sqrt{\frac{1}{3}l_1^2 + \frac{2EAu_2}{f_A}} - \frac{2}{3}l_1\right)$$

(17)

The bending moment is obtained by

$$M = EI\frac{d^2w}{dx^2}$$

(18)

where w stands for w_1 or w_2, and the bending stress σ_M is

$$\sigma_M = \frac{MD}{2I}$$

(19)

The maximum stress is

$$\sigma_m = \sigma_P + \sigma_{Mm}$$

(20)

where the stresses σ_P and σ_{Mm}, induced by axial force P and maximum bending moment M_m, respectively, are

$$\begin{cases} \sigma_P = \frac{P}{A} \\ \sigma_{Mm} = \left|\frac{M_mD}{2I}\right| \end{cases}$$

(21)

3. Results

The mathematical model is validated by comparing it with the analytical solution in [26], and the discrepancy between them is discussed. Next, the influence of μ_{brk}, v_{om} and μ_s is analysed. The results are obtained by employing the analytical formulation developed in Section 2 and taking the parameters in Table 1.

Table 1. Parameters.

Parameter	Value	Unit
External diameter D	323.9	mm
Wall thickness t	12.7	mm
Elastic modulus E	206	GPa
Steel density ρ	7850	kg/m^3
Coefficient of thermal expansion α	1.1×10^{-5}	°C
Axial friction coefficient μ_A	0.5	—

One should note that only the analytical solutions in Figures 4 and 5 come from [26], which is used to validate the numerical results obtained in this study. In [26], the lateral soil resistance is assumed to be constant, while in the present study, nonlinear lateral soil resistance is considered. Moreover, in [26], analytical solutions are obtained due to the assumption of constant lateral soil resistance. In the present study, because the lateral soil resistance is nonlinear, Equation (6) cannot be solved analytically. Thus, the shooting method is used to solve Equation (6) to obtain the numerical results.

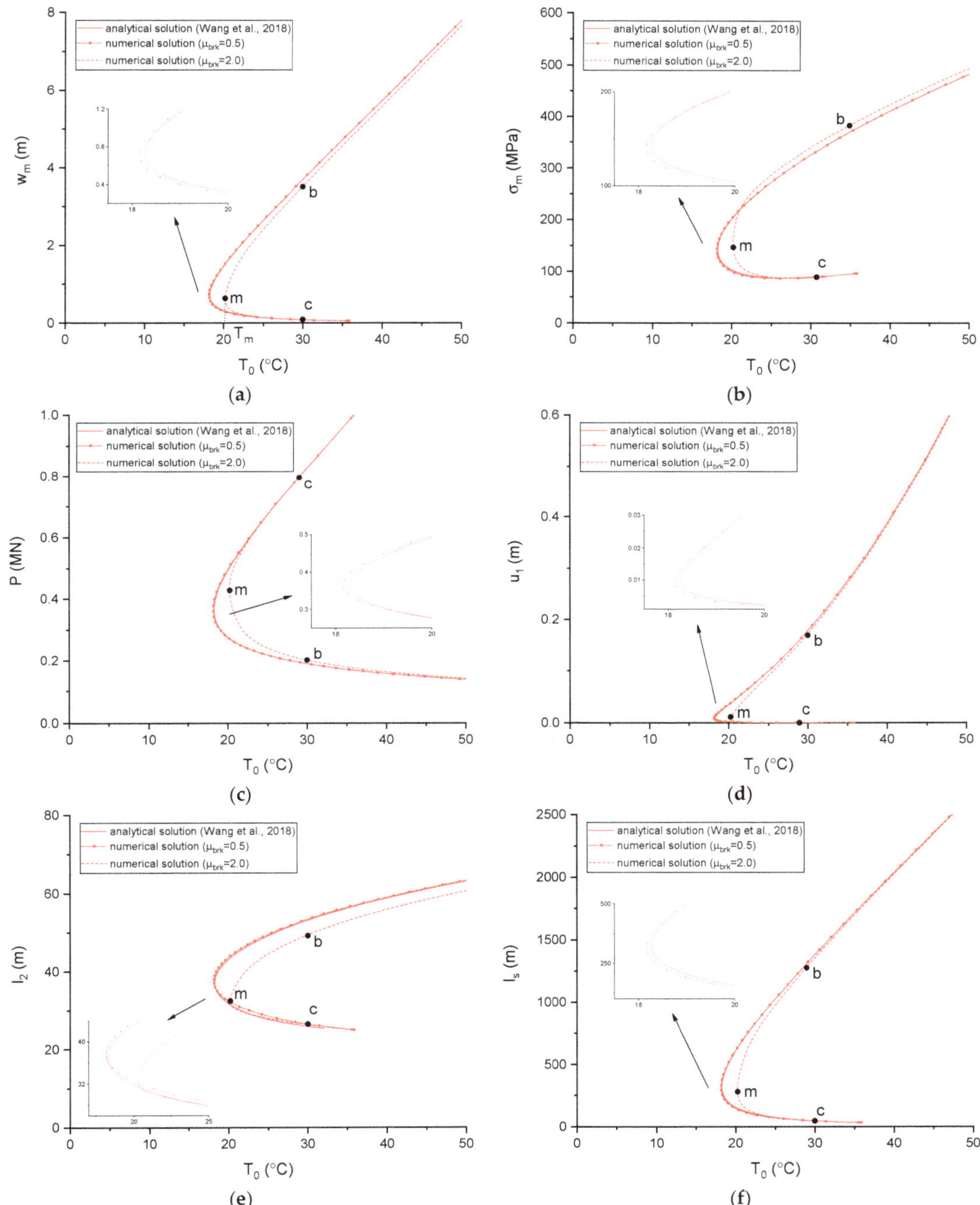

Figure 4. Validation. (**a**) Displacement amplitude w_m. (**b**) Maximum stress σ_m. (**c**) Axial force P. (**d**) Thermal expansion u_1. (**e**) Half-length of buckled region l_2. (**f**) Half-length of feed-in region l_s. ($v_{om} = 0.1$ m, $\mu_s = 0.1$) [26].

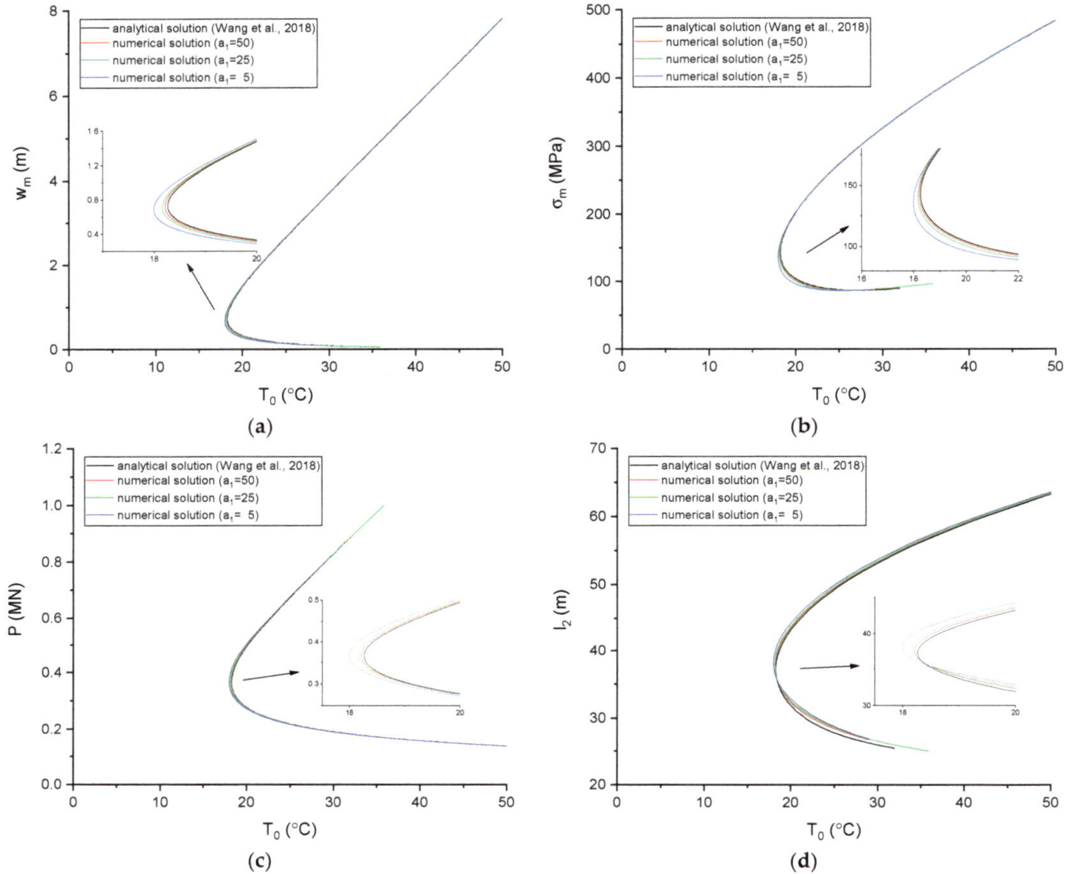

Figure 5. Error analysis. (**a**) w_m. (**b**) σ_m. (**c**) P. (**d**) l_2 ($\mu_{brk} = 0.5$, $v_{om} = 0.1$ m, $\mu_s = 0.1$) [26].

3.1. Validation

The solutions obtained in this study were validated by comparing them with the analytical solutions in [26], as shown in Figure 4. An error analysis is shown in Figure 5. In Figure 4, the analytical solutions are obtained by using the formulas derived in Wang et al. [26] with constant lateral soil resistance. To compare with the analytical solutions, the numerical solutions shown in Figure 4 are obtained by assuming $\mu_{brk} = 0.5$. For $\mu_{brk} = 0.5$, the nonlinear pipe–soil interaction model is reduced to elastic-plastic (see Figure 3). The numerical solutions for $\mu_{brk} = 2.0$ are also illustrated in Figure 4 to show the influence of the nonlinear pipe–soil interaction.

In Figure 4, there are two branches for each solution, which are denoted as **m-b** and **m-c**. The temperature difference at m, i.e., T_m, is called the minimum critical temperature difference, since solutions only exist for $T_0 > T_m$.

From Figure 4, the numerical solutions for $\mu_{brk} = 0.5$ are in good agreement with the analytical solutions, except that there is a slight discrepancy between them around T_m. This discrepancy comes from the difference in mobilization distance. For the rigid-plastic model, the resistance is always constant (see Figure 3). For the elastic-plastic model, the lateral soil resistance increases from zero to residual resistance gradually (see Figure 3).

In Figure 4, the discrepancy between the analytical and numerical solutions reduces as the temperature difference increases. The reason for this is that the displacement amplitude increases along with the temperature difference, so that more pipe sections fall into the region of constant lateral soil resistance for numerical solutions.

A more detailed error analysis is illustrated in Figure 5. Figure 6 shows that the mobilization distance is controlled by the parameter a_1. The mobilization distance is the distance that the lateral resistance reaches μ_{brk}. In Figure 6, $\mu_{brk} = \mu_{res} = 0.5$, so the mobilization distance in Figure 6 is the distance at which μ reaches 0.5. The elastic-plastic model approaches the rigid-plastic model for larger a_1, since the mobilization distance becomes smaller. In Figure 5, the discrepancy between the analytical and numerical solutions becomes smaller for larger a_1. In Figures 4 and 5, the discrepancy in half-buckled length l_2 is larger than in the other parameters. This is because the deflection at the ends of the buckled section is small, and it is affected by the mobilization distance. For the remaining parameters, the discrepancy between the analytical and numerical solutions is small enough.

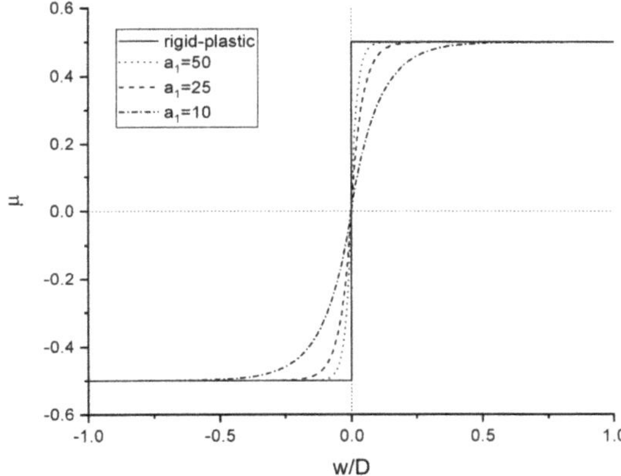

Figure 6. Pipe–soil interaction model with different a_1.

In Figure 4a, T_m becomes larger when the nonlinear pipe—soil interaction model is considered, which means that lateral buckling can only be triggered at higher temperature differences. At the same temperature difference, both w_m and l_2 become smaller when considering the nonlinear pipe–soil interaction model (see Figure 4a,e). Thus, the use of an additional pipe to create lateral deflection, which comes from thermal expansion, also reduces (see Figure 4d), so that l_s decreases, as shown in Figure 4f. However, P within the buckled section becomes larger due to the restriction of the breakout resistance. Moreover, at the same temperature difference, the maximum stress σ_m becomes larger when nonlinear pipe–soil interaction is considered. This means that the maximum stress is underestimated when assuming the lateral soil resistance to be constant.

3.2. Parametric Study

3.2.1. Influence of μ_{brk}

The influence of μ_{brk} on the buckled configuration, post-buckling behaviour, and minimum critical temperature difference T_m is shown in Figures 7–9, respectively.

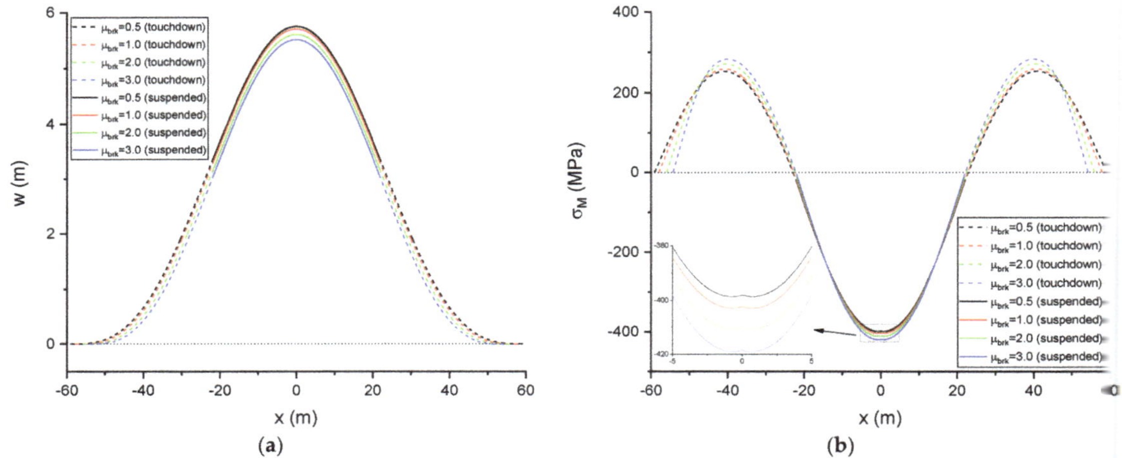

Figure 7. Influence of μ_{brk} on configurations. (**a**) Deformed shapes. (**b**) Bending stresses ($v_{om} = 0.1$ m, $\mu_s = 0.1$, $T_0 = 40\,°C$).

In Figure 7, the dashed curves are the pipe sections in contact with the seabed, and the solid curves are the pipe sections suspended due to the existence of the sleeper. In Figure 7a, both the touchdown and suspended pipe segments shrink with larger μ_{brk}. In Figure 7b, there are two extrema of bending stress in the positive direction. Another three extrema of bending stress occur around the sleeper. One local minimum (absolute value) of bending stress appears at the sleeper, while there are two other local maxima (absolute value) of bending stress close to the sleeper. The occurrence of the local minimum (absolute value) of bending stress at the sleeper is induced by the friction force between the pipeline and the sleeper. For each specific μ_{brk}, the maximum bending stress is located at the local maxima (absolute value) of bending stress close to the sleeper. For larger μ_{brk}, all the extrema of bending stress in both positive and negative directions become larger.

In Figure 8a, T_m is larger for larger μ_{brk}. A more detailed analysis on the influence of μ_{brk} on T_m is shown in Figure 9, which shows that T_m increases with increasing μ_{brk} for specific values of v_{om} and μ_s, and the increasing rate of T_m reduces with the increase in μ_{brk}. In Figure 9a, under the same μ_{brk}, T_m is larger for smaller v_{om}. The increasing rate of T_m with increasing μ_{brk} is also larger for smaller v_{om}. The reason is that since there are less length of suspended pipeline and larger length of touchdown pipeline with the smaller v_{om}, the breakout resistance has a larger influence on the initiation of lateral buckling. In Figure 9b, under the same μ_{brk}, T_m becomes larger for larger μ_s. The increase in the rate of T_m along with the increasing μ_{brk} remains almost the same for different values of μ_s. The reason for this is that the friction force between the pipeline and the sleeper becomes larger for larger μ_s; however, the value of μ_s has no influence on the lengths of the suspended or touchdown pipeline segments.

In Figure 8a,e, both the displacement amplitude w_m and the half-buckled length l_2 increase with the increasing T_0, and need more thermal expansion u_1 to form the buckled deflection (see Figure 8d). Thus, larger l_s is required for larger T_0, as shown in Figure 8f. The maximum stress also increases with increasing T_0 (see Figure 8b) since large deflection occurs; however, the axial force P reduces with increasing T_0 (see Figure 8c).

In Figure 8a,e, at a specific temperature difference, both w_m and l_2 become smaller for larger μ_{brk}. The reason for this is that since the breakout resistance is larger for larger μ_{brk}, the pipeline is subjected to greater lateral soil resistance. The deflection shrinks with larger μ_{brk}, as shown in Figure 7a. Therefore, both u_1 and l_s become smaller with larger μ_{brk} at the same temperature difference (see Figure 8d,f).

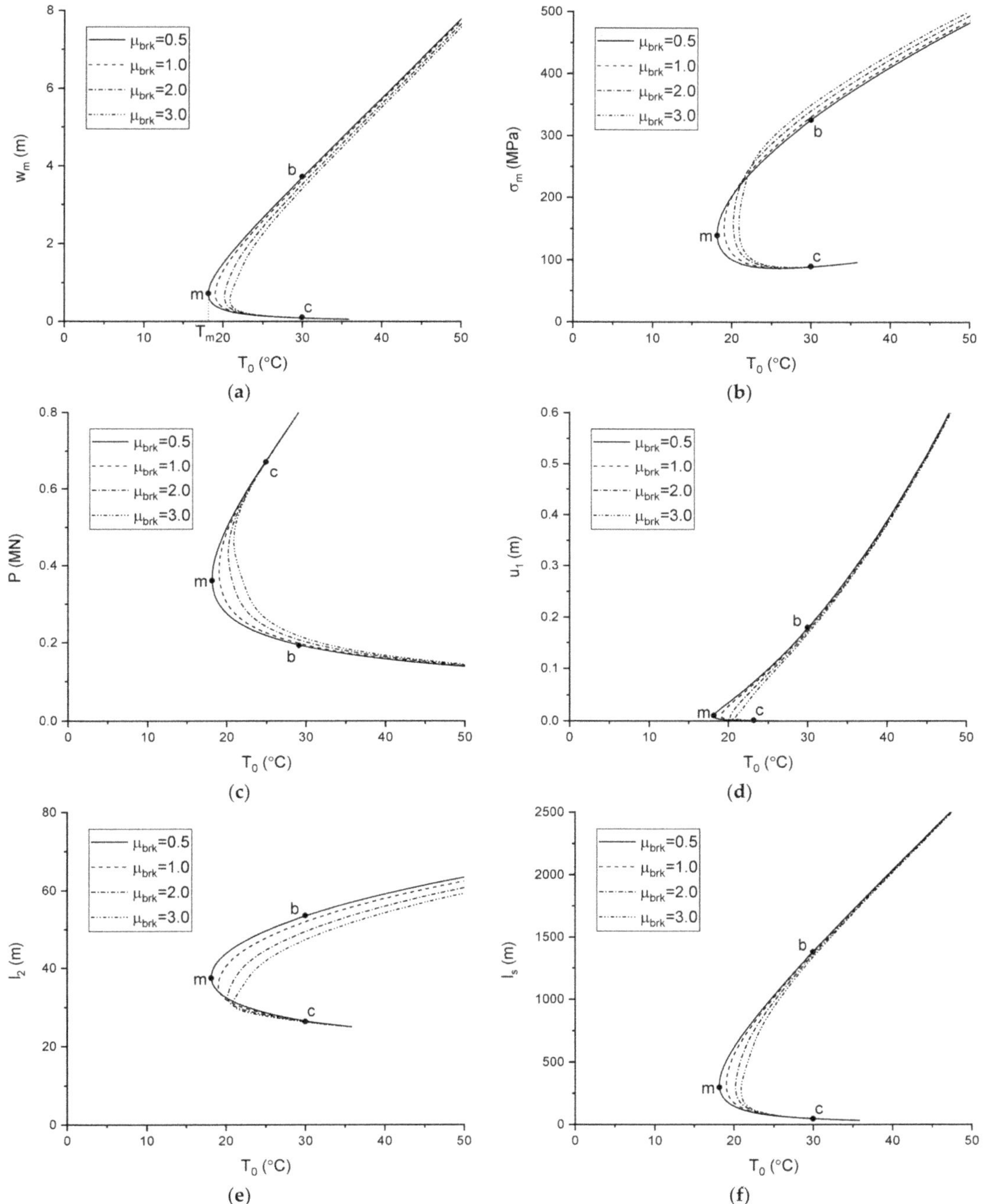

Figure 8. Influence of μ_{brk} on the buckling behaviour. (**a**) w_m. (**b**) σ_m. (**c**) P. (**d**) u_1. (**e**) l_2. (**f**) l_s ($v_{om} = 0.1$ m, $\mu_s = 0.1$).

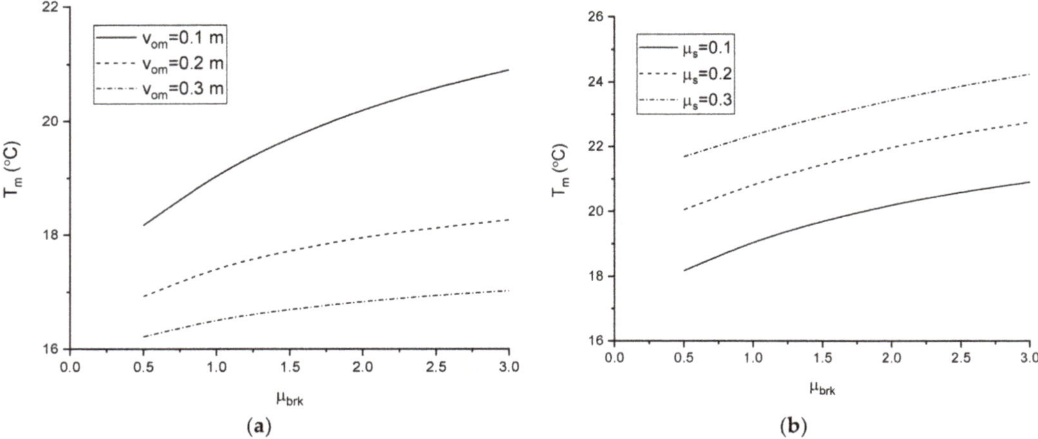

Figure 9. (**a**) Influence of μ_{brk} on T_m with different v_{om} ($\mu_s = 0.1$). (**b**) Influence of μ_{brk} on T_m with different μ_s. ($v_{om} = 0.1$ m).

However, P becomes larger with larger μ_{brk} (see Figure 8c). The reason for this is that the reduction in the axial force reduces, since the greater breakout resistance restricts the deflection of the buckled pipeline. Moreover, σ_m along the buckled pipeline becomes larger for larger μ_{brk} (see Figure 8b). After considering the nonlinear pipe–soil interaction model, both T_m and σ_m in the pipeline became larger. When the nonlinear pipe–soil interaction model is not included, lateral buckling may fail to be triggered by the sleeper, and the maximum stress along the buckled pipeline may exceed the allowable stress in the design.

3.2.2. Influence of v_{om}

The influence of v_{om} on the buckled configuration, post-buckling behaviour, and minimum critical temperature difference T_m are shown in Figures 10–12, respectively.

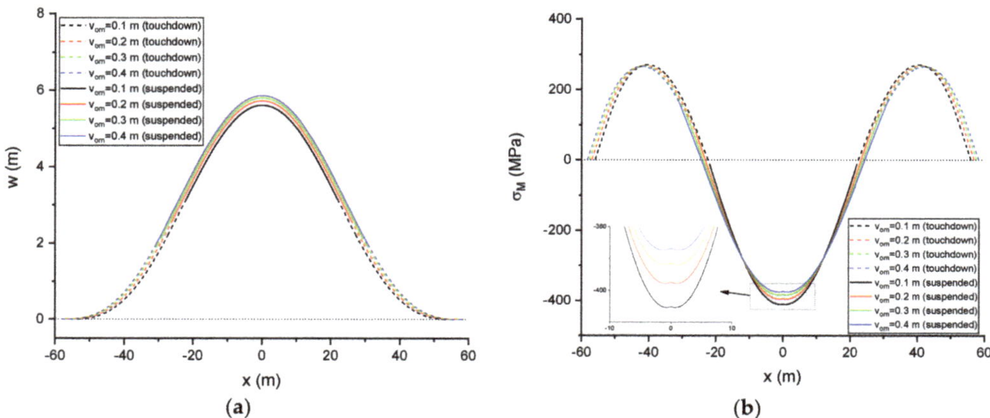

Figure 10. Influence of v_{om} on configurations. (**a**) Deformed shapes. (**b**) Bending stresses ($\mu_{brk} = 2.0$, $\mu_s = 0.1$, $T_0 = 40\,°\text{C}$).

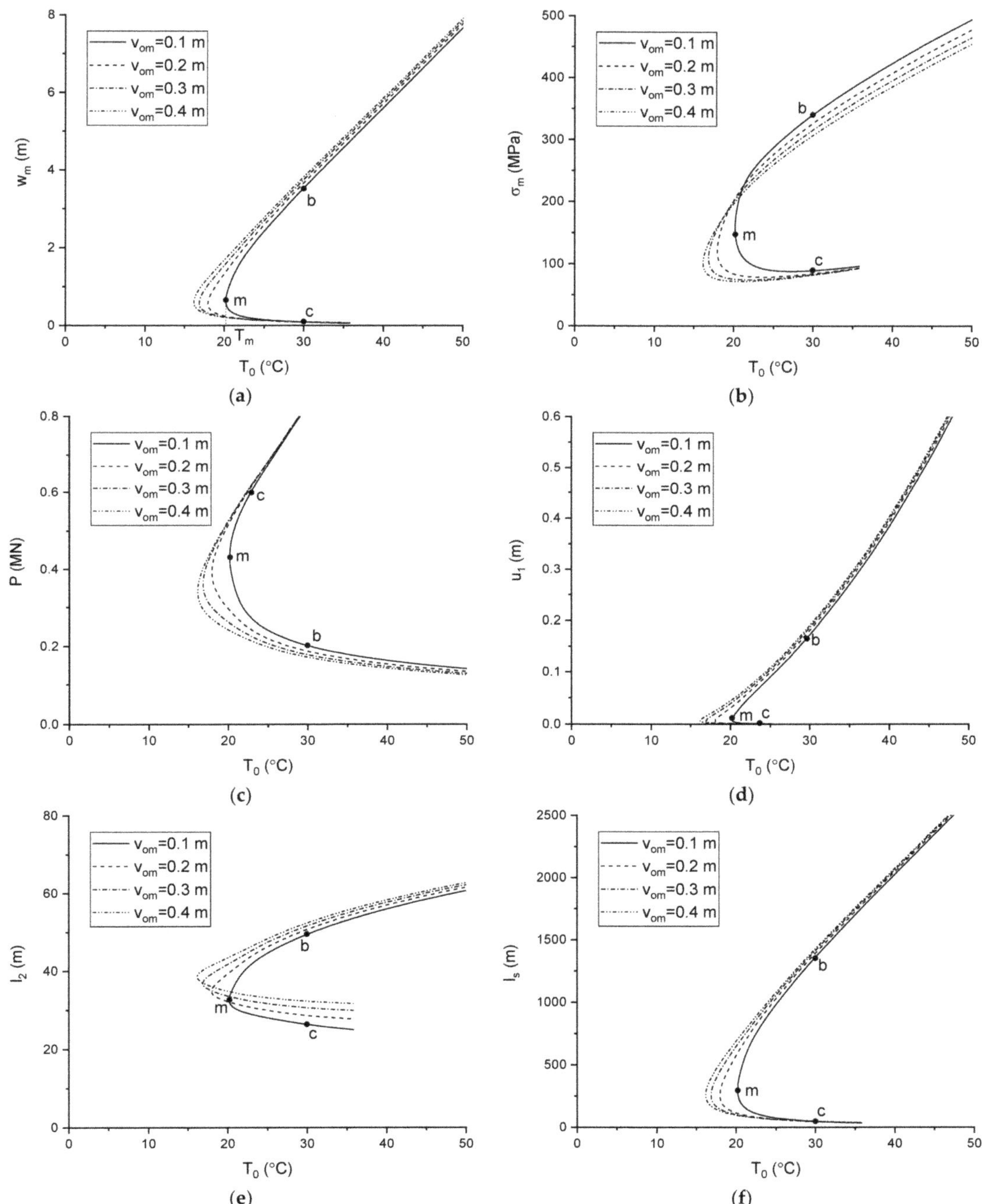

Figure 11. Influence of v_{om} on the buckling behaviour. (**a**) w_m. (**b**) σ_m. (**c**) P. (**d**) u_1. (**e**) l_2. (**f**) l_s ($\mu_{brk} = 2.0$, $\mu_s = 0.1$).

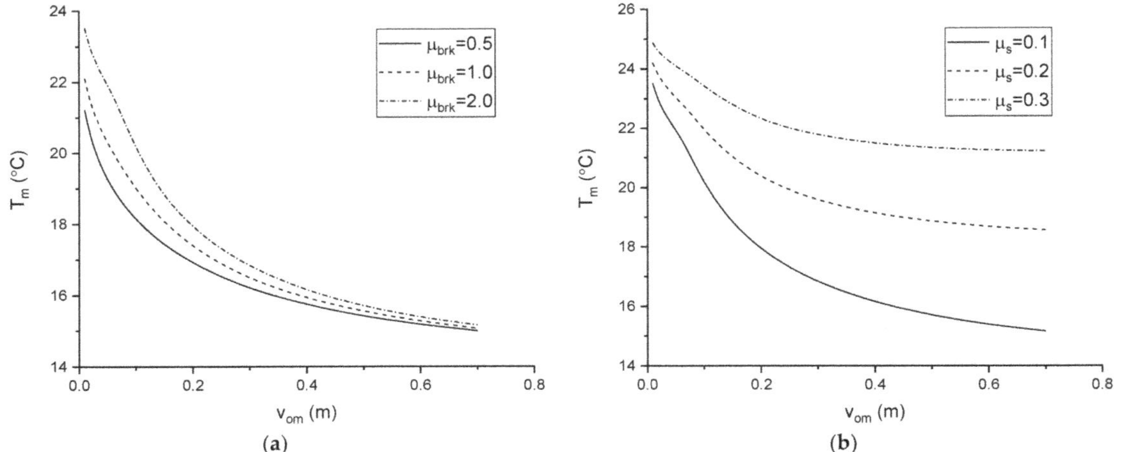

Figure 12. (**a**) Influence of v_{om} on T_m with different μ_{brk} ($\mu_s = 0.1$). (**b**) Influence of v_{om} on T_m with different μ_s ($\mu_{brk} = 2.0$).

From Figure 10a, it is clear that the length of the suspended pipe segment becomes larger with larger sleeper heights, v_{om}. The deflection of the buckled pipeline enlarges with larger v_{om}. Because the soil resistance for the suspended pipeline is zero, the buckled pipeline has less restriction from the seabed foundation with larger v_{om}. The deflection of the buckled pipeline enlarges with larger v_{om}, but both the local minimum and the local maximum (absolute value) of the bending stress become smaller with larger v_{om}, as shown in Figure 10b. This is because the deflection of the buckled pipeline is more benign with larger v_{om}.

In Figure 11a, T_m becomes smaller with larger v_{om}. Figure 12 illustrates the influence of v_{om} on T_m in detail. In Figure 12, T_m decreases with increasing v_{om} for specific values of μ_{brk} and μ_s, and the decreasing rate of T_m reduces with increasing v_{om}. In Figure 12a, under the same v_{om}, T_m becomes larger for larger μ_{brk}. The influence of μ_{brk} on T_m becomes smaller for larger v_{om}, since the length of suspended pipeline with zero soil resistance is greater. In Figure 12b, under the same v_{om}, T_m is larger with larger μ_s. The decreasing rate of T_m with increasing v_{om} becomes smaller with larger μ_s. The influence of μ_s on T_m is larger with larger v_{om}. The reason for this is that the concentrated contact force between the pipeline and the sleeper becomes larger with larger v_{om}, so that the friction force between the pipeline and the sleeper becomes larger with larger v_{om}. Therefore, an effective way to reduce T_m is to increase the sleeper height v_{om}; however, the corresponding weakness is that the suspended pipeline will be longer, which may lead to vortex-induced vibration.

In Figure 11a,e, at a specific temperature difference, both w_m and l_2 become larger with larger v_{om}. This is because the length of the suspended pipeline with zero soil resistance increases with increasing v_{om}, as shown in Figure 10a. There is less restriction from the seabed foundation with larger v_{om}. Due to the larger deflection with larger v_{om}, the requirement of additional pipes to feed into the buckled section increases, which creates the need for more thermal expansion (see Figure 11d) and a longer feed-in region (see Figure 11f). The axial force P becomes smaller with larger v_{om}, since a larger deflection occurs to release more axial force, as shown in Figure 11c. The maximum stress σ_m along the buckled pipeline reduces with larger μ_{brk} (see Figure 11b).

3.2.3. Influence of μ_s

The influence of μ_s on the buckled configuration, post-buckling behaviour, and minimum critical temperature difference T_m are shown in Figures 13–15, respectively.

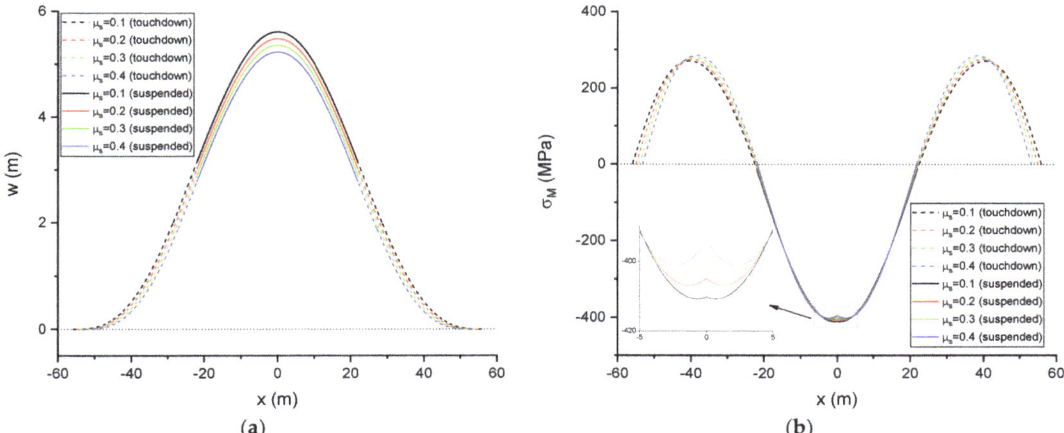

Figure 13. v_{om} on configurations. (**a**) Deformed shapes. (**b**) Bending stresses ($\mu_{brk} = 2.0$, $v_{om} = 0.1$ m, $T_0 = 40\,^{\circ}$C).

In Figure 13a, the deflection of the buckled pipeline shrinks with larger μ_s. This is because the friction force between the pipeline and the sleeper becomes larger with larger μ_s, which restricts the deflection of the buckled pipeline. In Figure 13b, the extrema of the bending stress in the positive direction becomes slightly larger with larger μ_s. However, both the local minimum and the local maximum (absolute value) of the bending stress close to the sleeper in the negative direction become smaller with larger μ_s, as shown in Figure 13b. With larger μ_s, the difference between the local minimum and the local maximum of the bending stress close to the sleeper becomes larger. Taking $\mu_s = 0.4$ as an example, it is clear that the local minimum (absolute value) of the bending stress is smaller than the local maximum (absolute value) of the bending stress.

In Figure 14a, T_m is larger with larger μ_s. The effect of μ_s on T_m is illustrated in Figure 15, with different values of μ_{brk} and v_{om}. In Figure 15, T_m increases with increasing μ_s for specific values of μ_{brk} and v_{om}, and the increasing rate of T_m slightly reduces with increasing μ_s. The friction force between the sleeper and the pipeline becomes larger with larger μ_s, which makes it more difficult to trigger the lateral buckling. In Figure 15a, at the same μ_s, T_m becomes larger with larger μ_{brk}. The increasing rate of T_m with increasing μ_s is similar for different values of μ_{brk}. In Figure 15b, under the same μ_s, T_m becomes smaller for larger v_{om}. The increasing rate of T_m with increasing μ_s is larger for larger v_{om}. The influence of v_{om} on T_m gradually reduces with increasing μ_s. Thus, the friction coefficient between the sleeper and the pipeline μ_s should be carefully controlled. When the value of μ_s is too large, such as $\mu_s = 0.6$, T_m is barely affected by the sleeper height v_{om}.

In Figure 14a,e, under a specific T_0, both w_m and l_2 reduce with larger μ_s. This is because, since the friction force between the sleeper and the pipeline becomes larger with larger μ_s, the deflection of the buckled pipeline is restricted by the larger resistance between the sleeper and the pipeline. Thus, the requirements of both u_1 and l_s decrease with larger μ_s, as shown in Figure 14d,f. Due to the restriction of the larger friction force between the sleeper and the pipeline, the axial force P increases with increasing μ_s, as shown in Figure 14c. However, the maximum stress σ_m along the buckled pipeline reduces with larger μ_s (see Figure 14b), which is induced by the decrease in the maximum bending stress (absolute value) along the buckled pipeline.

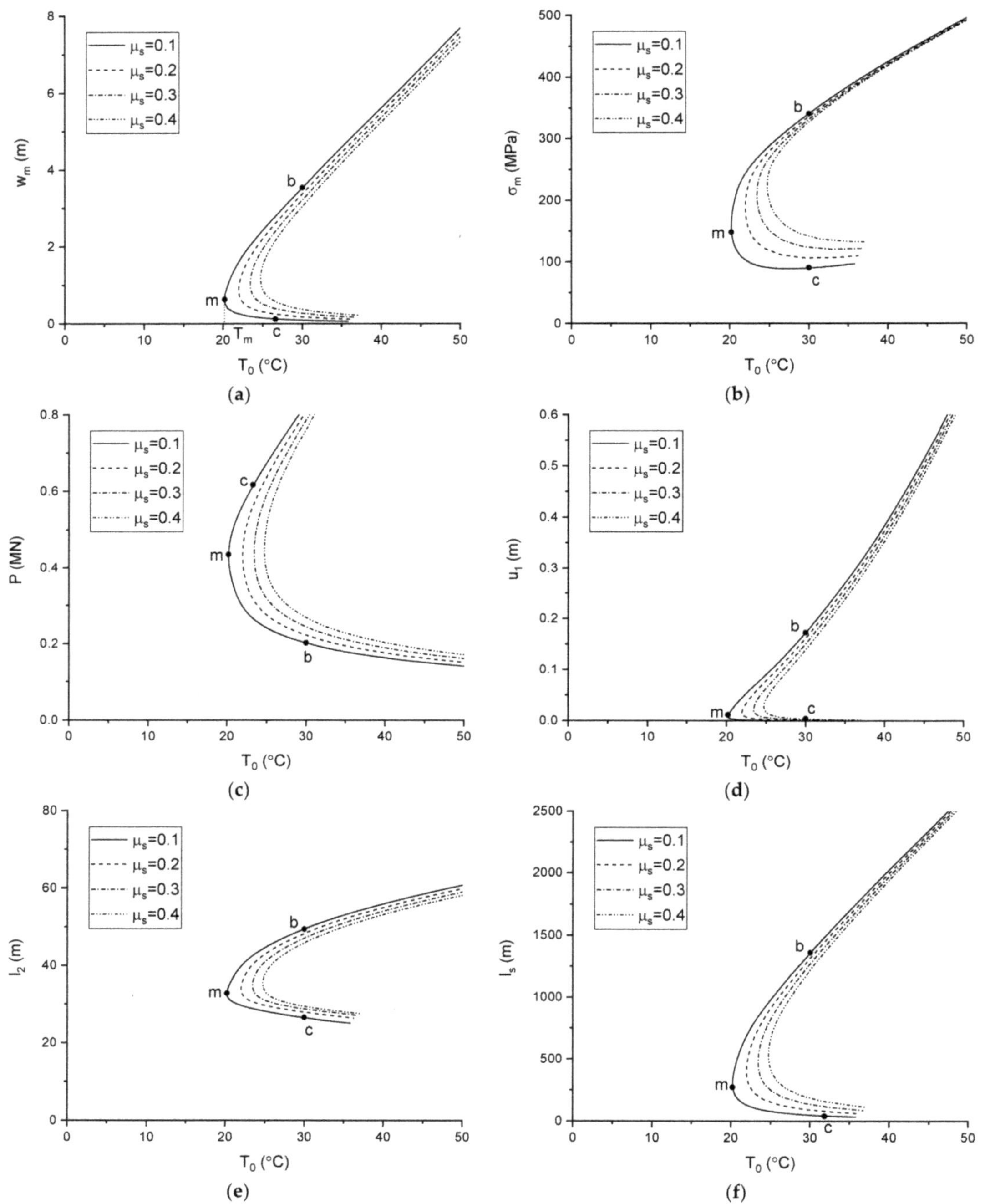

Figure 14. Influence of μ_s on the buckling behaviour. (**a**) w_m. (**b**) σ_m. (**c**) P. (**d**) u_1. (**e**) l_2. (**f**) l_s ($\mu_{brk} = 2.0$, $v_{om} = 0.1$ m).

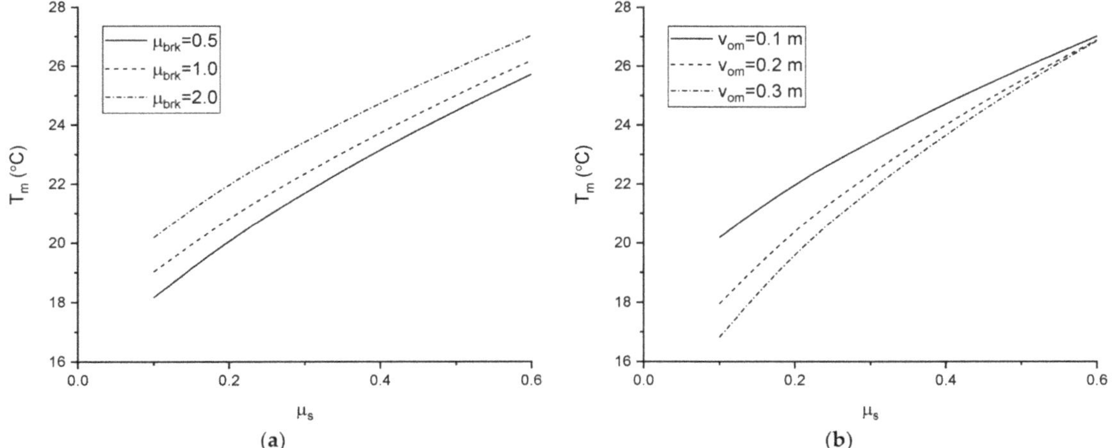

Figure 15. (**a**) Influence of μ_s on T_m with different μ_{brk} ($v_{om} = 0.1$ m). (**b**) Influence of μ_s on T_m with different v_{om} ($\mu_{brk} = 2.0$).

4. Conclusions

Through a consideration of the nonlinear pipe–soil interaction model, a mathematical model was proposed to simulate the lateral buckling of subsea pipelines triggered by a sleeper. The model was solved numerically and validated by comparing its predictions with the analytical solutions from [26]. The discrepancy between the numerical and analytical solutions was analysed through the discussion of the mobilization distance. A detailed parametric analysis was presented to show the effect of the breakout resistance, sleeper height, and sleeper friction coefficient on the buckling behaviour of a pipeline laid on a sleeper. The conclusions are:

(i)   The discrepancy between the numerical and analytical solutions comes from the difference between the elastic-plastic and rigid-plastic pipe–soil interaction models, which reduces with decreasing mobilization difference in the elastic-plastic pipe–soil interaction model.

(ii) When the nonlinear pipe–soil interaction model is taken into account, both the displacement amplitude and the buckled length reduce due to the occurrence of breakout resistance, which decreases further with increasing breakout resistance. However, both the axial force and the maximum stress, along with the buckled pipeline, increase, and increase further with increasing breakout resistance.

(iii) The deflection of the buckled pipeline enlarges as the sleeper height increases and shrinks as the sleeper friction coefficient increases. The axial force decreases with increasing sleeper height and increases with increasing sleeper friction coefficient. Moreover, the maximum stress along the buckled pipeline decreases with increasing sleeper height and with decreasing sleeper friction coefficient.

(iv) The minimum critical temperature difference increases with increasing breakout resistance and sleeper friction coefficient, and decreases with increasing sleeper height. The influence of the breakout resistance on the minimum critical temperature difference gradually reduces with increasing sleeper height. Moreover, the sleeper height has little effect on the minimum critical temperature difference when the sleeper friction coefficient is large enough.

In conclusion, it is better to incorporate the nonlinear pipe–soil interaction model into the mathematical model when simulating the lateral buckling of subsea pipelines triggered by a sleeper, since both the minimum critical temperature difference and the maximum stress increase. Moreover, both the sleeper height and the sleeper friction coefficient should be carefully selected and controlled.

Author Contributions: Conceptualization, Z.W. and C.G.S.; methodology, Z.W.; software, Z.W.; validation, Z.W.; investigation, Z.W.; writing—original draft preparation, Z.W.; writing—review and editing, C.G.S.; supervision, C.G.S.; project administration, C.G.S.; funding acquisition, C.C.S. All authors have read and agreed to the published version of the manuscript.

Funding: Zhenkui Wang would like to acknowledge that the work described in this paper was funded by the National Natural Science Foundation of China (grant number: 52001229). This work contributes to the Strategic Research Plan of the Centre for Marine Technology and Ocean Engineering (CENTEC), which is financed by the Portuguese Foundation for Science and Technology (Fundação para a Ciência e Tecnologia—FCT) under contract UIDB/UIDP/00134/2020.

Institutional Review Board Statement: Not applicable.

Informed Consent Statement: Not applicable.

Data Availability Statement: Not applicable.

Conflicts of Interest: The authors declare no conflict of interest.

References

1. *DNV-RP-F110*; Global Buckling of Submarine Pipelines. Det Norske Veritas: Oslo, Norway, 2019.
2. Hobbs, R.E. In-service buckling of heated pipelines. *J. Transp. Eng.* **1984**, *110*, 175–189. [CrossRef]
3. Taylor, N.; Gan, A.B. Submarine pipeline buckling-imperfection studies. *Thin-Walled Struct.* **1986**, *4*, 295–323. [CrossRef]
4. Croll, J.G.A. A simplified model of upheaval thermal buckling of subsea pipelines. *Thin-Walled Struct.* **1997**, *29*, 59–78. [CrossRef]
5. Karampour, H.; Albermani, F.; Veidt, M. Buckle interaction in deep subsea pipelines. *Thin-Walled Struct.* **2013**, *72*, 113–120. [CrossRef]
6. Liu, R.; Xiong, H.; Wu, X.; Yan, S. Numerical studies on global buckling of subsea pipelines. *Ocean Eng.* **2014**, *78*, 62–72. [CrossRef]
7. Hong, Z.; Liu, R.; Liu, W.; Yan, S. Study on lateral buckling characteristics of a submarine pipeline with a single arch symmetric initial imperfection. *Ocean Eng.* **2015**, *108*, 21–32. [CrossRef]
8. Liu, R.; Wang, X. Lateral global buckling high-order mode analysis of a submarine pipeline with imperfection. *Appl. Ocean. Res.* **2018**, *73*, 107–126. [CrossRef]
9. Konuk, I. Coupled lateral and axial soil-pipe interaction and lateral buckling Part II: Solutions. *Int. J. Solids Struct.* **2018**, *132–133*, 127–152. [CrossRef]
10. Konuk, I. Coupled lateral and axial soil-pipe interaction and lateral buckling Part I: Formulation. *Int. J. Solids Struct.* **2018**, *132–133*, 114–126. [CrossRef]
11. Zhang, X.; Duan, M. Prediction of the upheaval buckling critical force for imperfect submarine pipelines. *Ocean Eng.* **2015**, *109*, 330–343. [CrossRef]
12. Zhang, X.; Guedes Soares, C.; An, C.; Duan, M. An unified formula for the critical force of lateral buckling of imperfect submarine pipelines. *Ocean Eng.* **2018**, *166*, 324–335. [CrossRef]
13. Liu, R.; Li, C. Determinate dimension of numerical simulation model in submarine pipeline global buckling analysis. *Ocean Eng.* **2018**, *152*, 26–35. [CrossRef]
14. Zeng, X.; Duan, M. Mode localization in lateral buckling of partially embedded submarine pipelines. *Int. J. Solids Struct.* **2014**, *51*, 1991–1999. [CrossRef]
15. Chee, J.; Walker, A.; White, D. Effects of variability in lateral pipe-soil interaction and pipe initial out-of-straightness on controlled lateral buckling of pre-deformed pipeline. *Ocean Eng.* **2019**, *182*, 283–304. [CrossRef]
16. Zhang, X.; Guedes Soares, C. Lateral buckling analysis of subsea pipelines on nonlinear foundation. *Ocean Eng.* **2019**, *186*, 106085. [CrossRef]
17. Peek, R.; Yun, H. Flotation to trigger lateral buckles in pipelines on a flat seabed. *J. Eng. Mech.* **2007**, *4*, 442–451. [CrossRef]
18. Shi, R.; Wang, L. Single buoyancy load to trigger lateral buckles in pipelines on a soft seabed. *J. Eng. Mech.* **2015**, *141*, 1–7. [CrossRef]
19. Wang, Z.; Tang, Y. Antisymmetric thermal buckling triggered by dual distributed buoyancy sections. *Mar. Struct.* **2020**, *74*, 102811. [CrossRef]
20. Chee, J.; Walker, A.; White, D. Controlling lateral buckling of subsea pipeline with sinusoidal shape pre-deformation. *Ocean Eng.* **2018**, *151*, 170–190. [CrossRef]
21. Silva-Junior, H.C.; Cardoso, C.O.; Carmignotto, M.A.P.; Zanutto, J.C. Reduced Model Device of Solutions to Control Thermal Buckling Effects in HP-HT Subsea Pipelines (OMAE2008-57637). In Proceedings of the International Conference on Ocean, Offshore and Arctic Engineering, Estoril, Portugal, 15 June 2008.

22. de Oliveira Cardoso, C.; Solano, R.F. Performed of triggers to control thermal buckling of subsea pipelines using reduced scale model (ISOPE-I-15-445). In Proceedings of the International Offshore and Polar Engineering Conference, Honolulu, HI, USA, 21–26 June 2015.
23. Bai, Q.; Qi, X.; Brunner, M. Global buckle control with dual sleepers in HP/HT pipelines (OTC-19888-MS). In Proceedings of the Offshore Technology Conference, Houston, TX, USA, 4–7 May 2009.
24. Wang, Z.; Tang, Y. Analytical study on controlled lateral thermal buckling of antisymmetric mode for subsea pipelines triggered by sleepers. *Mar. Struct.* **2020**, *71*, 102728. [CrossRef]
25. Hong, Z.; Liu, W. Modelling the vertical lifting deformation for a deep-water pipeline laid on a sleeper. *Ocean Eng.* **2020**, *199*, 107042. [CrossRef]
26. Wang, Z.; Tang, Y.; van der Heijden, G.H.M. Analytical study of lateral thermal buckling for subsea pipelines with sleeper. *Thin-Walled Struct.* **2018**, *122*, 17–29. [CrossRef]
27. Lagrange, R.; Averbuch, D. Solution methods for the growth of a repeating imperfection in the line of a strut on a nonlinear foundation. *Int. J. Mech. Sci.* **2012**, *63*, 48–58. [CrossRef]
28. Wang, Z.; Tang, Y.; van der Heijden, G.H.M. Analytical study of distributed buoyancy sections to control lateral thermal buckling of subsea pipelines. *Mar. Struct.* **2018**, *58*, 199–222. [CrossRef]
29. Chatterjee, S.; White, D.J.; Randolph, M.F. Numerical simulations of pipe-soil interaction during large lateral movements on clay. *Géotechnique* **2012**, *62*, 693–705. [CrossRef]
30. *Wolfram Mathematica*, version 11.0; Wolfram Research, Inc.: Champaign, IL, USA, 2016.

Journal of
Marine Science and Engineering

Article

Design Equation of Buckle Propagation Pressure for Pipe-in-Pipe Systems

Ruoxuan Li, Bai-Qiao Chen and C. Guedes Soares *

Centre for Marine Technology and Ocean Engineering (CENTEC), Instituto Superior Técnico, Universidade de Lisboa, 1049-001 Lisboa, Portugal
* Correspondence: c.guedes.soares@centec.tecnico.ulisboa.pt

Abstract: This paper focuses on the buckle propagation pressure of a pipe-in-pipe system under uniform external pressure. After the validation of the finite element model used in this research, several effects were studied. At first, the effects of the initial imperfection of the inner pipe, including the ovality and eccentricity, were checked out. Second, the influence of the scantling of the inner and outer pipes was tested. It was found that the dominant factor which affects the buckle propagation mode is changeable. Then, the boundary value of the above factor was investigated by varying the ratio of the outer and inner diameters. Finally, after a summary of the calculation results, an empirical formula was proposed to illustrate the relationship between the buckle propagation pressure and the geometrical parameters of the pipe-in-pipe system. Utilizing this formula, the buckle propagation pressure of the pipe-in-pipe system can be estimated rapidly with good accuracy.

Keywords: subsea pipeline; pipe-in-pipe; external pressure; buckle propagation

1. Introduction

With the development of the offshore oil and gas industry, the deep-sea area becomes more and more important and valuable. As they constitute the main transmission unit, pipelines are required to be stronger to withstand higher water pressure. In an early work on the analytical solution of the collapse pressure of a single-walled pipe under external pressure, the significant influence of the ovality of the pipe cross-section was highlighted [1]. The pressure-carrying capacity is dependent on the D/t ratio of pipes under pure external pressure [2–6].

For subsea pipelines settled on the seabed with great water depth, thermal protection is still a task to be solved. Since the transmission distance of the pipeline system is up to several kilometres, the decrease in temperature will slow down the flow of the fluid, and even stop it. To avoid possible economic loss and environmental pollution in the aforementioned situation, pipe-in-pipe (PIP) systems are designed. The PIP contains two pipes: the outer pipe responsible for tolerating the hydrostatic pressure due to the subsea environment, and the inner pipe responsible for carrying the fluid gas or oil at high temperatures. Between the outer and inner pipes, the gap is expected to slow down the temperature dissipation of the fluid during the transmission over a long distance.

Alrsai et al. [7,8] figured out that the ultimate collapse strength of PIP is equal to that of a single-walled pipe with the same geometrical features as the outer pipe under external pressure. Generally, the two pipes of the PIP system are not very close to each other. Before the ultimate buckling state of the outer pipe, contact between the two pipes does not occur. Bhardwaj et al. [9] showed the uncertainty of the research of the PIP system.

In real engineering applications, the initial imperfection is always introduced onto the pipes during manufacture and installation procedures. When the pipes are subjected to external pressure, failure first occurs in the cross-section with the most severe initial imperfection, and then the buckle spreads rapidly along the axial direction [10]. Thus, in

Citation: Li, R.; Chen, B.-Q.; Guedes Soares, C. Design Equation of Buckle Propagation Pressure for Pipe-in-Pipe Systems. *J. Mar. Sci. Eng.* **2023**, *11*, 622. https://doi.org/10.3390/jmse11030622

Academic Editor: Erkan Oterkus

Received: 2 February 2023
Revised: 24 February 2023
Accepted: 7 March 2023
Published: 16 March 2023

addition to the ultimate collapse strength of pipelines, the buckle propagation pressure (P_p) is considered another key characteristic.

In general, the buckle propagation pressure is lower than the ultimate collapse strength. It is worth determining the value of the buckle propagation pressure. Netto and Estefen [10] performed experimental tests on the buckle propagation of single-walled pipes. Then, the dynamic propagation phenomenon and the formation of flip-flop was introduced by Lee et al. [11].

Based on the proposed buckle deformation mode for single-walled pipes [12], the analytical solutions of P_p are derived. Figure 1 shows the buckle propagation mode of the pipeline in the 2D scenario. Based on that, Kyriakides and Vogler [13] investigated the buckle propagation phenomenon in the PIP system. The cross-section of the pipeline contains four plastic hinges for each pipe, as shown in Figure 1a,b.

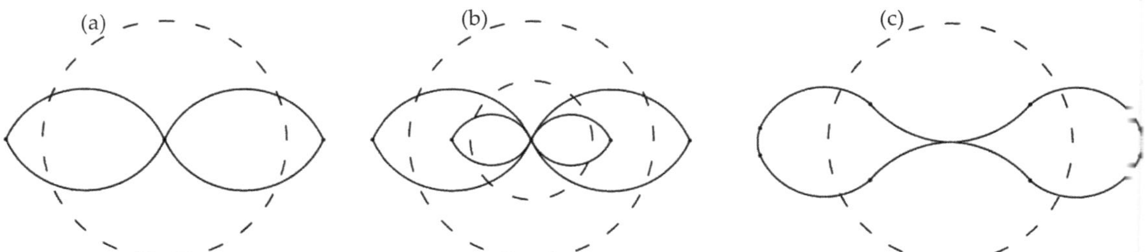

Figure 1. Buckle propagation mode in the 2D scenario: (**a**) single-walled pipe, (**b**) PIP, and (**c**) single-walled pipe.

The buckle propagation pressure of single-walled pipe proposed by Kyriakides et al. [12] is,

$$P_{Ps} = \frac{2\pi}{\sqrt{3}}\sigma_0\left(\frac{t}{D}\right)^2 \tag{1}$$

where, σ_0 is the yield stress,
    t is the thickness of the pipe,
    D is the diameter of the pipe.
    The buckle propagation pressure of PIP proposed by [13] is,

$$P_P = \frac{2\pi}{\sqrt{3}}\sigma_0\left(\frac{t}{D}\right)^2\left[1 + \frac{\sigma_{0i}}{\sigma_0}\left(\frac{t_i}{t}\right)^2\right] \tag{2}$$

where, σ_{0i} is the yield stress of the inner pipe,
    t_i is the thickness of the inner pipe.
    Wierzbicki and Bhat [14] applied the balance of the internal and external energy in the derivation of the buckle propagation pressure. The internal energy is contributed to by eight plastic hinges as shown in Figure 1c. External energy is the work of the hydrostatic external pressure. The buckle propagation pressure is as follows:

$$P_P \cong \left[3 + 12\left(\frac{1}{3}\frac{E}{\sigma_0}\frac{t}{D}\right)^{0.7}\right]\sigma_0\left(\frac{t}{D}\right)^2 \tag{3}$$

where, E is Young's modulus of the pipe.

Alrsai et al. [7] found two different buckle propagation modes in the numerical simulation of their study on a PIP system:

Mode A: The buckle propagation occurs after the contact between the upper and lower parts of the inner surface of the inner pipe;

Mode B: During buckle propagation, the inner surface of the inner pipe maintains a distance away from contact.

The difference between these two modes is illustrated in Figure 2. It was also concluded that the buckle propagation mode changes from Mode A to Mode B when the geometrical parameters satisfy the relationship of $D_o/t_o = 1.25D_i/t_i$ [7].

Mode A:

Mode B:

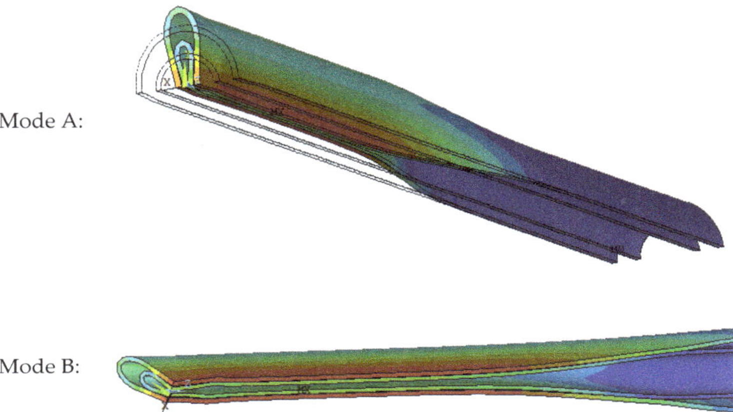

Figure 2. Two different buckle propagation modes in a PIP system.

Similar phenomena were also illustrated by Gong and Li [15]. They classified buckle propagation into four modes, which could be also summarised into two modes according to the criteria of Mode A and B from [7]. The research investigated the buckle propagation pressure experimentally and numerically, while empirical equations were derived.

In the previous research, the dominant factor influencing the buckle propagation mode is assumed to have a constant value. Moreover, in the parameter studies, the variables are considered to affect separately the carrying capacity of the PIP system, lacking the coupling effect between them.

Aiming at providing an efficient approach for the evaluation of the buckle propagation pressure in a PIP system, the current research performed a numerical study with the help of the finite element method. At first, the numerical simulation in ANSYS was verified by comparing the results with those from the model experiment in [14] in 3D scenarios. Then, the influence of several kinds of initial imperfection was checked. Third, the effects of the geometrical parameters of both inner and outer pipes were studied. Fourth, a critical parameter used for figuring out the buckle propagation mode was proposed. Finally, a semi-empirical formula was derived to express the buckle propagation pressure with the geometrical parameters of the PIP system.

2. FEM Validation

In this section, the validity of the numerical simulation is verified. The results from the finite element analyses in ANSYS are compared with those from the experiments performed in [15]. Then, the mesh sensitivity is checked by four models with different mesh divisions but with the same scantling.

2.1. Validation with Model Experiment

In model experiments, steel grade SS316 is applied [15]. The geometrical parameter and physical properties of the pipe are shown in Table 1. The ultimate collapse strength and buckle propagation pressure obtained in the model experiment were 29.54 MPa and 14.98 MPa, respectively.

Table 1. Geometrical parameters and physical properties of the pipe in the model experiment; the data are from [15].

	L (mm)	D (mm)	t (mm)	Δ_0 (%)	$\sigma_{0.5}$ (MPa)	E (GPa)
Outer pipe	1500	60	4	7.85	319.2	188.9
Inner pipe	1500	25	2	0	319.2	188.9

In this research, finite element analyses were performed with the help of ANSYS using the Solid element SOLID185. One case, entitled N40, was analysed to be verified against the model experiment, of which the mesh division is shown in Table 2. The strain-stress curve of the steel shown in Figure 3 is considered in the current analyses.

Table 2. Detail mesh division of N-series.

Case		Element Size (mm)			Mesh Number			Total Number
		θ	t	z	θ	t	z	
N20	outer pipe	9.42	2	10	20	2	150	13,800
	inner pipe	9.82	2	10	8	1	150	
N30	outer pipe	6.28	1.33	7.5	30	3	200	36,000
	inner pipe	6.54	1	7.5	12	2	200	
N40	outer pipe	4.71	1	5	40	4	300	84,000
	inner pipe	4.91	1	5	16	2	300	
N60	outer pipe	3.14	0.67	3.75	60	6	400	225,600
	inner pipe	3.27	0.67	3.75	24	3	400	

θ-circumferential direction; t-thickness direction; z-axial direction.

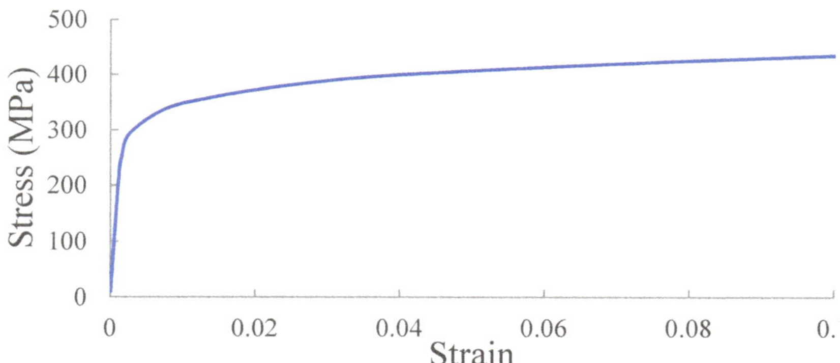

Figure 3. Strain–stress curve applied in numerical calculation.

Due to the symmetricity of the structure, only a quarter of the PIP is modelled, with half of the cross-section and half of the length, as shown in Figure 4. The geometrical parameters are identical to those in the model experiment (as shown in Table 1). The initial ovality is introduced only onto the outer pipe which locates in the middle span of the pipe. The total length area of this initial ovality is 60 mm which is equal to the diameter of the outer pipe. The introduction of the above-mentioned initial ovality is realized by changing the nodes' coordinates according to

$$R = R_0 + \Delta_0 \times R_0 \cos 2\theta \tag{4}$$

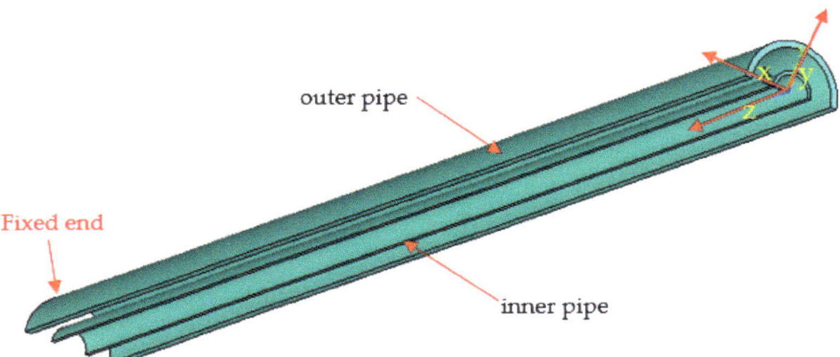

Figure 4. Geometrical model of the PIP in Case N40.

The definition of ovality is as follows:

$$\Delta_0 = \frac{D_{max} - D_{min}}{D_{max} + D_{min}} \tag{5}$$

An initial ovality of 7.85% is introduced into Case N40.

Contacts are defined between the inner surface of the outer pipe and the outer surface of the inner pipe, as well as the self-contact of the inner surface of the inner pipe. The fixed boundary condition is applied on nodes at $z = 750$. The external pressure is applied on the outer surface of the outer pipe.

The deformation and von Mises stress distribution during the buckle propagation procedure are shown in Figures 5 and 6, respectively. In such a dynamic procedure, five states are displayed in both figures. In the first state, local buckling occurs in the middle of the pipe where the initial ovality is introduced. After the second state when the outer pipe contacts the inner pipe, self-contact of the inner pipe is observed in the third state. Then in the last two states, the buckle propagates along the axial direction of the PIP system.

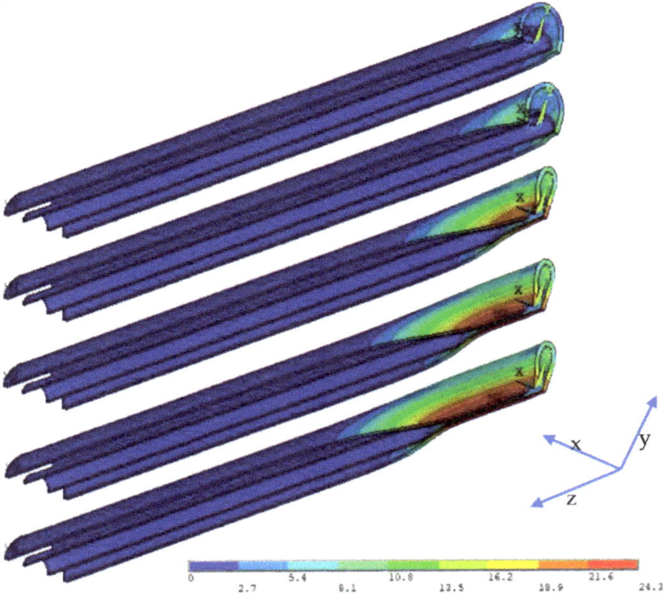

Figure 5. Deformation of PIP system during buckle propagation of Case N40 (unit: mm).

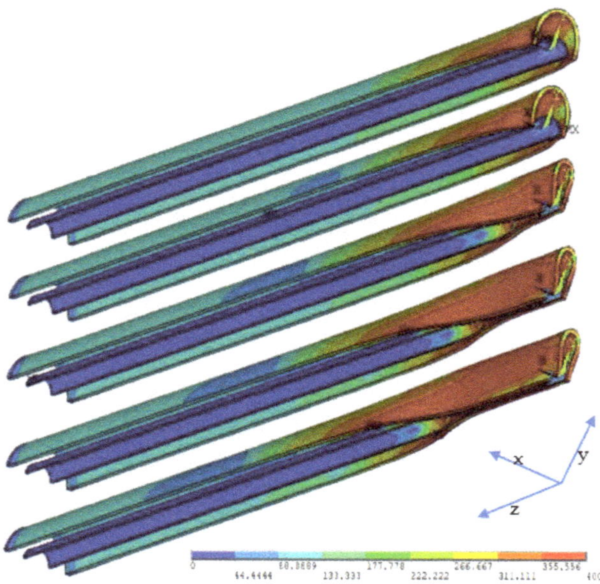

Figure 6. Distribution of von Mises stress during buckle propagation of Case N40 (unit: MPa).

The ultimate collapse pressure and buckle propagation pressures are 31.39 MPa and 16.30 MPa, respectively. After comparing with the results obtained from the model experiment, the errors are 6.27% and 8.81%, respectively, which indicates an acceptable accuracy provided by the finite element model. Thus, in the following numerical simulation of this research, similar parameter settings were adapted.

2.2. Effect of Mesh Sensitivity

To figure out the mesh division with good computing efficiency and simultaneously acceptable accuracy, four cases are further calculated. Considering the geometrical parameters shown in Table 1, different mesh divisions are presented in the new finite element models, as shown in Figure 7.

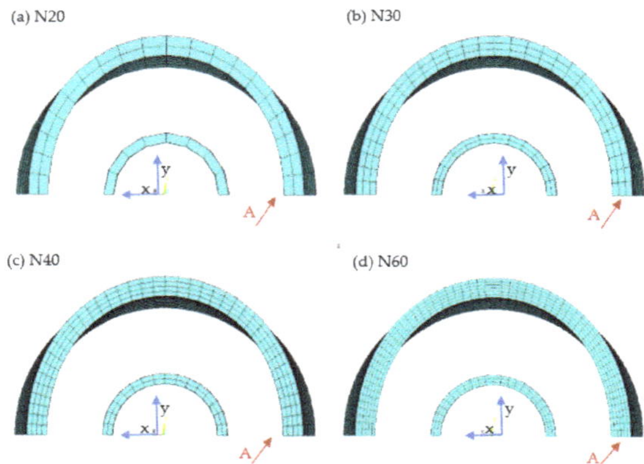

Figure 7. Mesh divisions of N-series cases from the back view: (**a**) Case N20, (**b**) N30, (**c**) N30, (**d**) N50.

The details of element number and size in the four models are listed in Table 2. These four cases are in the N-series named by the mesh number of the outer pipe in the circumferential direction. The comparison between the results obtained from the N-series cases and the experiments is listed in Table 3. The result of the external pressure with respect to the displacement in the cross-sections of the models is shown in Figure 8. The displacement of the cross-section is represented by the displacement in the *x*-direction of the node belonging to the outer surface of the outer pipe with the maximum initial ovality as shown in Figure 7.

Table 3. Comparison results between N-series and experiment.

Case	Ultimate Collapse Pressure (P_{co})			Buckle Propagation Pressure (P_p)		
	FEM (MPa)	Experiment (MPa)	Error (%)	FEM (MPa)	Experiment (MPa)	Error (%)
N20	34.12	29.54	15.50	19.50	14.98	30.17
N30	32.21	29.54	9.04	17.00	14.98	13.48
N40	31.39	29.54	6.27	16.30	14.98	8.81
N60	30.75	29.54	4.10	15.55	14.98	3.81

Error = (FEM − Experiment)/Experiment × 100%.

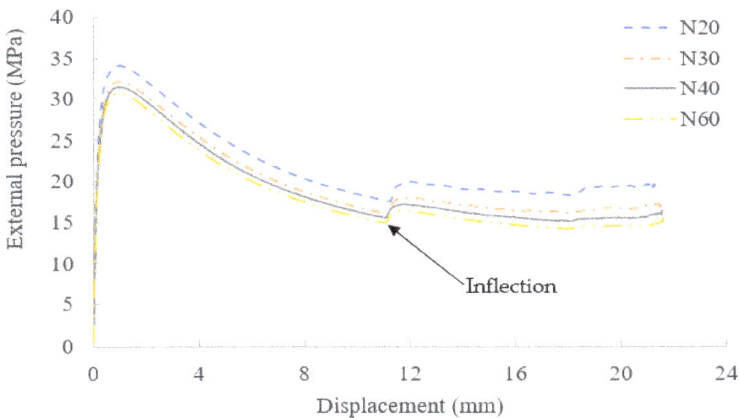

Figure 8. Relationship between external pressure and displacement of cross-section.

An inflexion exists in the curves in Figure 8 when the displacement is approximately 11 mm. At this moment, the contact between the outer and the inner pipes occurs. Consequently, the carrying capacity of PIP starts to increase because the inner pipe starts to participate in the resistance to structural deformation.

Despite the fact that Case N60 has the minimum errors compared with the experiment, the large element number of Case N60 indicates a great computational effort required. Case N40 utilizes 1/3 elements of those used in N60 but obtains similar accuracy. Thus, in the following calculations, the mesh division of Case N40 is applied.

Moreover, in terms of the initial dent area, the length is set to be 60 mm in the following cases.

3. Effect of Inner Pipe Imperfection

In the previous section, the initial ovality is merely introduced onto the outer pipe. The inner pipe is kept perfect, and concentric with the outer pipe. In this section, the effects of two imperfection patterns (ovality and eccentricity) of the inner pipe are checked.

3.1. The Effect of Inner Pipe Ovality

The direction of ovality is defined as the same as the minor axis of the oval. At first, the influence of the angles between directions of ovality of the outer and inner pipe is tested. Two cases are created based on Case N40, with the initial ovality of the inner pipe of a value of 7.85%. The length of the initial ovality area of the inner pipe is the same as that of the outer pipe and locates at the same position in the axial direction of the pipe. The only difference between these two cases is the direction of the inner pipe ovality. In one case, it is the same as that of the outer pipe, while in the other, it is vertical to that of the outer pipe, as shown in Figure 9.

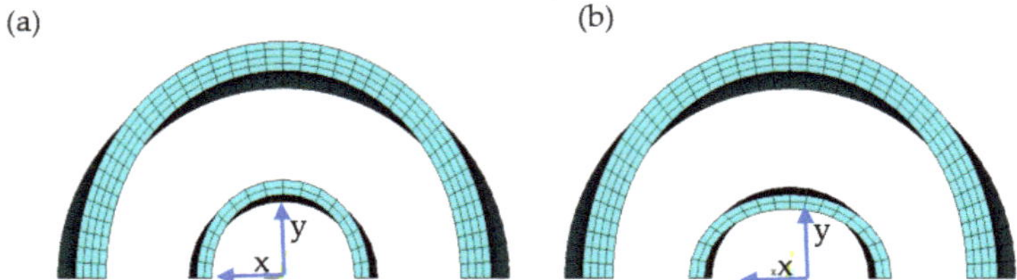

Figure 9. Directions of ovality of outer and inner pipe: (**a**) same and (**b**) vertical.

The comparison of pipe strength between both cases is listed in Table 4. In addition to the ultimate collapse pressure and buckle propagation pressure, another two featured pressures P_t and P_i are listed to represent the occurrence of the contact. P_t is the pressure under which the outer pipe contacts the inner pipe while P_i is the pressure under which the inner surface of the inner pipe contacts itself. As shown in Table 4, in both cases, the buckle propagation pressure is the same. The difference in the values of the pressure when contact occurs between both cases is due to the geometrical features. From Figure 8, the external pressure decreases after the ultimate collapse pressure is reached. A larger distance or the gap between two pipes results in a lower value of P_t. So, this pressure is smaller in the case of the same ovality direction. Additionally, the same reason leads to the difference in values of P_i. Nevertheless, the difference between these two cases is insignificant. Thus, it is confirmed that the direction of the ovality of the inner pipe does not influence the buckle propagation of PIP.

Table 4. Effect of ovality direction.

Direction of Ovality	P_{co} (MPa)	P_t (MPa)	P_i (MPa)	P_p (MPa)
a. same	31.39	16.29	15.77	16.30
b. vertical	31.39	16.95	15.46	16.30

P_t: pressure under which outer pipe touches inner pipe; P_i: pressure under which inner pipe contacts itself.

3.2. Effect of Eccentricity

The maximum eccentricity of the inner pipe locates in the middle of the PIP in the axial direction, the same as for the ovalities of the outer and inner pipes. The degree of eccentricity, e_0, is defined as follows:

$$e_0 = \frac{e_{max}}{(D_o - D_i)/2} \tag{6}$$

where, e_{max} is the maximum offset of the inner pipe centre, as shown in Figure 10.

The eccentricity is introduced as a sinusoidal wave during the axial direction of the pipe, from zero in the boundary to the maximum in the middle. The geometrical parameters follow the same settings in Case N40, with the details listed in Table 5. Slight differences are observed when introducing different combined types of the initial imperfection of the inner pipe. The finite element model of Case N40.03 is shown in Figure 10, considering e_0 as 0.5 which is far beyond the acceptable range in the manufacturing stage.

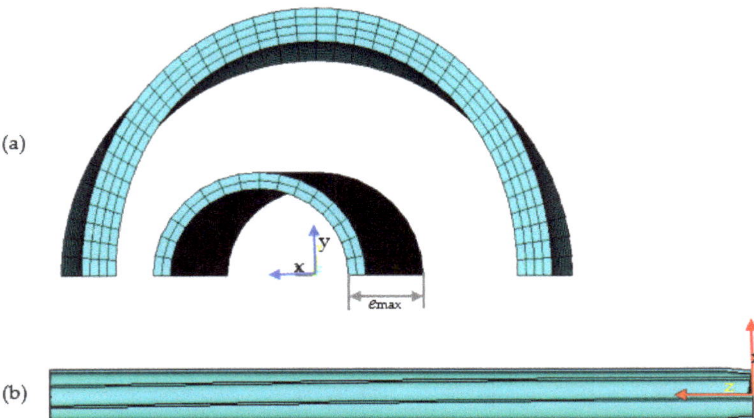

Figure 10. Model and mesh of Case N40.03: (**a**) back view and (**b**) bottom view.

Table 5. The initial imperfection of the inner pipe.

Case	Δ_i	e_0	P_{co} (MPa)	P_t (MPa)	P_i (MPa)	P_p (MPa)
N40	0	0	31.39	15.66	15.66	16.30
N40.01	0	0.1	31.39	15.95	15.68	16.33
N40.02	0	0.3	31.39	15.86	15.69	16.35
N40.03	0	0.5	31.39	15.79	15.69	16.35
N40.11	7.85	0.1	31.39	14.92	15.86	16.30
N40.12	7.85	0.3	31.39	15.28	15.84	16.35
N40.13	7.85	0.5	31.39	15.24	15.85	16.40

In the notation of the cases, the first number after the decimal sign shows two different levels of ovality: '0' for no ovality and '1' for an ovality of 7.85%. The second number after the decimal sign, which is also the last number in the notation, is used to identify the levels of eccentricity: '1', '2', and '3' stand for the eccentricity of 0.1, 0.3, and 0.5, respectively. An ovality with a value of 7.85% is also at a high level of imperfection [16–18]. In this regard, the initial imperfection introduced in this section considers the severe situation of the PIP system.

For the sake of further confirmation of the effects of these parameters, additional cases are created in a similar method as those shown in Table 5 by changing the diameter and thickness of the inner pipe. By checking the results, it is confirmed that the imperfection of the inner pipe, neither the ovality of the inner pipe nor the eccentricity, has an insignificant influence on the buckle propagation pressure, which was also verified in the case of sandwich pipes by Fu et al. [19]

After the validation in this section, in the following numerical simulation cases, the length of PIP is set to be 1500 mm, and the initial imperfection introduced onto PIP is only the ovality on the outer pipe with a value of 7.85%.

4. Effect of the Geometry of the Inner Pipe

When checking the influence of the initial imperfection of the inner pipe in the previous section, the geometrical parameters of the inner pipe varied. In this section, the names of cases are defined as 'D_o-t_o-D_i-t_i'. For instance, the name of the case in the first line of Table 6 is '60-4-10-2'.

Table 6. Cases with different geometrical parameters of the inner pipe.

D_o (mm)	t_o (mm)	D_i (mm)	t_i (mm)
60	4	10	2
60	4	15	2
60	4	20	2
60	4	30	2
60	4	35	2
60	4	25	1.43
60	4	25	1.67
60	4	25	2.5
60	4	25	3.33
60	4	25	5

In this section, besides Case '60-4-25-2' (N40), five more ratios of D_i/t_i are selected. Five cases with a constant D_i of 25 mm and another five cases with the same t_i as 2 mm are shown in Table 7. The ultimate collapse strength of PIP equals that of the outer pipe. Thus, because the outer pipes of cases listed in Table 7 share the same geometrical parameters, the ultimate collapse strength is not presented.

Table 7. Geometry and buckle propagation of PIP with the same outer pipe (D_o = 60 mm, t_o = 4 mm).

D_i (mm)	t_i (mm)	D_o/D_i	D_i/t_i	$(D_o/t_o)/(D_i/t_i)$	P_p (MPa)	Mode
25	2	2.4	12.50	1.20	16.30	A
10	2	6	5.00	3.00	15.80	B
15	2	4	7.50	2.00	15.90	A
20	2	3	10.00	1.50	16.10	A
30	2	2	15.00	1.00	16.50	A
35	2	1.71	17.50	0.86	17.10	A
25	1.43	2.4	17.48	0.86	15.75	A
25	1.67	2.4	14.97	1.00	15.95	A
25	2.5	2.4	10.00	1.50	17.00	A
25	3.33	2.4	7.51	2.00	18.50	B
25	5	2.4	5.00	3.00	18.70	B

Mode A: The buckle propagation occurs after the contact between the upper and lower parts of the inner surface of the inner pipe; Mode B: During buckle propagation, the inner surface of the inner pipe maintains a distance away from contact.

From the listed results in Table 7, it is noticed in Case '60-4-15-2' and Case '60-4-25-3.33', in which the value of the ratio $(D_o/t_o)/(D_i/t_i)$ is equal to 2 in both cases, the buckle propagation mode is different. It is supposed that the buckle propagation mode is influenced by the ratio of $(D_o/t_o)/(D_i/t_i)$, but the ratio is not a fixed value.

5. Effect of the Geometry of the Outer Pipe

In this section, the effect of the geometrical parameters of the outer pipe is investigated. First, the thickness of the outer pipe is changed in the range of 2–6 mm. The results of ultimate collapse pressure and buckle propagation pressure, as well as the buckle propagation mode, are listed in Table 8. It is observed that the buckle propagation mode switches at the thickness of the outer pipe in the range of 3–3.5 mm.

Then, to check the effect of the diameter of the outer pipe, new cases are created considering similar ratios of D_o/t_o as those shown in Table 8, in which Case N40 is included. These cases maintain the thickness of the outer pipe at the value of 4 mm, as listed in Table 9. By comparing the results listed in Tables 8 and 9, it is noticed that the buckling mode of Case '80-4-25-2' is Mode A, whereas that of Case '60-3-25-2' is Mode B, although the ratio of D_o/t_o is the same in both cases.

Table 8. Effect of thickness of the outer pipe.

Case	t_o (mm)	D_o/t_o	P_{co} (MPa)	P_p (MPa)	Mode
60-3-25-2	3.0	20	19.50	8.75	B
60-3.5-25-2	3.5	17.14	25.10	11.95	A
60-4-25-2 (N40)	4.0	15	31.39	16.30	A
60-4.5-25-2	4.5	13.33	37.80	21.10	A
60-5-25-2	5.0	12	44.83	27.00	A
60-5.5-25-2	5.5	10.91	52.09	33.70	A
60-6-25-2	6.0	10	59.76	41.20	A

Table 9. Effect of the diameter of the outer pipe.

Case	D_o (mm)	D_o/t_o	P_{co} (MPa)	P_p (MPa)	Mode
80-4-25-2	80	20	20.61	7.95	A
68.58-4-25-2	68.58	17.15	26.07	11.75	A
60-4-25-2 (N40)	60	15	31.39	16.30	A
53.4-4-25-2	53.34	13.34	37.02	22.10	A
48-4-25-2	48	12	44.26	31.00	A
43.6-4-25-2	43.6	10.9	50.15	39.40	A
40-4-25-2	40	10	57.32	51.00	A

6. Critical Parameter of Two Different Buckle Propagation Modes

The ratio of D_o/D_i for the PIP system is generally in the range of 1.3 to 2.2 [20–25]. Additional series of cases were created in the present study and calculated with outer pipe diameters of 32 mm and 36 mm. In both series, the other geometrical parameters were kept the same as in Case N40. To figure out the boundary between the two different buckle propagation modes, the thickness of the outer pipes of the cases listed in Table 9 is changed with an increment of 0.1 mm. The purpose is to illustrate the switch of the buckle propagation mode from one to another. The thickness of the outer pipe from both buckle propagation modes is summarized in Table 10.

Table 10. The thickness of the outer pipe from both buckle propagation modes of cases with D_o/D_i less than 2.2 (D_i = 25 mm, t_i = 2 mm).

D_o (mm)	D_o/D_i	t_o (mm)		$(D_o/t_o)/(D_i/t_i)$	
		Mode A	Mode B	Mode A	Mode B
53.34	2.13	3.1	3	1.38	1.42
48	1.92	2.7	2.6	1.42	1.48
43.6	1.74	2.3	2.2	1.52	1.59
40	1.60	2.0	1.9	1.60	1.68
36	1.44	1.8	1.7	1.60	1.69
32	1.28	1.5	1.4	1.71	1.83

It is confirmed that the boundary value of the ratio $(D_o/t_o)/(D_i/t_i)$ is indeed influenced by the ratio D_o/D_i. The relationship between $(D_o/t_o)/(D_i/t_i)$ ratio and the D_o/D_i ratio is shown in Figure 11, where a monotonically decreasing tendency is observed.

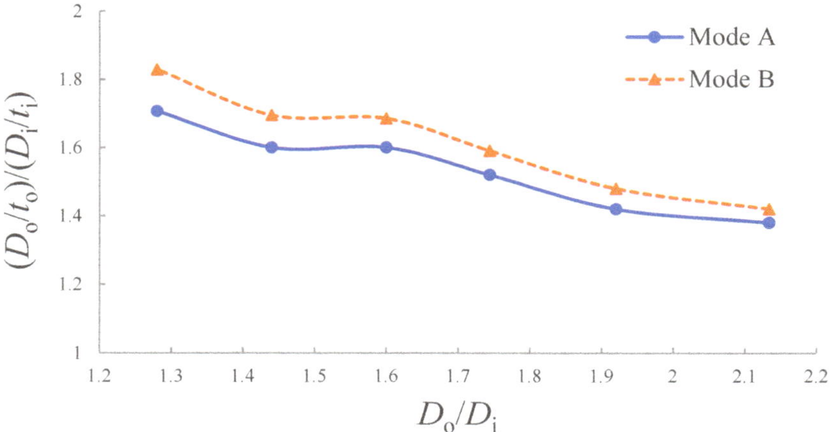

Figure 11. Critical $(D_o/t_o)/(D_i/t_i)$ ratio of Mode A and B.

7. Buckle Propagation Pressure of a PIP System

In the process of pointing out the boundary between two buckle propagation modes, a large number of numerical simulation cases was created and calculated. In this section, the buckle propagation pressure of the PIP system is studied. Both Alrsai et al. [7] and Gong and Li [15] derived empirical equations of buckle propagation pressure. Both equations are made up of a combination of separated parameters together. Based on the results of the present research, the semi-empirical formula is derived to describe the relationship between buckle propagation pressure and the geometrical parameters of the PIP system.

Both effects of the ratio D_o/t_o and D_i/t_i were investigated. For the case with diameters of the outer and inner pipes as 60 mm and 25 mm, respectively, the buckle propagation pressure is shown in Table 11.

Table 11. Buckle propagation pressure of PIP (D_o = 60 mm, D_i = 25 mm).

t_o (mm)	t_i (mm)	D_o/D_i	D_o/t_o	D_i/t_i	P_p (MPa)
4	2	2.4	15	12.5	16.30
3.2	2	2.4	18.75	12.5	9.95
3.1	2	2.4	19.35	12.5	9.35
3	2	2.4	20	12.5	8.75
2.7	2	2.4	22.22	12.5	7.20
4	1.43	2.4	15	17.48	15.75
4	1.67	2.4	15	14.97	15.95
4	2	2.4	15	12.5	16.30
4	2.5	2.4	15	10	17.00
4	3.33	2.4	15	7.51	18.50
4	5	2.4	15	5	18.60

The relationship between the buckle propagation pressure and the D/t ratios of outer and inner pipes are illustrated in Figure 12, in which both trendlines are in the style of power functions. So, it is assumed that the buckle propagation pressure can be expressed by the following:

$$P_P = \chi \left(\frac{D_o}{t_o}\right)^{\alpha} \left(\frac{D_i}{t_i}\right)^{\beta} \tag{7}$$

where, χ, α and β are related to the ratio of D_o/D_i, and will be explained later in this section.

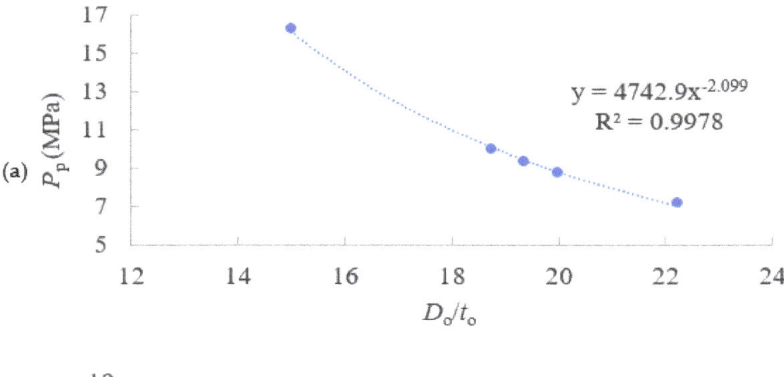

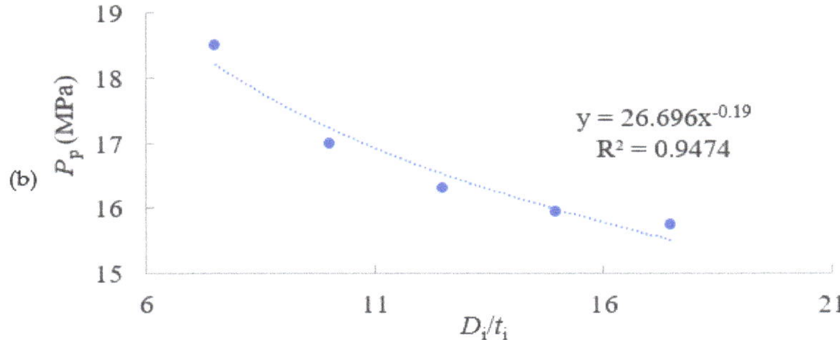

Figure 12. Relationship between buckle propagation pressure and D/t ratios: (**a**) outer pipe and (**b**) inner pipe (D_o = 60 mm, D_i = 25 mm).

To determine the expression of χ, α, and β, several new cases listed in Table 12 were analysed. The ratios of D_o/t_o and D_i/t_i are in the ranges of 10 to 33 and 7 to 50, respectively. The ratio of D_o/D_i is in the range of 1.28 to 3.2. As pointed out in the last case in Table 11, the buckle propagation pressure did not increase to a large extent when the thickness of the inner pipe changed from 3.33 mm to 5 mm. The same phenomenon is also found in other series of calculations. So, the results of these cases are excluded from the summary.

Table 12. Calculated series for the effect of D_o/D_i.

D_o (mm)	D_i (mm)	D_o/t_o	D_i/t_i
80	25	√	√
68.58	25	√	
60	25	√	√
53.34	25	√	
48	25	√	
43.6	25	√	√
40	25	√	
36	25	√	
32	25	√	√

After summarizing the results of 73 cases, the expression of χ, α, and β can be expressed as follows:

$$\chi = 1910.5\left(\frac{D_o}{D_i}\right)^2 - 7523.7\left(\frac{D_o}{D_i}\right) + 13281 \tag{8}$$

$$\alpha = -0.3371\left(\frac{D_o}{D_i}\right) - 1.1944 \tag{9}$$

$$\beta = 0.1625 \left(\frac{D_\text{o}}{D_\text{i}} \right) - 0.5984$$

The comparison between the calculation results and the results predicted by Equation (7) are shown in Figure 13, which leads to the conclusion that a good agreement is achieved.

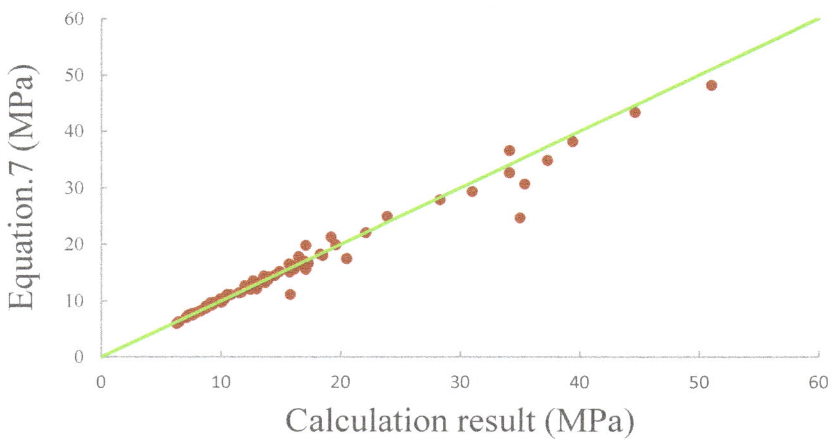

Figure 13. Comparison between the calculation results and results from Equation (7).

8. Conclusions

The phenomenon of buckle propagation in a PIP system under uniform external pressure was studied in this paper. After the validation of the numerical result by comparison with experiments, the buckle propagation pressures of a series of cases were obtained.

Still some limitations exist in the research of this paper. The physical properties applied in the numerical simulation, such as yield stress and Young's modulus, are not considered to be variables in deriving the empirical equation. Thus, further research is required on the effects of physical properties as another aspect.

Despite the above limitation, several conclusions based on the results were obtained as follows:

1. The initial imperfections of PIP, i.e., the ovality and eccentricity of the inner pipe, have an insignificant influence on the buckle propagation pressure
2. The switch of two buckle propagation modes is dependent on the ratio of $(D_\text{o}/t_\text{o})/(D_\text{i}/t_\text{i})$ and the critical value of the ratio is related to the ratio of D_o/D_i; instead of that in the previous literature, the critical value is considered to be constant. In practice situations of the ratio of D_o/D_i in the range of 1.2 to 2.2, a monotonically decreasing tendency of $(D_\text{o}/t_\text{o})/(D_\text{i}/t_\text{i})$ can be observed with respect to D_o/D_i
3. The buckle propagation pressure is highly influenced by the ratio of D_o/t_o and D_i/t_i. In other words, the strength of the outer pipe compared with the inner pipe not only results in the buckle propagation mode of PIP but also leads to the stability of the buckle propagation pressure. One phenomenon observed is that during the increase of thickness of the inner pipe, after it exceeds a certain value (such as 3.33 mm in Case 60-4-25-3.33), the buckle propagation pressure will not increase any longer in line with the increase of thickness
4. The relationship between the buckle propagation pressure and the ratio of D/t of pipes in PIP follows the format of power functions, which includes the coupling effect of the variables. Based on the results from the present analyses, a fitted formula (Equation (7)) is proposed for predicting the buckle propagation pressure of the PIP with good accuracy.

Author Contributions: R.L.: Methodology, Formal analysis, Writing—original draft, Visualization. B.-Q.C.: Methodology, Writing—original draft. C.G.S.: Conceptualization, Writing—review and editing, Supervision. All authors have read and agreed to the published version of the manuscript.

Funding: This work was developed in the scope of the project "Cementitious cork composites for improved thermal performance of pipelines for ultradeep waters"–SUPBSEAPIPE, with the reference n.o POCI-01-0145-FEDER-031011 funded by the European Regional Development Fund (FEDER) through COMPETE2020-Operational Program Competitiveness and Internationalization (POCI) and with financial support from the Portuguese Foundation for Science and Technology (Fundação para a Ciência e Tecnologia-FCT) under grant 02/SAICT/032108/2017. This study contributes to the Strategic Research Plan of the Centre for Marine Technology and Ocean Engineering, which is financed by FCT, under contract UIDB/UIDP/00134/2020.

Institutional Review Board Statement: Not applicable.

Informed Consent Statement: Not applicable.

Data Availability Statement: The data presented in this study are available on request from the corresponding author.

Conflicts of Interest: The authors declare no conflict of interest.

References

1. Bai, Y.; Igland, R.T.; Moan, T. Tube collapse under combined external pressure, tension and bending. *Mar. Struct.* **1997**, *10*, 389–410. [CrossRef]
2. Park, T.D.; Kyriakides, S. On the collapse of dented cylinders under external pressure. *Int. J. Mech. Sci.* **1996**, *38*, 557–578. [CrossRef]
3. Vasilikis, D.; Karamanos, S.A. Stability of confined thin-walled steel cylinders under external pressure. *Int. J. Mech. Sci.* **2009**, *51*, 21–32. [CrossRef]
4. He, T.; Duan, M.; An, C. Prediction of the collapse pressure for thick-walled pipes under external pressure. *Appl. Ocean Res.* **2014**, *47*, 199–203. [CrossRef]
5. Timoshenko, S.P.; Gere, J.M. *Theory of Elastic Stability*, 2nd ed.; Dover Publications: New York, NY, USA, 1961.
6. Liang, H.; Zhou, J.; Lin, J.; Jin, F.; Xia, F.; Xue, J.; Xu, J. Buckle propagation in steel pipes of ultra-high strength: Experiments, theories and numerical simulations. *Acta Mech. Solida Sin.* **2020**, *33*, 546–563. [CrossRef]
7. Alrsai, M.; Karampour, H.; Albermani, F. On collapse of the inner pipe of a pipe-in-pipe system under external pressure. *Eng. Struct.* **2018**, *172*, 614–628. [CrossRef]
8. Alrsai, M.; Karampour, H.; Albermani, F. Numerical study and parametric analysis of the propagation buckling behaviour of subsea pipe-in-pipe systems. *Thin-Walled Struct.* **2018**, *125*, 119–128. [CrossRef]
9. Bhardwaj, U.; Teixeira, A.P.; Guedes Soares, C. Reliability assessment of a subsea pipe-in-pipe system for major failure modes. *Int. J. Press. Vessels Pip.* **2020**, *188*, 104177. [CrossRef]
10. Netto, T.A.; Estefen, S.F. Buckle arrestors for deepwater pipelines. *Mar. Struct.* **1996**, *9*, 873–883. [CrossRef]
11. Lee, L.H.; Kyriakides, S.; Netto, T.A. On the flip-flop mode of dynamic buckle propagation in tubes under external pressure. *Extreme Mech. Lett.* **2021**, *48*, 101378. [CrossRef]
12. Kyriakides, S.; Babcock, C.D.; Elya, D. Initiation of propagating buckles from local pipeline damages. *J. Energy Resour. Technol.* **1984**, *106*, 79–87. [CrossRef]
13. Kyriakides, S.; Vogler, T.J. Buckle propagation in pipe-in-pipe systems: Part II. Analysis. *Int. J. Solids Struct.* **2002**, *39*, 367–392. [CrossRef]
14. Wierzbicki, T.; Bhat, S.U. Initiation and propagation of buckles in pipelines. *Int. J. Solids Struct.* **1986**, *22*, 985–1005. [CrossRef]
15. Gong, S.; Li, G. Buckle propagation of pipe-in-pipe systems under external pressure. *Eng. Struct.* **2015**, *84*, 207–222. [CrossRef]
16. Li, R.; Guedes Soares, C. Numerical study on the effects of multiple initial defects on the collapse strength of pipelines under external pressure. *Int. J. Press. Vessels Pip.* **2021**, *194*, 104484. [CrossRef]
17. Li, R.; Chen, B.Q.; Guedes Soares, C. Design equation for the effect of ovality on the collapse strength of sandwich pipes. *Ocean Eng.* **2021**, *235*, 109367. [CrossRef]
18. Li, R.; Chen, B.Q.; Guedes Soares, C. Effect of ovality length on collapse strength of imperfect sandwich pipes due to local buckling. *J. Mar. Sci. Eng.* **2022**, *10*, 12. [CrossRef]
19. Fu, G.; Li, M.; Yang, J.; Sun, B.; Shi, C.; Estefen, S.F. The effect of eccentricity on the collapse behaviour of sandwich pipes. *Appl. Ocean Res.* **2022**, *124*, 103190. [CrossRef]
20. Binazir, A.; Karampour, H.; Sadowski, A.J.; Gilbert, B.P. Pure bending of pipe-in-pipe systems. *Thin-Walled Struct.* **2019**, *145*, 106381. [CrossRef]
21. Jiwa, M.Z.; Kim, D.K.; Mustaffa, Z.; Choi, H.S. A systematic approach to pipe-in-pipe installation analysis. *Ocean Eng.* **2017**, *142*, 478–490. [CrossRef]
22. Karampour, H.; Alrsai, M.; Albermani, F.; Guan, H.; Jeng, D.S. Propagation buckling in subsea pipe-in-pipe systems. *J. Eng. Mec.* **2017**, *143*, 04017113. [CrossRef]

23. Sun, C.; Zheng, M.; Guedes Soares, C.; Duan, M.; Wang, Y.; Onuoha, M.D.U. Theoretical prediction model for indentation of pipe-in-pipe structures. *Appl. Ocean Res.* **2019**, *92*, 101940. [CrossRef]
24. Zhang, Z.; Liu, H.; Chen, Z. Lateral buckling theory and experimental study on pipe-in-pipe structure. *Metals* **2019**, *9*, 185. [CrossRef]
25. Zhang, X.; Duan, M.; Guedes Soares, C. Lateral buckling critical force for submarine pipe-in-pipe pipelines. *Appl. Ocean Res.* **2018**, *78*, 99–109. [CrossRef]

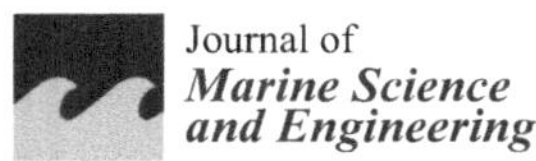
Journal of
Marine Science and Engineering

Article

Test System Development and Experimental Study on the Fatigue of a Full-Scale Steel Catenary Riser

Jianxing Yu [1,2,3], Fucheng Wang [1,2,*], Yang Yu [1,2], Xin Liu [1,2], Pengfei Liu [1,2] and Yefan Su [1,2]

[1] State Key Laboratory of Hydraulic Engineering Simulation and Safety, Tianjin University, Tianjin 300072, China
[2] Tianjin Key Laboratory of Port and Ocean Engineering, Tianjin University, Tianjin 300072, China
[3] College of Mechanical and Marine Engineering, Beibu Gulf University, Qinzhou 535011, China
* Correspondence: tjuwfc1993@163.com; Tel.: +86-188-4603-5521

Abstract: This paper presents a full-scale deep-water steel catenary riser fatigue test system. The proposed system can carry out fatigue tests on steel catenary risers, hoses, and subsea pipelines up to 21 m in length, ranging from 8 to 24 inches in diameter. The test system was realized by mechanical loading with loading control systems, and could carry out axial tension and compression, bending moment, torsion, and internal pressure to simulate all load types on deep-water steel catenary risers or subsea pipelines. The counterforce was sustained by a counterforce frame. Through mechanical simulation analysis, the authors determined the size of the counterforce frame and designed the connection form of the counterforce frame and loading system. According to the required loading capacity, the appropriate cylinder thickness and diameter were obtained through calculation. After the design and construction of the test system, the authors designed a fatigue test to confirm the loading capacity and accuracy of the test system. The authors performed full-scale testing to assess the fatigue performance of pipe-to-pipe mainline 5G girth welds fabricated to BS 7608. This test was designed according to the stress level of pipelines in the Lingshui 17-2 gas field, and the test results were compared with the calculation results of the S–N curve.

Keywords: full-scale riser; fatigue damage; test system; steel catenary risers; deep-water

Citation: Yu, J.; Wang, F.; Yu, Y.; Liu, X.; Liu, P.; Su, Y. Test System Development and Experimental Study on the Fatigue of a Full-Scale Steel Catenary Riser. *J. Mar. Sci. Eng.* **2022**, *10*, 1325. https://doi.org/10.3390/jmse10091325

Academic Editors: Baiqiao Chen and Carlos Guedes Soares

Received: 6 August 2022
Accepted: 10 September 2022
Published: 19 September 2022

Publisher's Note: MDPI stays neutral with regard to jurisdictional claims in published maps and institutional affiliations.

1. Introduction

A deep-water riser is the only channel connecting a subsea wellhead and surface floating facilities, and it is an important facility for the development of deep-water oil and gas fields. In contemporary riser construction, steel catenary risers are preferred for deep-water oil and gas development. In deep-water environments, due to the presence of wind, waves, currents, and pressure, both inside and outside the riser, the riser structure is subjected to complex loads. Under the effect of long-term loads, fatigue damage failure occurs in the riser, resulting in structural damage, and the consequences are very serious [1,2]. Low-cycle fatigue failure occurs rapidly in a short amount of time under extreme loads [3,4]. Therefore, the fatigue life of tubular structures has received more attention [5]. Particularly for steel tubular welded structures, fatigue failure is a very general failure mode. Nassiraei et al. [6] proposed a detailed fatigue calculation method for welded tubes, which was verified by an FE model, and finally validated according to the experimental data and UK DoE acceptance criteria.

Fatigue tests are the key method to solving the problem of riser fatigue and for obtaining the fatigue life of a riser. Especially in the weld structure of a riser, the toe of the weld can easily become the fatigue crack cracking site [7,8]. Moreover, high stress concentrations [9,10] and large residual stress [11–15] exist at the welding point under cyclic loading. Under the action of higher stress, the structure will undergo plastic deformation [16]. According to the research, even under the action of low-cycle fatigue, the specimen will have some structural plastic deformation [17], which makes the prediction of structural fatigue life more difficult. Therefore, fatigue life prediction using experimental methods is effective in fatigue research at present. For tubular welded structures, CT specimens are mostly used for material fatigue tests. A series of fatigue tests was carried out on CT specimens, considering corrosion [18]. However, the structural form will affect the stress concentration, and the fatigue life of the structure cannot be predicted by a CT sample test. Therefore a full-scale fatigue test is required to determine the fatigue life of the structure.

In this study, the design and construction of a full-scale riser fatigue test system were carried out, and the composition and layout of the test platform are described in detail, including the design of the reaction frame structure to withstand the load reaction force, the design of the core loading system of the test platform, and the design of the fatigue test. The experimental capability of the test platform was tested, and the experimental capability of the system was verified by comparing the theoretical calculation results with the loading test results. Finally, future functions of the test system and further improvements to the test ability of the system are proposed.

2. Development of the Counterforce Frame

The counterforce frame was developed to sustain a maximum axial tension/compression loading of 3000 kN, torque force of 200 kN, and bending moment of 1300 kN·m. Therefore, the reinforcement design was adopted for axial force, torsion force, and bending moment loading: the rest was trusswork. The external dimensions of the counterforce frame were 2.3 m× 2.3 m× 24 m (excluding the loading actuator). Its combined frame structure, which had a horizontal installation, was made of Q345 steel. Its overall structure is shown in Figure 1.

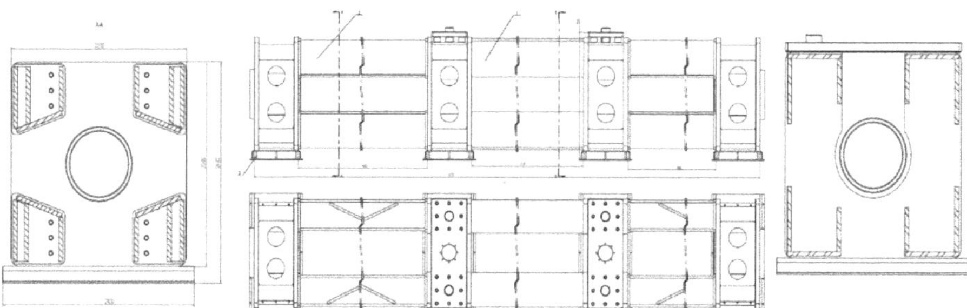

Figure 1. Overall structure of the counterforce frame.

2.1. Main Components of Counterforce Frame

The counterforce frame was composed of the end frame (i.e., tension/pressure and torsional loading counterforce frame), middle frame coupling (i.e., moment loading counterforce frame), stiffener coupling, and counterforce frame base, as shown in Figure 2.

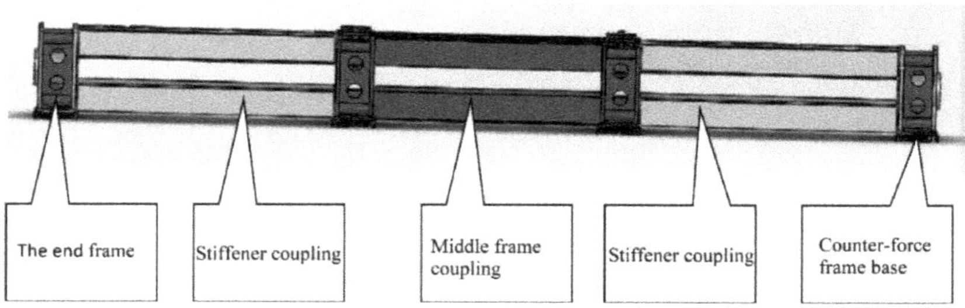

Figure 2. Structure of the counterforce frame.

The front and rear end frames were assembled using a middle frame coupling and stiffener couplings and securely connected by 24 M48*250 12.9 high-strength bolts. The connection form of the end frame and the stiffener coupling is shown in Figure 3.

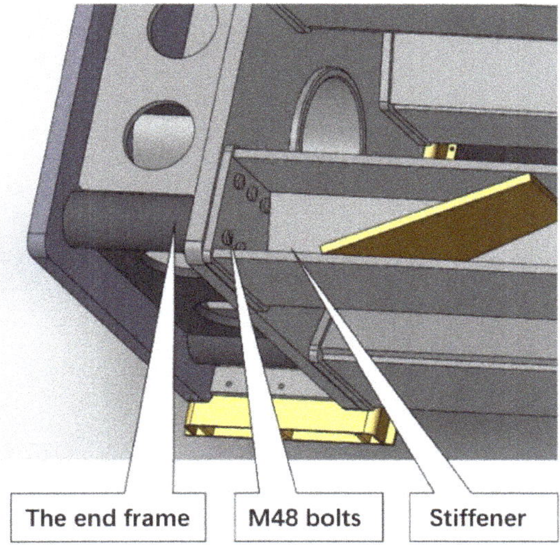

Figure 3. Connection form of the end frame and the middle frame.

The middle frame coupling was welded together from two moment loading counterforce frames and stiffener couplings. The moment cylinder mounting plate was installed on the upper part of the moment loading counterforce frame. The moment cylinder mounting plate could be rotated 90 degrees along the horizontal direction. The upper part of the moment loading reaction frame was slotted, and the U-shaped structure was convenient for lifting the riser specimen up and down. The connection form of the frame coupling is shown in Figure 4.

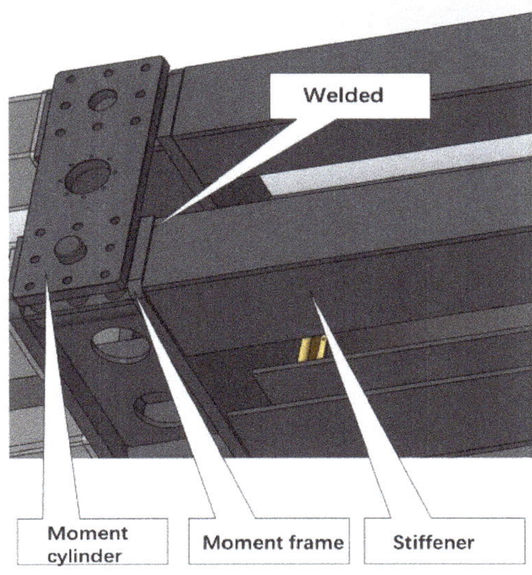

Figure 4. Connection form of the middle frame joint.

The thickness of the end frame and the middle frame coupling was 100 mm, and the thickness of the stiffener coupling was 50 mm.

2.2. Force Analysis of the Counterforce Frame

The model of the counterforce frame was established by SOLIDWORKS, and the strength and natural frequency were analyzed. The counterforce frame was used to sustain the reaction force of the test system, and the axial load and bending moment under the maximum loading capacity of the test platform could be loaded into the model in the form of the reaction force. An axial tension and compression of 3000 kN and an out-of-plane bending moment of 1300 kN·m were applied to the model, and a stress cloud diagram was obtained in Figures 5 and 6.

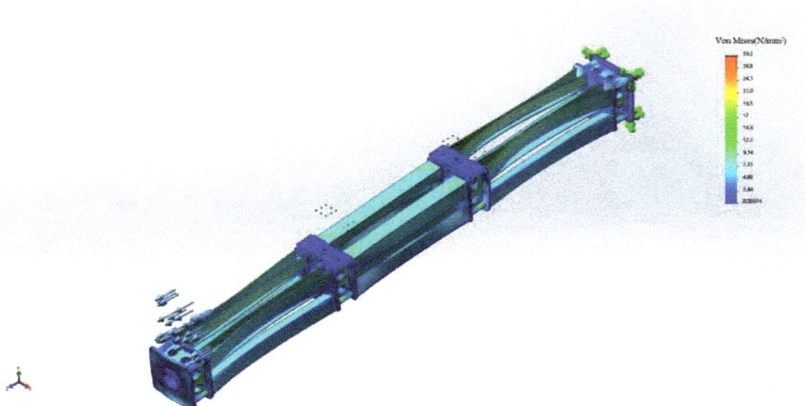

Figure 5. Stress cloud diagram under axial load.

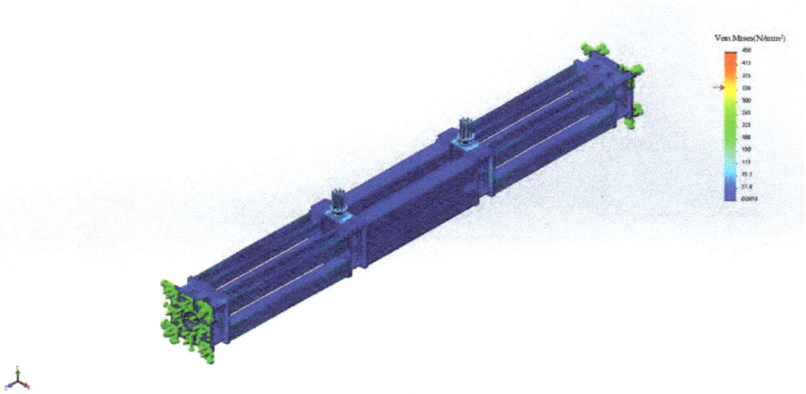

Figure 6. Stress cloud diagram under bending moment load.

The maximum tensile stress was 38 MPa, and the maximum compressive stress was 29.2 MPa, much lower than the yield stress of Q345.

In order to ensure the safety of the reaction frame structure during loading, the natural frequency of the counterforce frame was calculated to avoid loading resonance. The natural frequency of the counterforce frame was 3 Hz. Therefore, a loading frequency of 3 Hz was avoided when designing the test. The natural frequency analysis results of the structure are shown in Figure 7.

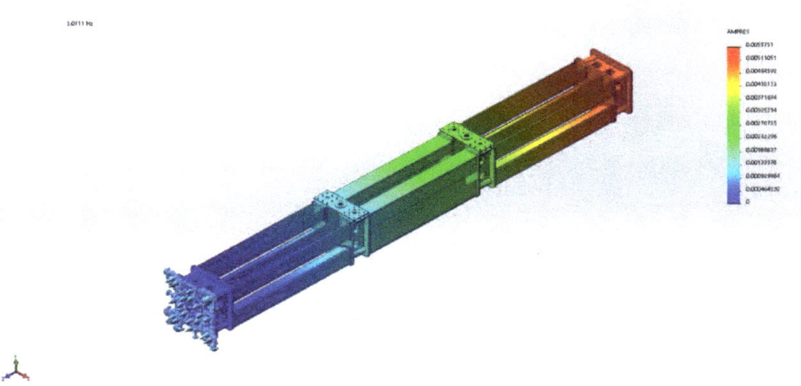

Figure 7. Natural frequency analysis.

3. Development of the Loading System

The loading system included an internal water pressure loading system, axial loading system, bending moment loading system, and torque loading system. Among them, the axial loading system and bending moment loading system can provide cyclic loading, equipped with a servo mechanism. The internal water pressure loading and torque loading were static loads. The axial loading mechanism and torque loading mechanism were combined, arranged at the end of the platform, and connected with the counterforce frame. The moment loading system was arranged at one-third and two-thirds of the counterforce frame to load the moment and in the form of a four-point bending moment. The internal pressure loading system was arranged on the side of the counterforce frame. The water pipe was connected to the flange at the end of the test pieces, and the water was injected into the pieces through the flange hole to provide internal pressure.

3.1. Internal Hydraulic Loading System

The maximum internal water pressure applied was 60 MPa, and the flow rate under the maximum pressure was 45 L/min. The maximum loading speed reached 10 MPa/min, and the loading control precision was less than or equal to 0.5 MPa, according to the maximum loading test pieces' size (the maximum riser diameter was 24 inches, and the length was 21 m) of the test system. The system can be used for single maximum linear loading or cyclic fatigue loading at low frequencies. The maximum loading power of the system was 55 KW. The specific technical parameters are shown in Table 1.

Table 1. Internal pressure loading system parameters.

Internal Pressure Loading Control System for the Specimen						
Loading Pressure Range	Maximum Loading Speed	Load Function	Pressure Control Accuracy		Pressure of the Pipeline and Valve	Environmental Conditions
0.5~60 MPa	10 MPa/min	Single linear loading or low-frequency cyclic fatigue loading	Lifting and lowering accuracy: less than ±0.2 MPa under 5 MPa;	Pressure load retention accuracy: less than ±0.5 MPa	≥70 MPa	Temperature: −5 °C~+40 °C; Medium: tap water or 3.5% saltwater

When the internal hydrostatic test was applied, one end of the riser specimen was fixed to the counterforce frame, and the other end was free to extend. The free end was connected to the axial force loading cylinder by the guide shaft. During the internal water pressure fatigue loading test, the axial force loading cylinder was in the free unloading state. A schematic diagram of the internal water pressure loading system is shown in Figure 8, and the electric automatic control valve is shown in Figure 9.

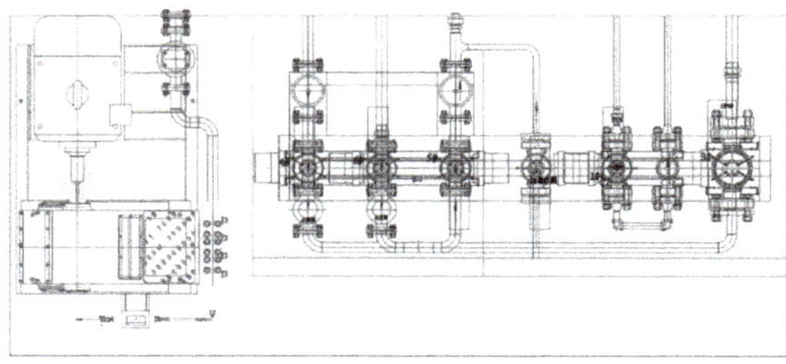

Figure 8. Internal hydraulic loading system.

Figure 9. Electric automatic control valve.

The internal water pressure loading system consisted of a water injection and drainage module, a high-pressure pipeline, an integrated valve, a pressure sensor and pressure gauge, an air compressor, etc. Among them, the water injection and drainage module were calculated according to the volume of the test pipe, meeting the maximum size of the test pipe (outer diameter: 24 inches; pipe length: 21 m) to complete the water pressure loading of 60 MPa within 1 min.

3.2. Servo Mechanism for Axial Tension/Compression Loading and Torsion Bidirectional Loading

In order to exert the axial force on the test specimen, an axial tension/compression loading torque servo mechanism and a bidirectional loading mechanism were developed and installed in the counterforce frame at one end. The test specimen, which had a perforated flange at the end, was connected to the combined loading system by the loading shaft through the torsion loading system. The flange at the other end of the specimen was connected to the end of the counterforce frame. The combined loading system was installed at the end of the counterforce frame, connected with the end frame through the anchor bolt. A disassembly of the combined loading system is shown in Figure 10, and its assembled form is shown in Figure 11.

The maximum applied axial tension force reached 3000 kN, and the effective tension/compression stroke was ±150 mm, with a maximum loading speed of 40 mm/s. Cyclic fatigue loading of the axial force was realized. The maximum applied bidirectional torque was 200 kN·m, with a loading angel of ±45 °, control accuracy ≤±2%, and maximum loading speed of 1°/s.

The tension/compression loading cylinder adopted frequency conversion speed regulation and proportional pressure valve control as well as hydraulic cylinder loading proportional tension/compression adjustment to achieve proportional and constant pressure loading. While unloading, the system was controlled by a proportional pressure valve. At the same time, a displacement sensor placed in the loading cylinder monitored the loading force and displacement changes of the test specimen in the process of tension/compression loading in real time.

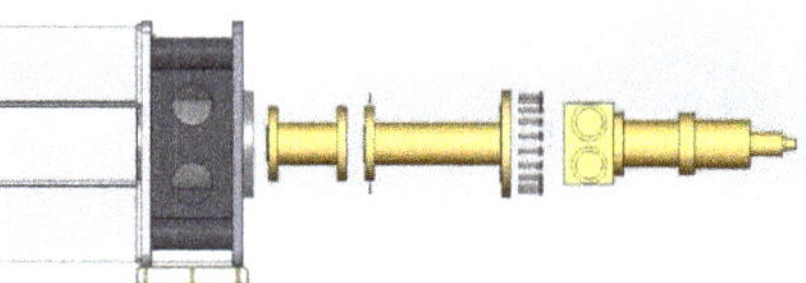

Figure 10. Disassembled diagram of the shaft end loading mechanism.

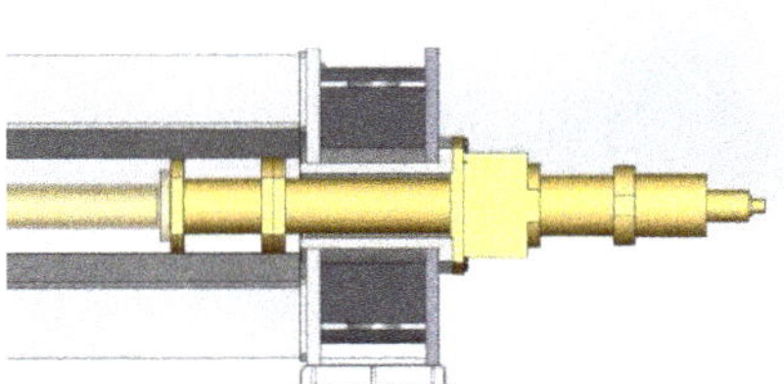

Figure 11. Combined load shaft.

The torsion loading cylinder was installed on the connecting cover flange of the tension loading cylinder. A rack-connecting rod was installed on the tension piston rod and the specimen, connected to the shaft of the axial tension loading cylinder. Bidirectional torsion loading was realized on the output shaft by loading the gear and rack in the swing cylinder.

As the water pressure and axial tension/compression changed, the output shaft of the specimen deformed outward while axial tension loaded, or inward while axial compression loaded; thus, the torsion could be normally loaded. The rack stroke under torsion load allowed for the horizontal movement distance of the specimen to be 150 mm.

3.3. Four-Point Bending Moment Bidirectional Loading Servo Mechanism

In order to generate a bending moment load on the specimen, a four-point moment loading servo mechanism perpendicular to the specimen was developed. It was installed in the middle of Section 3 of the counterforce frame, and it vertically and symmetrically applied the bending moment. The loading shaft end was a circular arc flange, holding the riser specimen and reciprocating the compression loading in the vertical direction of the specimen. The moment loading servo cylinder was symmetrically installed on the counterforce frame perpendicular to the specimen. The installation method was as follows: The lower bending moment loading mechanism was preinstalled at the bottom of the frame. The upper moment loading mechanism was installed on the upper flange. Before lifting the specimen, the upper flange was rotated 90 degrees towards the parallel direction of the reaction frame so as to facilitate the lifting of the specimen into the reaction frame. Then, the upper flange was rotated 90 degrees, reset, and tightened with bolts. Additionally, the loading shaft was put into the flange. In this way, the bending moment could be applied to the specimen. A schematic diagram of the bending moment loading system is shown in Figure 12.

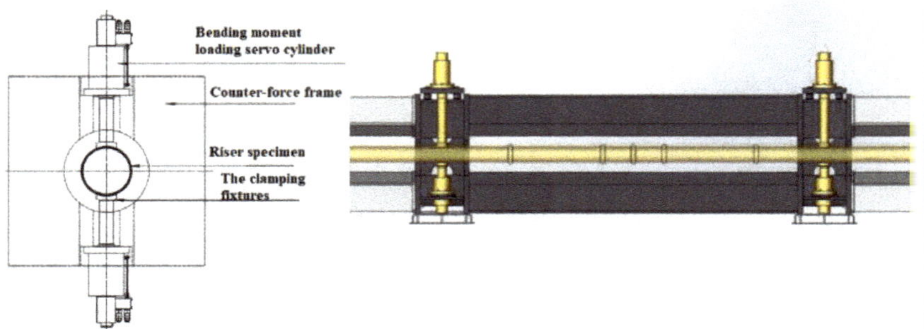

Figure 12. Moment loading device.

The loading device realized a maximum applied bending moment of 1300 KN·m. The effective stroke of the cylinder was ±150 mm, with a control accuracy of less than 1%. Cyclic fatigue loading was realized, with a maximum loading speed of 20 mm/s.

An MTS high-pressure magnetostrictive displacement sensor, external proportional servo valve, and pressure sensor were built into the moment loading cylinder, and displacement closed-loop loading or force closed-loop loading was realized. The bending moment loading cylinder is shown in Figure 13.

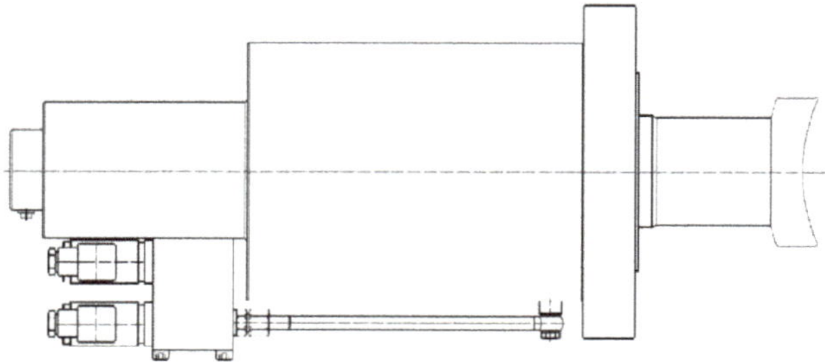

Figure 13. Bending moment loading cylinder.

The bending moment loading cylinder adopted frequency conversion speed regulation and proportional pressure reducing valve control to load or unload the hydraulic cylinder proportionally in order to realize torque proportional and constant loading. A proportional pressure reducing valve was used for unloading control. Meanwhile, the loading cylinder had a built-in sensor that monitored the loading force of the test pipe and the displacement of the specimen's deformation in the process of the bending moment loading. The moment loading cylinders on both sides could realize single-action/synchronous loading and unloading.

3.4. Hydraulic Loading Servo System

In order to meet the requirements of the compound loading mechanism, a hydraulic loading servo system was manufactured with a maximum power up to 400 KW. The hydraulic loading servo system was equipped with a variable frequency speed regulating motor, oil pump, servo valve, accumulator, sensor, and other hydraulic electrical components. Meanwhile, in order to meet the cyclic fatigue loading test, the system was equipped with an efficient cooling water tower. A schematic diagram of the hydraulic tank is shown in Figure 14. The servo system provides the power oil for the hydraulic cylinders.

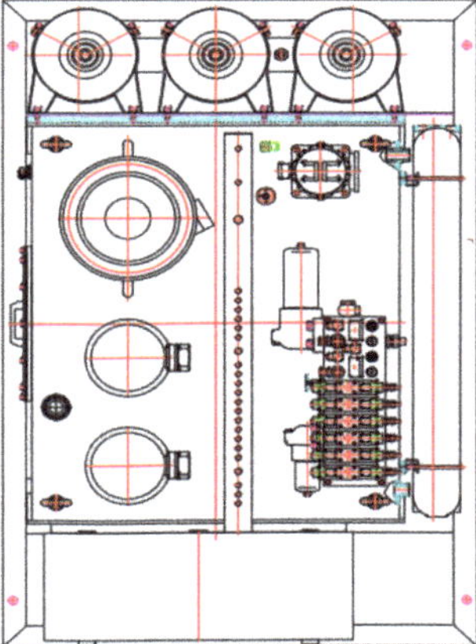

Figure 14. Hydraulic loading servo system.

The hydraulic pump station was equipped with a pump station electronic control cabinet. The electric control system for the pump station consisted of a sealed electric control box, a low-voltage electric pump controlled by a motor, a valve controlled by an amplifier, a servo valve real-time controller, a circuit drive board, and a connection interface with a control platform. The pumping station controller could realize basic distribution and water-cooling machine motor start, it could also realize onsite startup oil source system downtime, pressure, liquid level, temperature, and alarm signal collection and input. Meanwhile, TCP/PI ethernet interface communication with the remote central control center was also provided to realize remote monitoring control.

The test system was designed for mechanical loading. Its main characteristics were as follows:

(1) The full size of the test system (main body of the system) was 26 m, the size of the counterforce frame was 24 m, and the longest size of the test pipe section was 22 m;

(2) The maximum loading capacity of the test system was designed to be 3000 kN dynamic axial force, 1300 kN·m dynamic bending moment loading, 200 kN·m torque loading, and 60 MPa internal water pressure loading, with a loading frequency of 30 Hz.

The platform loading capability indicators were shown in Table 2.

Table 2. Loading capacity of a full-scale steel catenary riser fatigue test system.

Size of Specimen	Axial Force	Bending Moment	Torque	Internal Pressure MPa
L ≤ 22 m D < 609.6 mm	3000 kN	1300 kN·m Loading schedule ± 150 mm	200 kN·m	60

4. Full-Scale Riser Fatigue Test

In order to verify the performance of the fatigue test platform and test the fatigue strength of the steel catenary riser (SCR) in the Lingshui project, we designed and carried out riser fatigue tests. The SCR for the test was produced by Hengyang Valin Steel Tube Company and covered a 5G double jointing procedure. The test section was 5.6 m, with three weld joints arranged. The length of the connecting section was 11.2 m. The outer diameter of the test riser was 12 inches, and the thickness was 27 mm. Both ends were welded with perforated flanges and connected to the end of the test platform. The structure of the test riser section is shown in Figures 15 and 16.

Figure 15. Schematic diagram of the loading test riser specimen.

Figure 16. Full-size SCR test specimen.

The purpose of this experiment was to measure the fatigue life of the riser and weld under high-stress conditions in the South China Sea. Based on the measured stress in high-stress environments in the South China Sea, a stress cycle of 172 MPa (±86 MPa) was carried out on the basis of 138 MPa of stress.

The position and weld number of the test pipe section on the platform were as follows. The left side was the fixed end, and the right side was connected to the end of the extension section. The loading mechanism acted on the other end of the connecting section. From the left side, the welds were numbered G1, G2, and G3, as shown in Figure 17.

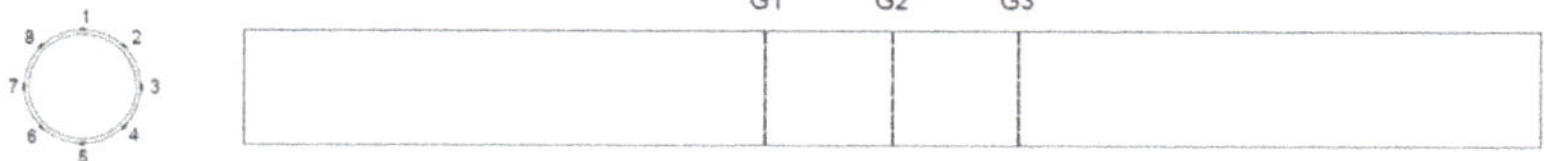

Figure 17. Weld diagram.

The section number of the weld is shown below, and the direction of the section Figure 17 is from left to right.

4.1. The Test Process

Before the fatigue test, the system and parameters of the specimen were tested, and the loading method was determined. The test analysis was as follows:

1. The first-order frequency of the filling riser was 3.4 Hz, and the effect of the rising water pressure on the first-order frequency could be ignored;
2. Under a loading of 48 MPa internal pressure, the axial displacement changed by 8.9 mm, and 732 kN axial tension continued to be applied. The overall axial displacement changed by 2.3 mm. The total change in pipeline displacement was 11.2 mm.
3. On the basis of 48 MPa internal pressure and 732 kN axial tension, when the axial tension was applied at 2900 kN, the axial displacement increased by 7 mm, and the

pipeline displacement reached 18.2 mm. The stress value reached the maximum stress value in the high-stress experiment;

4. On the basis of 48 MPa internal pressure and 732 kN axial tension, when the axial pressure was applied at -1500 kN, the axial displacement decreased by 7 mm, and the pipeline displacement reached 4.2 mm. The stress value reached the minimum stress value in the high-stress experiment.

Therefore, the loading scheme of this test was set as follows:

(1) An internal pressure of 48 MPa was applied;
(2) A fixed axial tension of 732 kN was applied, and the average stress reached 138 MPa;
(3) A cyclic axial force between -1500 and 2900 kN was applied. The cyclic range of axial displacement was ± 7 mm, meeting 172 MPa (± 86 MPa) of cyclic stress.

The stress measurement values of eight measured points for each three welding points within 15 s were intercepted, and the curves during stress generation were drawn as shown in Figure 18.

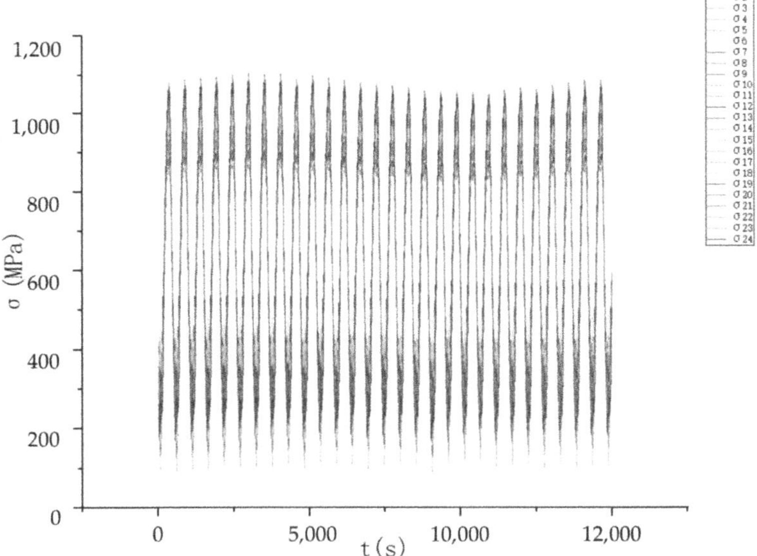

Figure 18. Stress changes at each weld measuring point.

Through the stress–time history curves of the 3 welding joints and the change in the stress–time history curves of the 24 measuring points, it can be seen that the stress cycle range of the riser test reached 172 MPa, and the stress cycle value was relatively stable during the test process. Based on the number of cycles calculated by the S–N curve, the number of test settings was increased and set at 2.85 million cycles. After the completion of the test, nondestructive testing was carried out on the weld of the test pipe to check the test results.

4.2. Weld Test Results

After the test, ultrasonic weld inspection of the specimen was conducted. The results of the inspection are shown in the Table 3.

Table 3. Weld inspection of the full-scale domestic steel catenary riser in a high-stress fatigue test.

The Weld Number	Defect Wave Reflection Region	Defect Location (mm)			Defect Indication Length (mm)	Defect Levels	Types of Crack
		L1	L2	Depth			
G1	III	57	318	3.1~27	261	III	
	III	531	855	Through-wall crack	324		
G2	III	114	327	6.2~27	213	III	Weld fatigue stress crack
	III	605	797	5.6~27	192	III	
G3	III	24	270	Through-wall crack	246	III	
	III	449	735	2.3~27	286	III	
	III	897	989	7.3~27	92	III	

It can be seen from the testing results that there were cracks caused by different degrees of fatigue stress in the three weld positions of the test riser's section, and through-wall cracks appeared in the specimen.

4.3. Fatigue Analysis Based on BS7608

In most cases, potential fatigue cracks will occur at the stress concentration of the base metal. In welded structures, fatigue failure mainly occurs in welded joints. Microcracks appeared near the weld as a result of welding. Under the action of alternating loads, stress concentration will appear around the microcrack, leading to crack propagation. When the length of the crack reaches a critical point, the member will suddenly fracture, causing the structure to fail. Reducing the stress concentration near the weld is the only way to solve the problem of fatigue crack in the design stage. The fatigue behavior of welded structures is a very complex phenomenon, because it depends on many factors affecting the stress/strain field at the point where the final fracture occurs. Obviously, structural fatigue, including welded joints, is much more complex than simple material fatigue.

In the British Standard 7608 (BS7608): Fatigue design and assessment of steel structures, the fatigue S–N curves of different welding forms and load types are given. Fatigue of welding joints is different from that of ordinary materials. As for the fatigue of welded joints, the applicable yield strength is between 200 MPa and 960 MPa. For each structural detail, there is a reference value for the fatigue strength limit. BS7608 provides different calculation methods for different welding types, and appropriate methods can be selected for calculation. By determining the specific form of the structure and selecting and using the appropriate S–N curve for the welding joints, the stress spectrum in the loading process is established, and the fatigue life is calculated by the stress spectrum. The S–N curve also considers the size, shape, residual stress, and crack shape in order to calculate the fatigue life of the structure more accurately. In specification BS7608, the S–N curve is calibrated according to nominal stress. However, in practical engineering, nominal stress does not strictly exist for welded joints with complex geometric shapes or under complex loads. Therefore, the generalized nominal stress is introduced. For numerical calculation, the regional stress whose stress gradient is close to zero is defined as the generalized nominal stress. Based on the above definition, BS7608 can be used to solve practical engineering problems.

In most cases, the underlying fatigue crack is located near a stress concentration in the base metal such as at the welding toes or bolt holes. It is assumed that the direction of principal stress does not change significantly during the stress cycling process. Therefore, the maximum cyclic range of the principal stress in the stress cycling process is taken as the cyclic stress range used in fatigue calculation, and the principal stress in any position near the crack on the base metal is correspondingly within this maximum cyclic stress range. It is assumed that the tensile stress is positive, whereas the compressive stress is negative. In practice, the stress component throughout the thickness has little effect and is usually negligible. When the principal stress direction changes periodically, the magnitude of the cyclic stress can be calculated by calculating the two extreme values in the process of stress change, i.e., the difference between the peak and trough of the wave. The peaks and troughs, here, are the value of the peaks and troughs in the main plane.

The S–N curve of the range of the cyclic stress and the number of cycles required to achieve fatigue is as follows:

$$\log N = \log C_0 - d\sigma - m \log S_r \tag{1}$$

where C_0 is the correlation constant of the average S–N curve, D is the standard deviation below the mean, σ is the relative standard deviation of N, and m is the reverse slope of the S–N curve under a double logarithm.

In fatigue calculation, the influence of the material's thickness on fatigue life should be considered. The non-joint class for which the basic S–N curve applies (i.e., corresponding the weld class B-G) requires a material thickness of no more than 16 mm. For joints of other thicknesses, the stress range of the fatigue strength should be modified by the following formula:

$$S = S_B \left(\frac{t_B}{t} \right)^{1/4} \tag{2}$$

where S is the fatigue strength equivalent to the stress of the specimen, S_B is the equivalent stress of the joint fatigue strength using the basic curve, t is the actual plate thickness when the thickness is greater than 16 mm, and t_B is the maximum plate thickness corresponding to the basic S–N curve, which is 16 mm.

The fatigue strength of the S–N curve can be increased by 30% for welded joints with initial cracks at the welding toes by local machining or grinding the toes.

The S–N curve is selected as a D curve for butt welding. According to the basic curve parameters, the expression of the S–N curve is as follows:

$$\log N = 12.6007 - 3 \log S_r \tag{3}$$

According to the D curve formula of the BS7608 specification and the data from the test, the S–N curve is drawn as Figure 19.

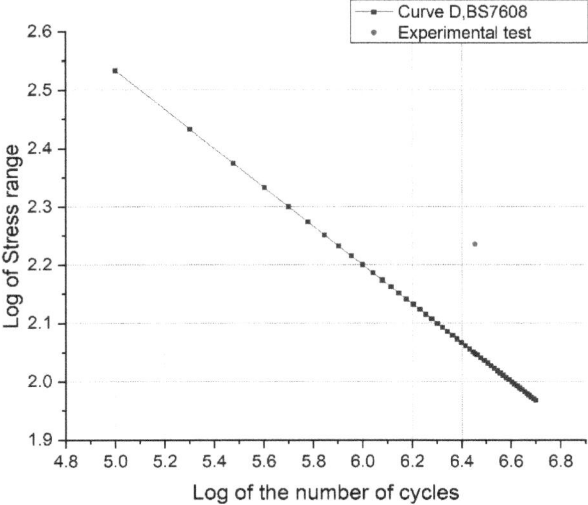

Figure 19. S–N curve in the specification.

It can be seen that the measured point is above the S–N curve, indicating that the test riser met the fatigue requirements of BS7608. At the same time, the thickness of the specimen needs to be corrected. According to the thickness stress correction in Formula (2), it was calculated that for a specimen pipe with a diameter under 27 mm, the actual stress value to be substituted by the S–N curve was 150.91 MPa. The stress value was substituted into (3), and the required number of cycles was 1.16 million. After 2.85 million cycles, cracks appeared in the specimen, indicating that the specimen and welding can meet the fatigue strength requirements of BS7608.

4.4. S–N Curve Selection and Riser Thickness Correction

The test in this section was designed according to the BS7608 fatigue specification, and the index of cycles should have been 1.16 million if calculated according to the specification. In the actual test, it took 2.85 million cycles for the welded riser to break. The actual cycle times increased by more than twofold compared with the theoretical cycle times. An increase in the actual cycle time is beneficial to the safety of the structure, but it will also increase the amount of steel used in engineering, resulting in a great increase in the cost. Therefore, it is necessary to redesign the thickness of the riser based on the test results and give the minimum thickness scheme of the riser to meet the requirements of the fatigue design. This can ensure that the structure meets the fatigue safety requirements and, at the same time, obtain the maximum economic benefits.

In general, under the same sea state, the stress response of the structure will change with different structural forms and thickness. Different cyclic stresses will cause different fatigue cycle indexes. Therefore, the S–N curve corresponds the cycle index with the stress value. For a certain S–N curve, the corresponding cycle index is also fixed under a certain stress value. However, for the same stress value but with a different thickness of the riser structure, the cycle index of the complete damage will be different. From the perspective of crack growth, the time of penetration crack in the thick-walled pipe with the same crack growth rate will be correspondingly longer. This is also the reason for the thickness modification in the BS7608 specification above. In addition, because of the different welding quality, the fatigue life will be increased. Therefore, different calculation formulas are defined for different welding methods in the specification. For whole pipe structure welding, according to the BS7608 specification, the test riser in this paper is more suitable for a D curve. However, according to the inspection of the test pipe in Table 2, it was obvious that the cracks were all generated at a depth of 27 mm, i.e., the welding toe of the riser. This indicates that the riser was fractured from the internal welding toe. Therefore, a C curve is more suitable for fatigue analysis. The C curve in the specification is as follows:

$$\log N = 14.0342 - 3.5 \log S_r \tag{4}$$

The C curve and D curve of the BS7608 specification and the test results are drawn in the Figure 20.

It is obvious that the C curve was closer to the actual test situation. Therefore, it was more appropriate to choose the C curve in the analysis. According to the modified formula of the BS7608 specification (i.e., Formula (2)), it can be determined that the thicker the riser, the smaller the equivalent stress value that should be substituted into the S–N curve, and the calculated index of the cycles under the corresponding also increased to a certain extent. In contrast, the thinner the riser, the larger the equivalent stress, and the smaller the corresponding cycle index value.

It can be concluded that when the number of cycles was 2.85 million (the response value was 6.45484), the corresponding C curve cyclic stress logarithm value was 2.166, and the cyclic stress was 146.55 MPa. By substituting the cyclic stress into the thickness correction formula, it can be calculated that under the welding condition of the test riser, its thickness was equivalent to 30.36 mm of the standard riser. The S–N curve was based on the linear accumulation of fatigue through the Palmgren–Miner rule; therefore, it can be approximated that the propagation of fatigue cracks along the thickness direction of the structure was linear. Thus, a linear formula for thickness correction is proposed based on the results of this test.

$$\frac{t_A}{t} = \frac{t'}{t_B} \tag{5}$$

where t_A is the standard thickness. Here, the thickness was 27 mm, which was selected for the experiment, as well as the reference thickness t_B of this test. t' is the standard thickness, i.e., the number of test cycles substituted into the S–N curve to obtain the corresponding cyclic stress value and corresponding the stress to the thickness correction formula to obtain the standard thickness t'.

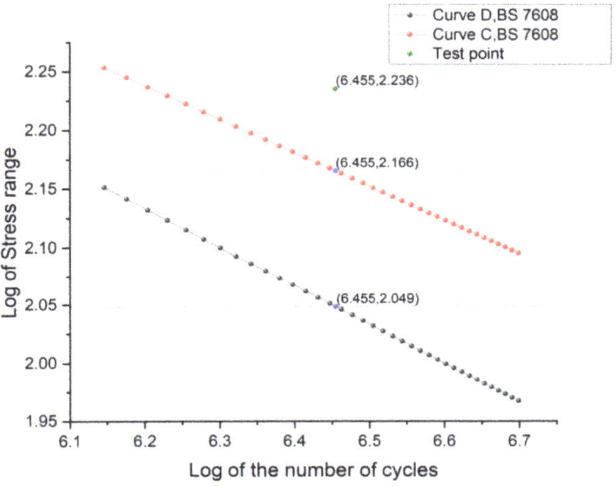

Figure 20. Calculation points in the S–N curve.

By substituting the test results, the riser thickness t that actually met the fatigue requirements was 24.016 mm. The safety factors of the riser's structural design are defined as follows:

$$\beta = \frac{t_d}{t_r} \tag{6}$$

where t_d is the designed thickness of the riser structure and t_r is the calculated actual thickness of the pipe. Accordingly, the safety factor of the test riser in this paper was 1.124. It can be concluded that the thickness design of the test pipe in this paper fully considered the economic benefit of steel quantity on the basis of ensuring the fatigue strength requirements.

5. Conclusions

In this paper, the full-scale riser fatigue test system and its main component have been described in detail. The system was capable of conducting fatigue tests for full-scale risers, flexible pipes, and seabed pipeline fatigue. The test system could complete a fatigue test on risers under a severe sea state and complex loads.

The test designed in this paper was a full-size riser high-stress loading test to determine the maximum loading capacity of the test platform. Thus, the axial loading method was used for loading. For the test with high stress, the platform adopted axial cyclic loading to meet the high-frequency cycle and shorten the test time. The bending moment loading could meet the maximum loading capacity. However, with the increase in the bending moment cylinder stroke, the loading frequency decreased, thus prolonging the test time. Therefore, at present, when conducting the fatigue test with high cyclic stress, the test was designed as an axial cyclic load test. The moment cyclic load can be used in tests with low stress levels. In the future, the moment loading capacity of the test system will be upgraded in order to realize high-stress fatigue tests under cyclic bending moment load.

Due to the high cost of full-scale riser tests, only fatigue tests at high stress levels were performed this time. In a subsequent study, several more sets of tests will be conducted to draw the S–N curve of the structure, which will be compared with the specification.

In the test, the fatigue behavior of a 27 mm thick welded riser was tested. The test results showed that the fatigue performance of the welded pipe exceeded the prediction of the S–N curve in the code. The thickness design was reasonable, and the economic benefits were fully considered.

This test was a fatigue test in an air environment; however, the environment of the South China Sea is complex. Therefore, corrosion fatigue correction tests will be designed in a future study to build a complete fatigue life prediction test system.

Author Contributions: Conceptualization, J.Y. and F.W.; methodology, F.W.; software, Y.Y.; validation, F.W., P.L. and X.L.; formal analysis, Y.S.; investigation, F.W.; resources, Y.S.; data curation, F.W.; writing—original draft preparation, F.W.; writing—review and editing, F.W.; visualization, X.L.; supervision, P.L.; project administration, Y.S.; funding acquisition, J.Y. All authors have read and agreed to the published version of the manuscript.

Funding: This research was funded by The National Natural Science Foundation of China (Grant No. 51879189) and National Natural Science Foundation of China (Grant No. 52071234).

Institutional Review Board Statement: Not applicable.

Informed Consent Statement: Not applicable.

Data Availability Statement: The data presented in this study are available upon request from the corresponding author.

Conflicts of Interest: The authors declare no conflict of interest.

References

1. Xu, J.; Jesudasen, A.S.; Fang, J.; Else, M. Wave Loading Fatigue Performance of Steel Catenary Risers (SCRs) in Ultradeepwater Applications. In Proceedings of the Offshore Technology Conference, Houston, TX, USA, 8 April 2006.
2. Yu, J.; Xu, W.; Yu, Y.; Fu, F.; Wang, H.; Xu, S.; Wu, S. CFRP Strengthening and Rehabilitation of Inner Corroded Steel Pipelines under External Pressure. *J. Mar. Sci. Eng.* **2022**, *10*, 589. [CrossRef]
3. Chatziioannou, K.; Karamanos, S.A.; Huang, Y. Ultra Low-Cycle Fatigue Performance of S420 and S700 Steel Welded Tubular X-Joints. *Int. J. Fatigue* **2019**, *129*, 105221. [CrossRef]
4. Song, W.; Liu, X.; Berto, F.; Razavi, S.M.J. Low-Cycle Fatigue Behavior of 10CrNi3MoV High Strength Steel and Its Undermatched Welds. *Materials* **2018**, *11*, 661. [CrossRef]
5. Feng, L.; Qian, X. Enhanced Crack Sizing and Life Estimation for Welded Tubular Joints under Low Cycle Actions. *Int. J. Fatigue* **2020**, *137*, 105670. [CrossRef]
6. Nassiraei, H.; Rezadoost, P. Stress Concentration Factors in Tubular T/Y-Joints Strengthened with FRP Subjected to Compressive Load in Offshore Structures. *Int. J. Fatigue* **2020**, *140*, 105719. [CrossRef]
7. Teng, T.-L.; Fung, C.-P.; Chang, P.-H. Effect of Weld Geometry and Residual Stresses on Fatigue in Butt-Welded Joints. *Int. J. Press. Vessel. Pip.* **2002**, *79*, 467–482. [CrossRef]
8. Lee, C.-H.; Chang, K.-H.; van Do, V.N. Modeling the High Cycle Fatigue Behavior of T-Joint Fillet Welds Considering Weld-Induced Residual Stresses Based on Continuum Damage Mechanics. *Eng. Struct.* **2016**, *125*, 205–216. [CrossRef]
9. Vieira Ávila, B.; Correia, J.; Carvalho, H.; Fantuzzi, N.; de Jesus, A.; Berto, F. Numerical Analysis and Discussion on the Hot-Spot Stress Concept Applied to Welded Tubular KT Joints. *Eng. Fail. Anal.* **2022**, *135*, 106092. [CrossRef]
10. da Silva, A.L.L.; Correia, J.A.F.O.; de Jesus, A.M.P.; Lesiuk, G.; Fernandes, A.A.; Calçada, R.; Berto, F. Influence of Fillet End Geometry on Fatigue Behaviour of Welded Joints. *Int. J. Fatigue* **2019**, *123*, 196–212. [CrossRef]
11. Jiang, W.; Chen, W.; Woo, W.; Tu, S.-T.; Zhang, X.-C.; Em, V. Effects of Low-Temperature Transformation and Transformation-Induced Plasticity on Weld Residual Stresses: Numerical Study and Neutron Diffraction Measurement. *Mater. Des.* **2018**, *147*, 65–79. [CrossRef]
12. Jin, Q.; Jiang, W.; Gu, W.; Wang, J.; Li, G.; Pan, X.; Song, M.; Zhang, K.; Wu, A.; Tu, S.-T. A Primary plus Secondary Local PWHT Method for Mitigating Weld Residual Stresses in Pressure Vessels. *Int. J. Press. Vessel. Pip.* **2021**, *192*, 104431. [CrossRef]
13. Peng, W.; Jiang, W.; Sun, G.; Yang, B.; Shao, X.; Tu, S.-T. Biaxial Residual Stress Measurement by Indentation Energy Difference Method: Theoretical and Experimental Study. *Int. J. Press. Vessel. Pip.* **2022**, *195*, 104573. [CrossRef]
14. Xin, H.; Correia, J.A.F.O.; Veljkovic, M.; Berto, F.; Manuel, L. Residual Stress Effects on Fatigue Life Prediction Using Hardness Measurements for Butt-Welded Joints Made of High Strength Steels. *Int. J. Fatigue* **2021**, *147*, 106175. [CrossRef]

15. Song, S.; Pei, X.; Dong, P. An Analytical Interpretation of Welding Linear Heat Input for 2D Residual Stress Models. In Proceedings of the ASME 2015 Pressure Vessels and Piping Conference, Boston, MA, USA, 19–23 July 2015.
16. Kang, G.; Luo, H. Review on Fatigue Life Prediction Models of Welded Joint. *Acta Mech. Sin.* **2020**, *36*, 701–726. [CrossRef]
17. Yang, H.; Qian, H.; Wang, P.; Dong, P. Analysis of Fatigue Behavior of Welded Joints in Orthotropic Bridge Deck Using Traction Structural Stress. *Adv. Mech. Eng.* **2019**, *11*, 1–14. [CrossRef]
18. Wang, H.; Yu, Y.; Yu, J.; Xu, W.; Chen, H.; Wang, Z.; Han, M. Effect of Pitting Defects on the Buckling Strength of Thick-Wall Cylinder under Axial Compression. *Constr. Build. Mater.* **2019**, *224*, 226–241. [CrossRef]

Journal of
Marine Science and Engineering

Article

Fatigue Strength Assessment of Single-Sided Girth Welds in Offshore Pipelines Subjected to Start-Up and Shut-Down Cycles

Yan Dong [1,2,3], Guanglei Ji [1], Lin Fang [4] and Xin Liu [1,*]

[1] Yantai Research Institute of Harbin Engineering University, Harbin Engineering University, Yantai 264000, China
[2] College of Shipbuilding Engineering, Harbin Engineering University, Harbin 150000, China
[3] HEU-UL International Joint Laboratory of Naval Architecture and Offshore Technology, Harbin Engineering University, Harbin 150000, China
[4] COOEC Subsea Technology Co., Ltd., Shenzhen 518000, China
* Correspondence: xin.liu@hrbeu.edu.cn

Abstract: During the service life of offshore pipelines, many start-up and shut-down cycles take place, possibly leading to significant cyclic loads. Fatigue failure may occur, resulting in serious environmental pollution and loss of property. The study aims to assess the fatigue strength of single-sided girth welds in offshore pipelines under these specific fatigue loads. The longitudinal stress range caused by the variation of the pipeline's internal pressure and temperature is calculated. The effective notch strain approach is used to assess the fatigue strength of welds. The plastic behaviour of the weld root is investigated for a study case to justify the use of low-cycle fatigue assessment approaches. The effect of weld root geometry on the notch stress factor is studied to identify the dominant geometrical parameters. The fatigue strength of the study case is assessed, and some limitations of the assessment are discussed. The results show that the plastic behaviour of the weld root is only significant for severe local stress concentrations, which is mainly governed by the axial misalignment, weld root angle and the weld root bead width. If the fatigue damage at failure is 0.1, a limited number of start-up and shut-down cycles are allowed during the service life of the pipeline for the study case, indicating the necessity of fatigue strength assessment.

Keywords: fatigue strength; offshore pipelines; girth welds; low cycle fatigue

Citation: Dong, Y.; Ji, G.; Fang, L.; Liu, X. Fatigue Strength Assessment of Single-Sided Girth Welds in Offshore Pipelines Subjected to Start-Up and Shut-Down Cycles. *J. Mar. Sci. Eng.* **2022**, *10*, 1879. https://doi.org/10.3390/jmse10121879

Academic Editor: Bruno Brunone

Received: 25 October 2022
Accepted: 20 November 2022
Published: 3 December 2022

Publisher's Note: MDPI stays neutral with regard to jurisdictional claims in published maps and institutional affiliations.

1. Introduction

Subsea pipelines are subjected to fatigue loading in operation conditions, which include those induced by motions of floating platforms, free span vortex-induced vibrations, and thermal cycles. These fatigue loadings may result in significant fluctuating stresses acting on the girth welds of pipelines, leading to fatigue failure. The fatigue strength of subsea pipelines has been a major concern.

An increasing number of high-pressure and high-temperature (HPHT) offshore pipelines have been applied in recent years because more and more HPHT oil and gas fields are being developed. HPHT applications of 150 °C and 68.95 MPa are common nowadays, and more severe HPHT operating conditions also exist [1]. Subsea pipelines under HPHT operating conditions have to face various challenges. The global lateral buckling may happen because of the high compressive axial force when the pipelines are laid on the seabed and heated, but the axial extensions are restrained by the soil [2]. The pipelines are subjected to a number of cycles of start-up and shut-down through the design life of an offshore field. The pipeline walking, i.e., global axial movement can be triggered by the start-up and shut-down cycles in some cases, leading to some problems that may endanger the pipeline system [3]. The HPHT conditions can increase the corrosion rate of the pipeline significantly [4]. Relative large stress ranges may take place due to the start-up and shut-down process, resulting in low cycle fatigue (LCF) of the pipeline [5].

In the present study, the focus is given to the fatigue damage of single-sided girth welds in offshore pipelines caused by the start-up and shut-down cycles. Since the loading cycles may be significant, the LCF approaches instead of high cycle fatigue (HCF) approaches are recommended. Compared with the HCF approaches, the LCF approach needs more computational efforts. The LCF approaches are generally classified into two types, one based on S-N curves and linear damage rule, and another relies on fatigue crack propagation analyses. The strain parameters or the elastic–plastic fracture mechanics are used in these approaches. The first type of approach includes the pseudo hotspot stress approach [6,7] structural strain approach [8,9], notch strain approach [10,11] and effective notch strain approach [12,13]. The approaches belonging to the second type usually use the cyclic J-integral [14–17] and the cyclic crack tip opening displacement [18,19] as the driving forces of crack growth. The first type of approach is easier to use in practice because a lower number of parameters is usually involved, and the calculation is more efficient, while the second type of approach is closer to the physical fatigue process.

Although many approaches have been developed for the LCF problem, the study on fatigue strength assessment of offshore pipelines subjected to start-up and shut-down cycles is rare. One reason may be that the special fatigue problem can be coupled with global bucking, which complicates the problem. When the pipeline is heated, global buckling may happen and can lead to a release of the global compressive force. Furthermore, the associated bending of the pipeline can either increase or decrease the local stress depending on the location of interest. In the present study, it is assumed that the pipeline remains straight during start-up and shut-down cycles, and the effects of global buckling are ignored.

The study on the fatigue strength assessment under these specific cyclic loads may contribute to improving the safety of subsea pipelines. The allowed number of cycles can be obtained from the analyses, which is valuable for the management of offshore oil and gas fields.

In the present study, the fatigue strength assessment of single-sided girth welds in offshore pipelines subjected to start-up and shut-down cycles is performed. The calculation of the longitudinal stress range due to the start-up and shut-down cycle and the effective notch strain approach is introduced. For a specific study case, the plastic behaviour of the weld root is investigated to justify the use of the LCF approach, and the effect of weld root geometry on the notch stress factor is studied to identify the dominant geometrical parameters. The fatigue strength of the study case is assessed, and some limitations of the assessment are discussed.

2. Methods

The methods involved in the fatigue strength assessment are introduced in this section. The fatigue loads acting on the welds are determined analytically. The effective notch strain approach is used for the fatigue strength assessment. The finite element method (FEM) and the analytical method are combined to calculate the effective notch strain.

2.1. Start-Up and Shut-Down Load Cycles

The stress components in pipelines are longitudinal (axial) stress, hoop (circumferential) stress and radial stress. Since fatigue cracks are usually initiated at girth welds and propagate along the direction of the weld line and the thickness, the stress component normal to the potential crack plane, i.e., the longitudinal stress, is of interest.

In the present study, the pipeline is assumed to be fully constrained. In other words, there is no longitudinal movement when the internal pressure or temperature is increased. The assumption can result in more conservative longitudinal stresses.

The longitudinal stress for the installation condition $\sigma_{l,ins}$ can be calculated by [1]:

$$\sigma_{l,ins} = \frac{F_{res} + P_{i,ins} A_i - P_e A_e}{A_s} \tag{1}$$

where F_{res} is the installation residual lay tension, $P_{i,ins}$ is the internal pressure of the pipe during installation, A_i is the internal bore area of the pipe, P_e is the external pressure of the pipe, A_e is the external area of the pipe, A_s is the cross-sectional area of the pipe.

When the temperature and pressure are increased to the operational condition, the pipeline tends to expand longitudinally. However, the pipeline is fully restrained, and thus, the compressive longitudinal stress is developed. For a fully restrained pipeline, the longitudinal stress is caused by the Poisson's ratio effect of hoop and radial stresses and the thermal effect [1]:

$$\sigma_l = \frac{2v(P_i A_i - P_e A_e)}{A_s} - E\alpha\left(T_{op} - T_a\right) \tag{2}$$

where σ_l is the longitudinal stress, E is Young's modulus, v is the Poisson's ratio, P_i is the operational internal pressure, α is the thermal expansion coefficient of pipe material, T_{op} is the operational temperature and T_a is the ambient temperature at installation.

Taking the longitudinal stress for the installation condition as the initial condition, the longitudinal stress for the operational condition can be calculated by:

$$\sigma_{l,op} = \sigma_{l,ins} + \frac{2v(P_i - P_{i,ins}) A_i}{A_s} - E\alpha\left(T_{op} - T_a\right) \tag{3}$$

The longitudinal stress range caused by the start-up and shut-down load cycle is

$$\Delta\sigma_l = \left|\sigma_{l,op} - \sigma_{l,ins}\right| = \left|\frac{2v(P_i - P_{i,ins}) A_i}{A_s} - E\alpha\left(T_{op} - T_a\right)\right| \tag{4}$$

For the fully restrained condition, increasing the operating internal pressure and increasing the operating temperature has the opposite effect on the longitudinal stress range. Increasing the operating internal pressure means increasing the hoop stress, leading to tensile longitudinal stresses due to Poisson's ratio effect for the fully restrained condition. However, increasing the operating temperature results in compressive longitudinal stress because the extension is restrained.

The assumption of a fully restrained condition is conservative, and the fatigue strength assessment based on this assumption is representative. If the pipeline is free at the end, the variation of longitudinal stress is limited around the end and continuously increases toward the centre of the pipeline until the fully restrained condition is developed. In the present study, only the fully restrained condition is considered.

Note that significant compressive stress can be developed in the operational condition, implying the stress variation due to the start-up and shut-down load cycle is mainly in the regime of the compressive stress. It does not mean the fatigue problem is less important because the local welding-induced residual stresses around the welds may approach the yield stress of the material, which can significantly improve the local mean stress.

2.2. Fatigue Strength Assessment

The effective notch strain approach is extended from the effective notch stress approach from the HCF regime. It can be used to deal with LCF and HCF problems and is employed in the present study. Compared with the notch train approach, the effective notch strain approach is easier to be applied in practice [13]. Detailed information on local geometry and residual stresses is not required, and it can result in the total fatigue life instead of the crack initiation life.

The method for the estimation of the effective notch strain, which is proposed by Dong et al. [13], is used. The method is simpler than the elastic–plastic finite element method. It can be divided into two steps. The first step is to calculate the effective notch stress as recommended by the International Institute of Welding [20]. The weld toe and root are rounded by a fictitious notch whose radius is $\rho_f = 1$ mm for welded steel structures with a thickness larger than 5 mm, as shown in Figure 1. The FEM is usually used. The maximum first principal stress along the notch is taken as the effective notch stress.

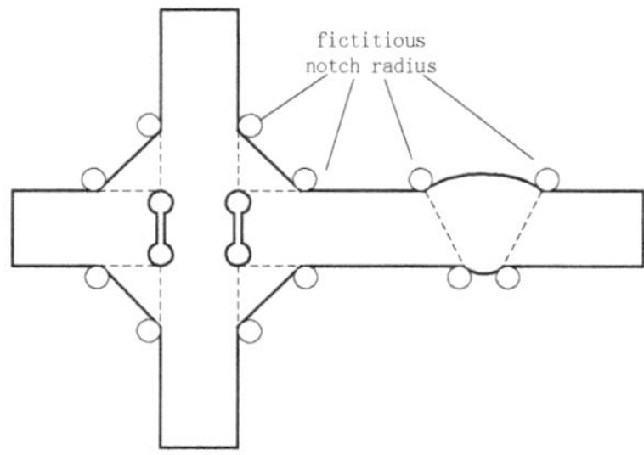

Figure 1. Fictitious notch rounding [20].

The second step is to convert the elastic stress to the elastic–plastic strain. The step is mainly based on the plane strain equivalent strain energy density approach [21,22]. The plane strain state is assumed, implying that the strain component along the notch is zero and a biaxial stress state exists at the notch tip when the notched component is remotely loaded. The uniaxial cyclic stress–strain curve of the material is assumed to follow the Ramberg–Osgood relation:

$$\varepsilon = \frac{\sigma}{E} + \left(\frac{\sigma}{K'}\right)^{\frac{1}{n'}} \tag{3}$$

where K' is the cyclic strength coefficient, n' is the cyclic strain hardening exponent and E is the elastic modulus. According to [23], the above curve should be transformed into the plane strain curve by using:

$$\begin{cases} \sigma_1 = \dfrac{\sigma}{\sqrt{1-\mu+\mu^2}} \\[2mm] \varepsilon_1 = \dfrac{\varepsilon(1-\mu^2)}{\sqrt{1-\mu+\mu^2}} \\[2mm] \mu = \dfrac{1}{2} - \left(\dfrac{1}{2}-\nu\right)\dfrac{\sigma}{E\varepsilon} \end{cases} \tag{4}$$

where stress and strain with a subscript 1 represent the first principal quantities, ν is the Poisson's ratio and μ is the generalised Poisson's ratio. The data points in the uniaxial cyclic stress–strain curve can be transformed into the data points of (σ_1, ε_1). The plane strain curve can be fitted based on these data points, and the plane strain curve is in the form of:

$$\varepsilon_1 = \frac{\sigma_1}{E}\left(1-\nu^2\right) + \left(\frac{\sigma_1}{K_1'}\right)^{\frac{1}{n_1'}} \tag{5}$$

where K_1' and n_1' are the new material properties. The plane strain equivalent strain energy density approach can be expressed by:

$$\frac{\sigma_{eff}^2\left(1-\nu^2\right)}{2E} = \int_0^{\varepsilon_1} \sigma_1 d\varepsilon_1 \tag{6}$$

where σ_{eff} is the effective notch stress. The first principal notch strain ε_1 can be solved by combining Equations (3) and (4). The generalised Poisson's ratio μ can also be obtained. According to Equation (6), the effective notch strain can be estimated by:

$$\varepsilon_{eff} = \frac{\varepsilon_1\sqrt{1-\mu+\mu^2}}{(1-\mu^2)} \tag{9}$$

Generally, the effective notch strain range is of interest in fatigue strength assessment. The effective notch strain amplitude ε_{eff} is obtained by solving Equations (3)–(5), in which the σ_{eff} is the effective notch stress amplitude. The effective notch strain range is twice the effective notch strain amplitude.

The curves of effective notch strain range $\Delta\varepsilon_{eff}$ vs. fatigue life N_f for a survival probability of 97.7% can be expressed by:

$$\begin{cases} N_f(\Delta\varepsilon_{\text{eff}})^{m_1} = C_1 & N_f < 10^4 \\ N_f(\Delta\varepsilon_{\text{eff}})^{m_2} = C_2 & N_f \geq 10^4 \end{cases} \tag{10}$$

The segment of $N_f < 10^4$ is derived from the LCF test data [12], and the segment of $N_f \geq 10^4$ is derived from the S-N curve of FAT225 for the effective notch stress approach [20]. The relationship between the effective notch strain range and effective notch stress range in the elastic domain is:

$$\Delta\varepsilon_{eff} = \frac{\Delta\sigma_{eff}\sqrt{1 - v + v^2}}{E} \tag{11}$$

The values of S-N curve parameters are listed in Table 1, and the S-N curves are shown in Figure 2.

Table 1. S-N curve parameters [13].

S-N Curve Parameter	Value
C_1	1.25×10^{-3}
m_1	3.195
C_2	1.83×10^{-3}
m_2	3

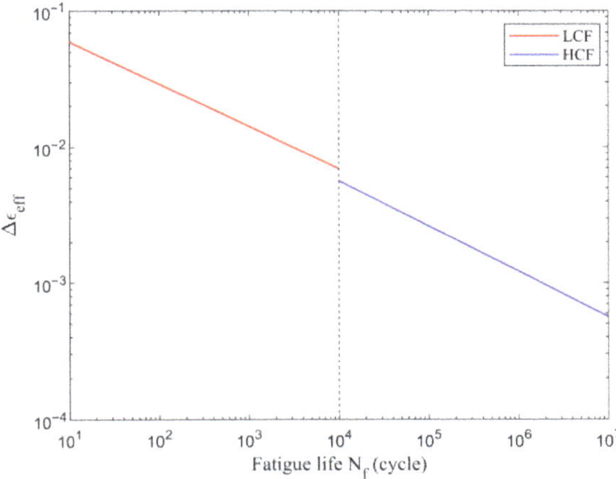

Figure 2. S-N curves for the effective notch strain approach [13].

3. Case Study

The fatigue strength of a specific case is investigated in the present study. The data of the example pipeline are presented in Table 2.

Table 2. Information for the example pipeline.

Modelling Parameters	Unit	Value
Pipeline outer diameter, D_e	mm	355.6
Pipeline wall thickness, t	mm	19.8
Area of steel pipeline's cross-section, A_s	m^2	0.0209
Internal pressure of installation/shut-down, $P_{i,ins}$	MPa	2
Operating internal pressure, P_i	MPa	20
Operating temperature, T_{op}	°C	120
Seabed ambient temperature, T_a	°C	12
Coefficient of thermal expansion, α	$°C^{-1}$	1.3×10^{-5}
Young's modulus, E	MPa	2.06×10^5
Poisson's ratio, ν	-	0.3
Cyclic strength coefficient, K'	MPa	923
Cyclic strain hardening exponent, n'	-	0.118

According to Equation (4), the longitudinal stress range due to start-up and shut-down cycles is approximately 248.7 MPa. This stress range is considered the nominal stress range $\Delta\sigma_n$ and used in the following fatigue strength assessment.

In the present study, the cyclic mechanical properties of API 5L X65 pipeline steel are used. Detailed information on the material and its mechanical properties can be found in [24]. In fact, the mechanical properties of heat affected zone are more relevant because fatigue cracks are usually initiated from this location. However, the cyclic mechanical properties of heat affected zone are not available. The use of mechanical properties of parent material usually results in the conservative estimation of local strains [25]. Therefore the mechanical properties of the parent material are still used in the present study.

For offshore pipelines, single-sided girth welds are more relevant in the industry. According to fatigue tests, the fatigue cracks most likely originate from the inside, i.e., the weld root, due to a possible poor weld root profile [26–28]. It has been shown that weld geometry plays a significant role in fatigue strength. It is assumed that the weld has a flush ground weld toe, and only the idealised geometry of the weld root is considered. The weld root geometry is characterised by four parameters, as shown in Figure 3. They are the weld root bead width W_i, weld root bead height h, weld root angle θ and axial misalignment δ. The weld toe bead width W_e is $W_i + 1.15t$, where t is the wall thickness.

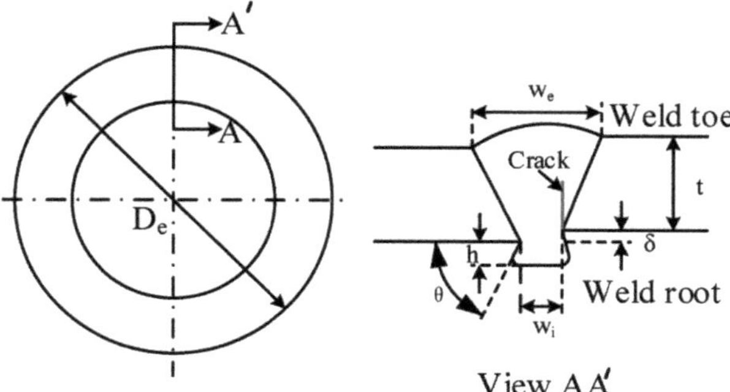

Figure 3. Weld root geometry.

To investigate the effect of weld geometry on the fatigue strength of single-sided girth welds, the weld geometry parameters are varied, as outlined in Table 3.

Table 3. Weld geometry of the single-sided girth weld.

Geometrical Parameters	Unit	Value
Weld root bead width, W_i	mm	4, 6, 8, 10, 12, 14, 16
Weld root bead height, h	mm	2, 3, 4, 5
Weld root angle, θ	°	50, 65, 97.5, 130, 145
Axial misalignment, δ	mm	0, 1, 2, 3

The finite element analyses for the estimation of the effective notch stress are performed using ANSYS [29]. Linear elastic analyses are carried out. The plane strain cross-sectional models, which are usually employed in plate-welded structures [13], are used. A circular notch with a radius of 1 mm is placed at the weld root. The element type of PLANE183 is chosen, and the maximum element size at the circular arc is less than 0.02 mm, which satisfies the requirement of [20]. In the recommendation, the maximum mesh size at the circular arc should be less than 0.25 mm to obtain convergent results. The mesh and boundary conditions of the finite element models are shown in Figure 4. The mesh conditions for $\theta < 90°$ or $\theta \geq 90°$ are both shown in the figure. All the nodes on the right side of the model are fixed, and the nominal stress is applied on the left side of the model. The first principal stress around the notch area is effective notch stress. The total number of nodes and elements varies with the geometrical parameters. The total number of elements is about 17,000, and the total number of nodes is about 50,000.

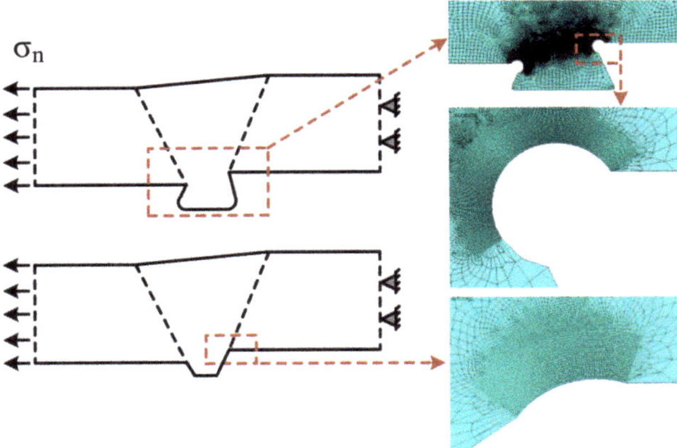

Figure 4. Finite element model for the estimation of the effective notch stress.

4. Results

In this section, the plastic behaviour of the weld root is investigated to justify the use of the LCF approach, and the effect of weld root geometry on the notch stress factor is studied to identify the dominant geometrical parameters. The fatigue strength of the welds is assessed, and the limitations of the assessment are discussed.

4.1. Plastic Behaviour

The effective notch strain approach is extended from the effective notch stress approach to consider the elastic–plastic behaviour. It is suitable for both LCF and HCF. However, the procedure for effective notch strain estimation is still complex, even though a simplified analytical method is used. For the study case, one may be interested in whether it is necessary to use only the effective notch stress approach instead of the effective notch strain approach.

Various notch stress factors K_f is assumed, which is defined by the effective notch stress divided by the nominal stress. For each K_f, the elastic–plastic effective notch strain range

$\Delta\varepsilon_{eff}$ and the effective notch strain calculated using Equation (11) $\Delta\varepsilon_{eff,e}$ are determined, respectively. The latter quantity is determined under the assumption of the elastic material. The ratio between the two quantities is used to represent the effect of plasticity:

$$r_p = \frac{\Delta\varepsilon_{eff}}{\Delta\varepsilon_{eff,e}} \tag{12}$$

The results of r_p are shown in Figure 5. The r_p increases nonlinearly with K_f. For the study case, if K_f is less than 4, the r_p is close to 1, indicating that the elastic behaviour dominates. With a further increase in K_f, plastic behaviour becomes more and more important. The result indicates that if the local stress concentration is kept at a low level, the HCF approach can also be used, and the use of the LCF approach is not necessary.

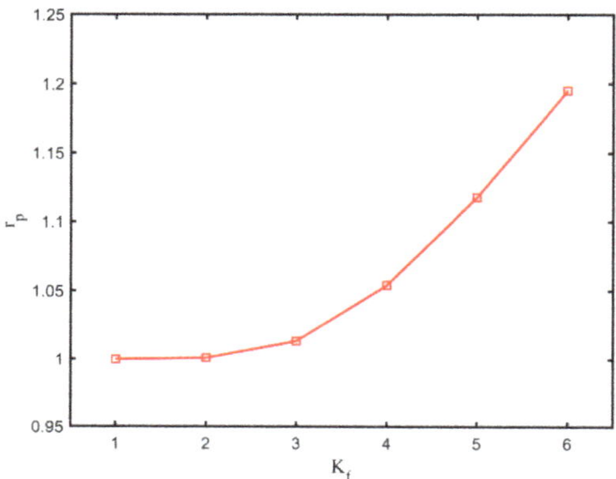

Figure 5. Effect of plasticity for $\Delta\sigma_n$ = 248.7 MPa, K' = 923 MPa, n' = 0.118.

4.2. Effect of Weld Geometry

The weld geometry has a significant impact on K_f. Some results of K_f for different weld geometries are illustrated in this section. The K_f is calculated using FEM. An example of the first principal stress contour is shown in Figure 6.

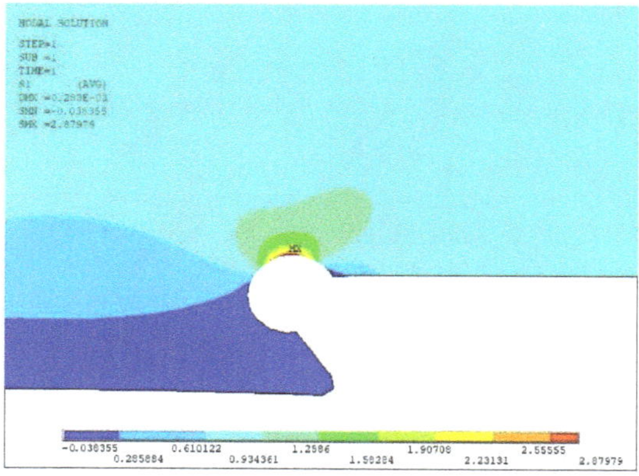

Figure 6. An example of the first principal stress contour (nominal stress is 1 MPa).

The effect h and W_i on K_f is shown in Figures 7 and 8. The effect of h is not as significant as the effect of W_i. The K_f is almost unchanged with the increase of h. The K_f increases rapidly with W_i when Wi is small but tends to approach a plateau value with a larger W_i.

The h has a minor effect on the plateau value. The lower h results in a slightly lower plateau value. The effect of W_i may be explained by the development of the eccentricity around the weld. Because of the unsymmetrical weld profile between the weld root and toe, the eccentricity of the neutral line exists around the weld [30]. With the increase of W_i, the eccentricity effect, i.e., local bending, is gradually developed. The fully developed eccentricity effect is similar to the problem of axial misalignment [31].

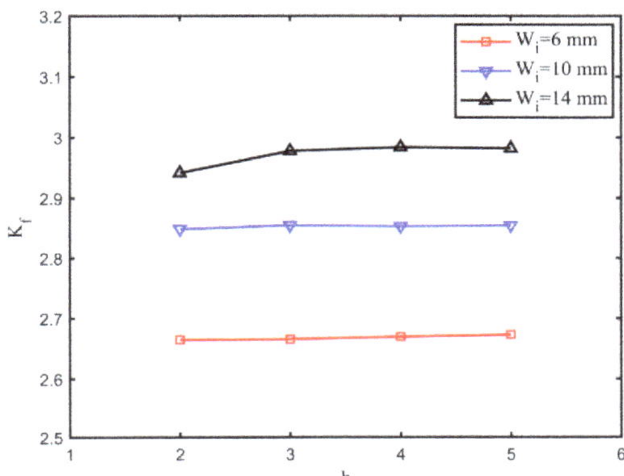

Figure 7. The effect of weld root bead height h ($\delta = 0$, $\theta = 50°$).

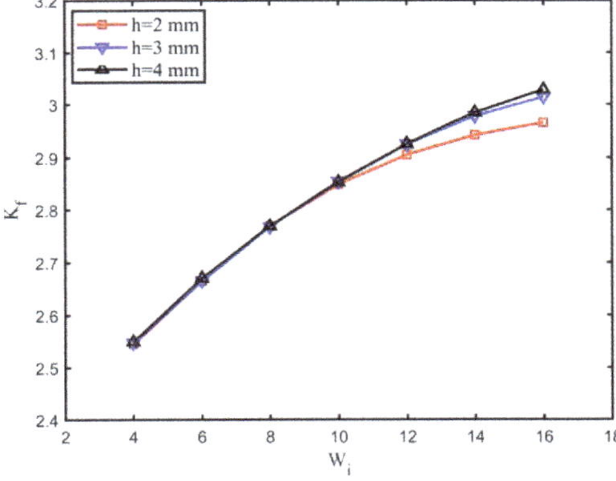

Figure 8. The effect of weld root bead width Wi ($\delta = 0$, $\theta = 50°$).

The effects of weld root angle and axial misalignment are more significant, as shown in Figures 9 and 10. The K_f decreases linearly with θ and increases linearly with δ. The effect of δ is the most significant among the four geometrical parameters. The K_f can be higher than four if δ is higher than 2 mm. The dangerous conditions are those with a low θ and a high δ.

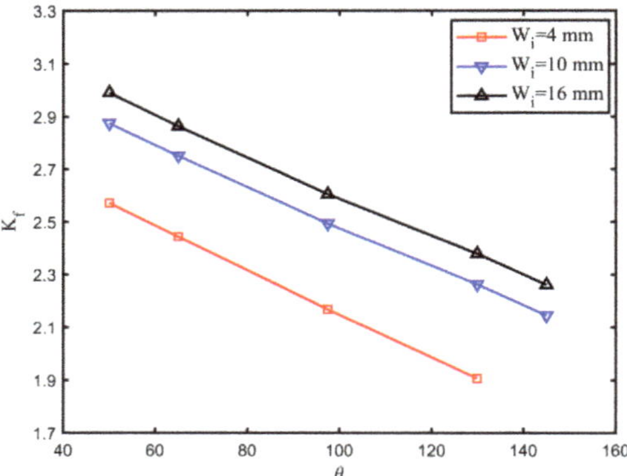

Figure 9. The effect of weld root angle ($\delta = 0$, h = 2 mm).

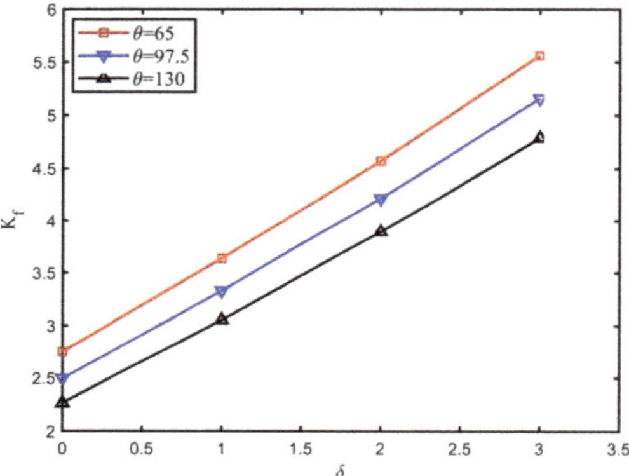

Figure 10. The effect of axial misalignment (h = 2 mm W_i = 10 mm).

4.3. Fatigue Strength

There are still some problems to be addressed to assess the fatigue strength of single-sided girth welds in offshore pipelines subjected to start-up and shut-down cycles. Firstly, the representative number of cycles should be assumed. The number of these stress cycles is usually up to several hundred during the design life of the offshore field [5]. However, knowledge of the exact number is not available. The number may be subjected to significant uncertainty. Statistical analyses should be performed based on the relevant data collected from various offshore fields. A representative number of cycles, which corresponds to a low level of probability of exceedance, may be employed.

Secondly, representative values of weld geometrical parameters should be assumed. The weld geometry is subjected to significant uncertainty. The representative values of some geometrical parameters can be assumed based on the worst-case scenario or the acceptance tolerance of fabrication. The notch radius of 1 mm is based on the worst-case scenario [20]. The fabrication tolerance for δ is less than 0.1t or a maximum 3 mm (t = wall thickness) [32]. However, the values of other parameters are not available. For other parameters that have a significant effect on the stress concentration, a statistical analysis of the data obtained from the measurement of the weld geometry is needed.

Thirdly, the effect of corrosion and the environment should be considered. The wall thickness, weld geometry and material properties can be changed during the service life due to corrosion. Using the values associated with the initial stage of the service life seems inappropriate. Additionally, the stress cycles are associated with significant thermal cycles, which may invalidate the traditional approaches for fatigue strength assessment because these approaches are usually developed under room temperatures with minor variation.

For the study case, the following assumptions on the values of geometrical parameters are made: W_i = 10 mm, h = 2 mm, θ = 90° and δ = 2 mm. The value of δ is equal to the tolerance for fabrication. The effective notch strain approach introduced in Section 2.2 is still used despite possible invalidation of the approach. The corrosion effects are ignored. The fatigue damages for different numbers of start-up and shut-down cycles N_{ss} are shown in Figure 11.

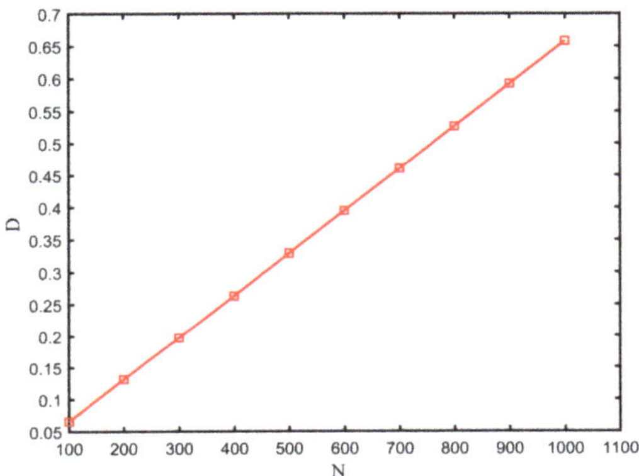

Figure 11. The fatigue damage for various numbers of start-up and shut-down cycles.

If the fatigue damage at failure is 1, the fatigue damage is acceptable even for N_{ss} = 1000. However, if the fatigue damage at failure is 0.1, which is usually used in offshore pipelines and risers, only about 150 start-up and shut-down cycles are allowed during the service life of the pipeline, indicating the necessity of fatigue strength assessment.

5. Conclusions

The following conclusions may be established:

1. For the fully constrained condition, the longitudinal stress range due to start-up and shut-down cycles depends on the variation of internal pressure and temperature. Increasing the operating internal pressure and increasing the operating temperature has the opposite effect on the longitudinal stress range.
2. For the study case, the plastic behaviour of the weld root is only significant for severe local stress concentrations. If the local stress concentration is kept at a low level, the HCF approach for fatigue strength assessment can also be used.
3. For single-sided girth welds, the axial misalignment, weld root angle, and weld root bead width are the main geometrical parameters influencing the notch stress factor of the weld root.
4. If the fatigue damage at failure is 0.1, a limited number of start-up and shut-down cycles are allowed during the service life of the pipeline for the study case, indicating the necessity of fatigue strength assessment.
5. There still exist some unsolved problems for the fatigue strength assessment of single-sided girth welds subjected to start-up and shut-down cycles. Some assumptions are made to simplify the fatigue problem and may be used in engineering practice. Further investigations are still required.

Author Contributions: Software, G.J.; formal analysis, G.J.; investigation, X.L.; data curation, G.J.; writing—original draft preparation, Y.D.; writing—review and editing, Y.D. and L.F.; visualization, X.L.; supervision, X.L. All authors have read and agreed to the published version of the manuscript.

Funding: The present work is supported by the National Natural Science Foundation of China (Grant No. 52101350).

Conflicts of Interest: The authors declare no conflict of interest.

Nomenclature

A_e	External area of the pipe
A_i	Internal bore area of the pipe
A_s	Cross-sectional area of the pipe
C_1, C_2	S-N curve parameters
E	Young's modulus
FEM	Finite element method
F_{res}	Installation residual lay tension
HCF	High cycle fatigue
HPHT	High-pressure and high-temperature
h	Weld root bead height
K'	Cyclic strength coefficient
K_f	Notch stress factor
LCF	Low cycle fatigue
m_1, m_2	S-N curve parameters
N_f	Fatigue life
n'	Cyclic strain hardening exponent
P_e	External pressure of the pipe
P_i	Operational internal pressure
$P_{i,ins}$	Internal pressure of the pipe during installation
r_p	Ratio between $\Delta\varepsilon_{eff}$ and $\Delta\varepsilon_{eff,e}$
T_a	Ambient temperature
T_{op}	Operational temperature
t	Wall thickness
W_e	Weld toe bead width
W_i	Weld root bead width
α	Thermal expansion coefficient
μ	Generalised Poisson's ratio
ν	Poisson's ratio
$\Delta\varepsilon_{eff}$	Effective notch strain range
$\Delta\varepsilon_{eff,e}$	Elastic effective notch strain range
$\Delta\sigma_l$	Longitudinal stress range
$\Delta\sigma_n$	Nominal stress range
δ	Axial misalignment
θ	Weld root angle
ρ_f	Fictitious notch radius
σ_{eff}	Effective notch stress
σ_l	Longitudinal stress
$\sigma_{l,ins}$	Longitudinal stress for the installation condition
$\sigma_{l,op}$	Longitudinal stress for the operational condition

References

1. Bai, Q.; Bai, Y. *Subsea Pipeline Design, Analysis, and Installation*; Gulf Professional Publishing: Houston, TX, USA, 2014.
2. Hong, Z.; Liu, R.; Liu, W.; Yan, S. A lateral global buckling failure envelope for a high temperature and high pressure (HT/HP) submarine pipeline. *Appl. Ocean Res.* **2015**, *51*, 117–128. [CrossRef]
3. Carr, M.; Sinclair, F.; Bruton, D. Pipeline walking-understanding the field layout challenges, and analytical solutions developed for the SAFEBUCK JIP. In Proceedings of the Offshore Technology Conference, Houston, TX, USA, 1–4 May 2006.
4. Hasan, S.; Sweet, L.; Hults, J.; Valbuena, G.; Singh, B. Corrosion risk-based subsea pipeline design. *Int. J. Press. Vessel. Pip* **2018**, *159*, 1–14. [CrossRef]
5. Bai, Q.; Pan, J. Pipeline Fatigue and Fracture Design. In *Encyclopedia of Maritime and Offshore Engineering*; John Wiley & Sons Ltd.: Hoboken, NJ, USA, 2017; pp. 1–18.

6.	Heo, J.; Kang, J.; Kim, Y.; Yoo, I.; Kim, K.; Urm, H. A study on the design guidance for low cycle fatigue in ship structures. In Proceedings of the 9th Symposium of Practical Design of Ships and Other Floating Structures, Luebeck-Travemuende, Germany, 12–17 September 2004.
7.	Wang, X.; Kang, J.-K.; Kim, Y.; Wirsching, P.H. Low cycle fatigue analysis of marine structures. In Proceedings of the ASME 2006 25th International Conference on Offshore Mechanics and Arctic Engineering, Hamburg, Germany, 4–9 June 2006; pp. 523–527.
8.	Dong, P.; Pei, X.; Xing, S.; Kim, M. A structural strain method for low-cycle fatigue evaluation of welded components. *Int. J. Press. Vessel. Pip.* **2014**, *119*, 39–51. [CrossRef]
9.	Pei, X.; Dong, P.; Xing, S. A structural strain parameter for a unified treatment of fatigue behaviors of welded components. *Int. J. Fatigue* **2019**, *124*, 444–460. [CrossRef]
10.	Lawrence, F.; Ho, N.; Mazumdar, P.K. Predicting the fatigue resistance of welds. *Annu. Rev. Mater. Sci.* **1981**, *11*, 401–425. [CrossRef]
11.	Dong, Y.; Garbatov, Y.; Guedes Soares, C. Strain-based fatigue reliability assessment of welded joints in ship structures. *Mar. Struct.* **2021**, *75*, 102878. [CrossRef]
12.	Saiprasertkit, K.; Hanji, T.; Miki, C. Fatigue strength assessment of load-carrying cruciform joints with material mismatching in low-and high-cycle fatigue regions based on the effective notch concept. *Int. J. Fatigue* **2012**, *40*, 120–128. [CrossRef]
13.	Dong, Y.; Garbatov, Y.; Guedes Soares, C. Improved effective notch strain approach for fatigue reliability assessment of load-carrying fillet welded cruciform joints in low and high cycle fatigue. *Mar. Struct.* **2021**, *75*, 102849. [CrossRef]
14.	Ngoula, D.T.; Beier, H.T.; Vormwald, M. Fatigue crack growth in cruciform welded joints: Influence of residual stresses and of the weld toe geometry. *Int. J. Fatigue* **2017**, *101*, 253–262. [CrossRef]
15.	Hutař, P.; Poduška, J.; Šmíd, M.; Kuběna, I.; Chlupová, A.; Náhlík, L.; Polák, J.; Kruml, T. Short fatigue crack behaviour under low cycle fatigue regime. *Int. J. Fatigue* **2017**, *103*, 207–215. [CrossRef]
16.	Cheng, A.; Chen, N.-Z.; Pu, Y.; Yu, J. Fatigue crack growth prediction for small-scale yielding (SSY) and non-SSY conditions. *Int. J. Fatigue* **2020**, *139*, 105768. [CrossRef]
17.	Cheng, A.; Chen, N.-Z. Structural integrity assessment for deep-water subsea pipelines. *Int. J. Press. Vessel. Pip.* **2022**, *199*, 104711. [CrossRef]
18.	Schweizer, C.; Seifert, T.; Nieweg, B.; Von Hartrott, P.; Riedel, H. Mechanisms and modelling of fatigue crack growth under combined low and high cycle fatigue loading. *Int. J. Fatigue* **2011**, *33*, 194–202. [CrossRef]
19.	Deng, J.; Yang, P.; Dong, Q.; Wang, D. Research on CTOD for low-cycle fatigue analysis of central-through cracked plates considering accumulative plastic strain. *Eng. Fract. Mech.* **2016**, *154*, 128–139. [CrossRef]
20.	Fricke, W. *IIW Recommendations for the Fatigue Assessment of Welded Structures by Notch Stress Analysis: IIW-2006-09*; Woodhead Publishing: Shaston, UK, 2012.
21.	Glinka, G. Calculation of inelastic notch-tip strain-stress histories under cyclic loading. *Eng Fract Mech* **1985**, *22*, 839–854. [CrossRef]
22.	Glinka, G. Energy density approach to calculation of inelastic strain-stress near notches and cracks. *Eng. Fract. Mech.* **1985**, *22*, 485–508. [CrossRef]
23.	Dowling, N.E. *Mechanical Behavior of Materials: Engineering Methods for Deformation, Fracture, and Fatigue*; Pearson: London, UK, 2013.
24.	Fatoba, O.; Akid, R. Low cycle fatigue behaviour of API 5L X65 pipeline steel at room temperature. *Procedia Eng.* **2014**, *74*, 279–286. [CrossRef]
25.	Dong, Y.; Garbatov, Y.; Guedes Soares, C. Fatigue strength assessment of an annealed butt welded joint accounting for material inhomogeneity. In *Progress in the Analysis and Design of Marine Structures*; Guedes Soares, C., Garbatov, Y., Eds.; Taylor & Francis Group: London, UK, 2017; pp. 349–359.
26.	Horn, A.M.; Lotsberg, I.; Orjaseater, O. The rationale for update of SN curves for single sided girth welds for risers and pipelines in DNV GL RP C-203 based on fatigue performance of more than 1700 full scale fatigue test results. In Proceedings of the International Conference on Ocean, Offshore and Arctic Engineering (OMAE), Madrid, Spain, 17–22 June 2018.
27.	Zhang, Y.-H.; London, T.; DeBono, D. Developing Mk Solutions for Fatigue Crack Growth Assessment of Flaws at Weld Root Toes in Girth Welds. In Proceedings of the International Conference on Offshore Mechanics and Arctic Engineering, Madrid, Spain, 17–22 June 2018.
28.	Dong, Y.; Kong, X.; An, G.; Kang, J. Fatigue reliability of single-sided girth welds in offshore pipelines and risers accounting for non-destructive inspection. *Mar. Struct.* **2022**, *86*, 103268. [CrossRef]
29.	ANSYS. Online Manuals. 2012.
30.	Dong, Y.; Teixeira, A.P.; Guedes Soares, C. Fatigue reliability analysis of butt welded joints with misalignments based on hotspot stress approach. *Mar. Struct.* **2019**, *65*, 215–228. [CrossRef]
31.	Hobbacher, A. *Recommendations for Fatigue Design of Welded Joints and Components*; Springer: Berlin/Heidelberg, Germany, 2015.
32.	DNVGL. *Fatigue Design of Offshore Steel Structures*; Det Norske Veritas: Hovik, Norway, 2016; Volume DNVGL-RP-C203.

Journal of
Marine Science and Engineering

Article

Fatigue Reliability Analysis of Submarine Pipelines Using the Bayesian Approach

Arman Kakaie [1], C. Guedes Soares [1,*], Ahmad Kamal Ariffin [2] and Wonsiri Punurai [3]

[1]   Centre for Marine Technology and Ocean Engineering (CENTEC), Instituto Superior Técnico, Universidade de Lisboa, 1049-001 Lisbon, Portugal
[2]   Computational Mechanics Laboratory, Department of Mechanical and Manufacturing Engineering, Faculty of Engineering and Built Environment, Universiti Kebangsaan Malaysia, Bangi 43600, Malaysia
[3]   Civil Engineering Group, Department of Civil and Environmental Engineering, Faculty of Engineering, Mahidol University, Nakhon Pathom 73170, Thailand
*   Correspondence: c.guedes.soares@centec.tecnico.ulisboa.pt

Abstract: A fracture mechanics-based fatigue reliability analysis of a submarine pipeline is investigated using the Bayesian approach. The proposed framework enables the estimation of the reliability level of submarine pipelines based on limited experimental data. Bayesian updating method and Markov Chain Monte Carlo simulation are used to estimate the posterior distribution of the parameters of a fracture mechanics-based fatigue model regarding different sources of uncertainties. Failure load cycle distribution and the reliability-based performance assessment of API 5L X56 submarine pipelines as a case study are estimated for three different cases. In addition, the impact of different parameters, including the stress ratio, maximum load, uncertainties of stress range and initial crack size, corrosion-enhanced factor, and also the correlation between material parameters on the reliability of the investigated submarine pipeline has been indicated through a sensitivity study. The applied approach in this study may be used for uncertainty modelling and fatigue reliability-based performance assessment of different types of submarine pipelines for maintenance and periodic inspection planning.

Keywords: fatigue reliability; submarine pipeline; Bayesian approach; fracture mechanics

Citation: Kakaie, A.; Guedes Soares, C.; Ariffin, A.K.; Punurai, W. Fatigue Reliability Analysis of Submarine Pipelines Using the Bayesian Approach. *J. Mar. Sci. Eng.* **2023**, *11*, 580. https://doi.org/10.3390/jmse11030580

Academic Editor: Bruno Brunone

Received: 2 February 2023
Revised: 1 March 2023
Accepted: 6 March 2023
Published: 8 March 2023

1. Introduction

Submarine pipelines are a key element in any offshore oil field development [1] and they are regarded as the main structures for the transportation of oil and gas on the seabed. These structures are exposed to a harsh seawater environment, which may lead to the deterioration of structural properties [2]. Corrosion on the pipe surface after a long time of servicing in the seawater environment is one of the common deterioration mechanisms in pipelines [3].

In addition, due to the erosion and unevenness of the seabed, a gap is often formed between some segments of the pipeline and the surface of the seabed. Periodic vibration may occur when the frequency of the vortex generated by current flows is close to the frequency of the free-spanning pipeline [4,5]. The long-term vibration may lead to fatigue failure of the pipeline. Also, motions of floating platforms, thermal cycles, and start-up and shut-down cycles have been regarded as another source of fatigue [6].

Additionally, the corrosive environment significantly affects the crack nucleation stage and reduces fatigue strength in dry air conditions [7]. Corrosion may enhance fatigue crack growth in the presence of cyclic loads, once cracks are initiated [8,9]. Therefore, corrosion fatigue crack growth is one of the most frequent phenomena which leads to the cracking of subsea pipelines resulting in a reduction of the resistance capacity and loss of asset integrity during their service life [5]. Although submarine pipelines are usually associated with the use of high-quality materials and innovative technology, the failure of these structures may result in serious environmental and economic consequences [10].

Fatigue cracking is affected by different kinds of uncertainties, including material properties, model choice, and its parameters, measurement data such as the number of cycles and crack length, and also simplification and idealization in analytical and numerical evaluations [11,12]. Therefore, the fatigue process should be considered a stochastic problem taking into account different sources of variability and uncertainty [13]. Probabilistic models related to structural deterioration are needed for planning structural maintenance and risk-based decision for marine structures [14–17]. Reliability methods provide a framework to consider the uncertainties and stochastic behaviour of the random variables on safety assessment and service-life prediction of structures which are important steps for risk-based maintenance and periodic inspection.

Limit state-based and data-based are the two main approaches in reliability analysis. The former approach is based on structural reliability theory combined with degradation models of structural resistance. In this approach, statistical information related to the basic variables of the limit state function is required. Ultimate, serviceability, and fatigue are major common limit states in structures. Ultimate limit state-based has been largely used in the reliability assessment of pipelines in recent years [18–22]. Also, a few studies have investigated the reliability level of pipelines regarding fatigue limit state. Shabani et al. [23] estimated the fatigue failure probability of subsea pipelines due to the vortex-induced vibration. He and Zhou [24] investigated the fatigue reliability of dented pipelines. They applied the S-N curve to define the limit state function. Also, Pinheiro et al. [25] proposed a new fatigue life assessment methodology for steel pipelines containing plain dents. They defined analytical expressions to estimate stress concentration factors for different types of dent shapes.

Another approach for estimating the reliability of structures is based on the statistical analysis of the failure data. The accuracy of this approach depends on the quality and quantity of the experimental data [13]. Garbatov and Guedes Soares [14] used historical data for structural maintenance planning of corroded deck plates of tankers. They also used the data-based approach to fatigue reliability assessment of dented pipelines by estimating the distribution of the number of load cycles to achieve crack initiation. They assumed Weibull distribution for the failure load cycles that fit the limited experimental data. Dong et al. [26] investigated the fatigue reliability of single-sided girth welds in offshore pipelines and risers based on the calibration of the crack propagation analyses to the S–N data.

S-N curve-based damage mechanics and fracture mechanics are the two main approaches to estimating the crack initiation and crack propagation of marine structures due to cyclic loads, respectively [27]. Most of the previous research on fatigue reliability-based assessment of pipelines concentrated on the S-N curve based on damage mechanics [14,24]. The S-N curve predicts the strength of non-cracked structures based on the crack initiation of a critical section as a function of the number of load cycles [28]. The curve is depicted by applying cyclic loads on smooth and non-cracked specimens [2]. However, mechanical damages in the form of cracks and defects are inevitable in the pipeline due to the operation activities, fabrication errors, and also corrosive seawater environment [29]. The fracture mechanics approaches take into account the initial cracks in fatigue life assessment and can be used in risk and reliability analysis of cracked structural components.

The main purpose of this study is to investigate the fracture mechanics-based fatigue reliability analysis of submarine pipelines in a corrosive environment. The main challenge in the fatigue reliability assessment of cracked pipelines is the lack of sufficient experimental or field data since the production of experimental fatigue data to construct a distribution is difficult and time-consuming. With such limited data, the Bayesian approach can be suitable. This approach has been applied for reliability assessment and maintenance and inspection planning of structures [30–33].

In this study, the Bayesian method is used to incorporate limited experimental data into a fracture mechanics-based model of fatigue to estimate the reliability of pipelines. The proposed methodology to estimate the fatigue failure probability has three main steps. At first, the probabilistic distribution of the parameters of the fatigue model is obtained using the Bayesian updating approach regarding associated uncertainties. Then, the distribution of load cycles at the critical size is estimated based on the updated model and defined limit state function. Markov Chain Monte Carlo (MCMC) simulation is used for draws the samples of the distribution. Finally, having a large number of generated failure data, the fatigue failure probability of pipelines can be calculated using the data-based reliability approach. The approach applied in this study is based on the limited experimental data related to the corrosion fatigue crack growth of serviced API 5L X56 submarine pipelines.

This study is organized as follows: In Section 2, a fatigue crack growth model based on Paris' law is introduced. Available experimental data for API 5L X56 submarine pipelines is presented in Section 3. In Section 4, probabilistic modelling of fatigue crack growth is developed based on the Bayesian approach for submarine pipeline, and the updated distribution of model parameters is obtained based on the experimental data and physic-based fatigue model. Finally, the reliability level of a submarine pipeline is evaluated, and the effect of different parameters on reliability analysis is investigated through a sensitivity study.

2. Fatigue Crack Growth Model

In general, the fatigue crack growth process includes three regions: Region I, which represents the early development of a fatigue crack with a small value of the crack growth rate, region II which represents the intermediate crack propagation with stable crack growth and region III, with a rapid fatigue crack growth. In the intermediate stage, the crack growth rate can be expressed based on the Paris law as follow [34]:

$$\frac{da}{dN} = C\Delta K^m \tag{1}$$

where a is the crack length, N is the number of load cycles, C and m are material constants. The stress range intensity factor, ΔK is proportional to stress range $\Delta\sigma$, geometry function $Y(a)$, and the square root of the crack length as follows:

$$\Delta K = Y(a)\Delta\sigma\sqrt{\pi a} \tag{2}$$

By integration of Equation (1), the number of cycles N which leads to the crack length a_N can be obtained as:

$$N = \int_{a_0}^{a_N} \frac{da}{C\left(Y(a)\Delta\sigma\sqrt{\pi a}\right)^m} \tag{3}$$

In the fatigue analysis of marine structures, the stress range is usually assumed to follow the Weibull distribution [35,36]. Regarding the Weibull distribution for the stress range and geometry factor as a power function [15,37], $Y(a) = Aa^B$, an analytical solution for Equation (3) can be obtained as:

$$a_N = \left\{ NC_{cr}C\left[1 - \left(\frac{m}{2} + mB\right)\right]\left(Au\sqrt{\pi}\right)^m \Gamma\left(1 + \frac{m}{\alpha}\right) + a_0^{1-\left(\frac{m}{2}+mB\right)} \right\}^{\frac{1}{1-\left(\frac{m}{2}+mB\right)}} \tag{4}$$

where u and α are scale and shape parameters of the Weibull distribution of stress range and Γ is the Gamma function. Since the investigated pipelines are located in a corrosive environment, the interaction between corrosion and fatigue needs to be considered in the physics-based fatigue model. Corrosion-enhance fatigue crack growth can be modelled by the production of a correction factor, C_{cr}, to the material parameters C [38], as indicated in the Equation (4). Assuming the failure limit state function as $g(X) = a_{cr} - a_N$, the number of cycles to failure can be calculated based on Equation (4) regarding $a_{cr} = a_N$.

$$N = \frac{a_{cr}^{1-\left(\frac{m}{2}+mB\right)} - a_0^{1-\left(\frac{m}{2}+mB\right)}}{C_{cr}C\left[1-\left(\frac{m}{2}+mB\right)\right]\left(Au\sqrt{\pi}\right)^m \Gamma\left(1+\frac{m}{\alpha}\right)} \tag{5}$$

However, the statistical information of the basic random variables vector is required to obtain the probabilistic distribution function of load cycles to failure. Regarding fracture mechanic-based fatigue reliability analysis of submarine pipelines, such information is not available. Instead, the limited experimental data related to the fatigue crack growth of pipelines (a-N curve) are available in the literature. In the first step of this study, the Bayesian approach has been applied to obtain the probabilistic model of fatigue crack growth parameters, which is explained in Section 4.

3. Experimental Data

The experimental data used in this paper is related to corrosion fatigue tests of API 5L X56 pipe-in-pipe pipelines serviced for 15 years, which is reported in [5]. Seven standard compact tensile (CT) specimens were extracted from the outer pipe of the submarine pipe in-pipe pipeline with dimensions of 219 × 12.2 mm. The designed CT specimens were in the hoop direction of the pipe and the thickness and width of them are 8 and 40 mm respectively. To simulate a corrosive seawater environment, a seawater circulation system was designed. In Figure 1, the corrosion fatigue test setup is shown. More details of the setup of the corrosion fatigue test can be found in the reference [5].

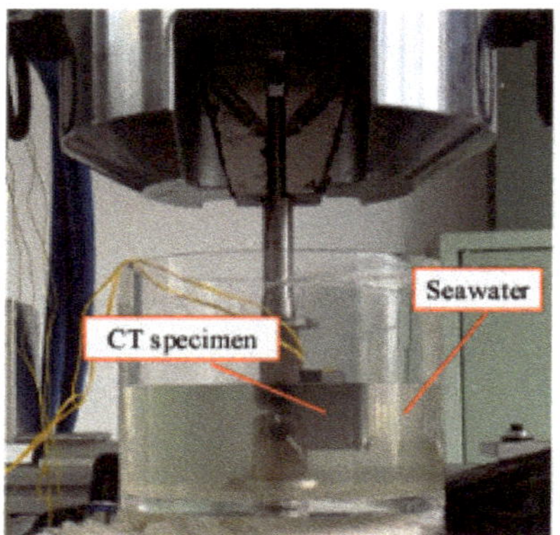

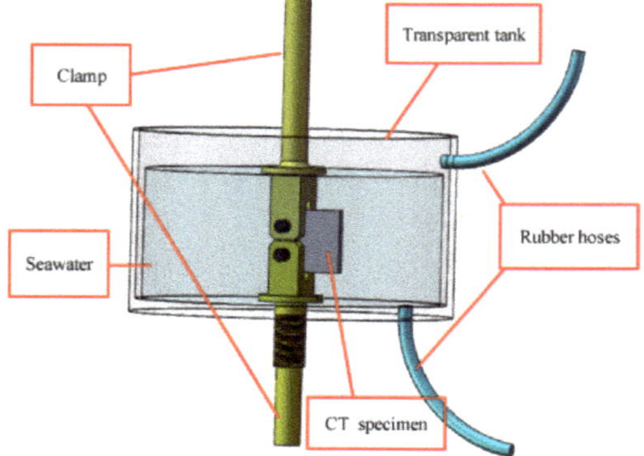

Figure 1. Corrosion fatigue test setup [5].

In pipe-in-pipe systems, the inner pipe is exposed to internal pressure and the outer pipe is designed to resist external pressure. Because of the minor amplitude of the internal pressure, fatigue failure of the inner pipe rarely occurred. However, the outer pipe is exposed to longitudinal tension caused by the ocean current in the free-spanning sections. Therefore fatigue failure is more probable on the outer surface of the outer pipe due to ocean current lateral pressure [5].

Fatigue tests were carried out by applying a sinusoidal tensile stress with constant amplitude. The fatigue crack growth of the CT specimens was monitored based on the back-face strain (BFS) method. The crack length-load cycles data and also the relationship between fatigue crack growth rate and the stress intensity factor in the Paris law function were calculated. In this study, the experimental data related to the three cases (specimens S_1, S_2 and S_3) are used to obtain probabilistic modelling of the fatigue crack growth and finally reliability analysis of the API 5L X56 submarine pipeline. In Table 1 and Figure 2 the details of the fatigue test and crack length-load cycles (a-N) curve for the investigated cases are shown.

Table 1. Parameters and results of fatigue crack growth test. Data adopted from [5].

Test ID	R	f (Hz)	P_{max} (KN)	a_0 (mm)	a_f (mm)
S_1	0.1	0.5	11	12.51	24.77
S_2	0.1	0.5	9	15.38	25.98
S_3	0.4	0.5	11	12.64	26.10

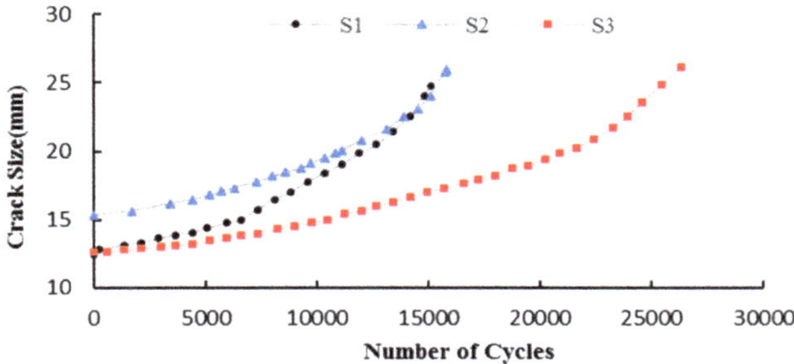

Figure 2. a-N fatigue test result extracted from [5].

It should be mentioned that the crack growth rate—stress intensity factor curve for investigated cases can be found in [5]. Based on the experimental fatigue crack growth results and using Equation (2) the geometry functions of the cracks in specimens can be calculated. Figure 3 presents the scatter of geometry function $Y(a)$ and fitted power function $Y(a) = Aa^B$ for three investigated cases using the least square method. R is the least square regression factor.

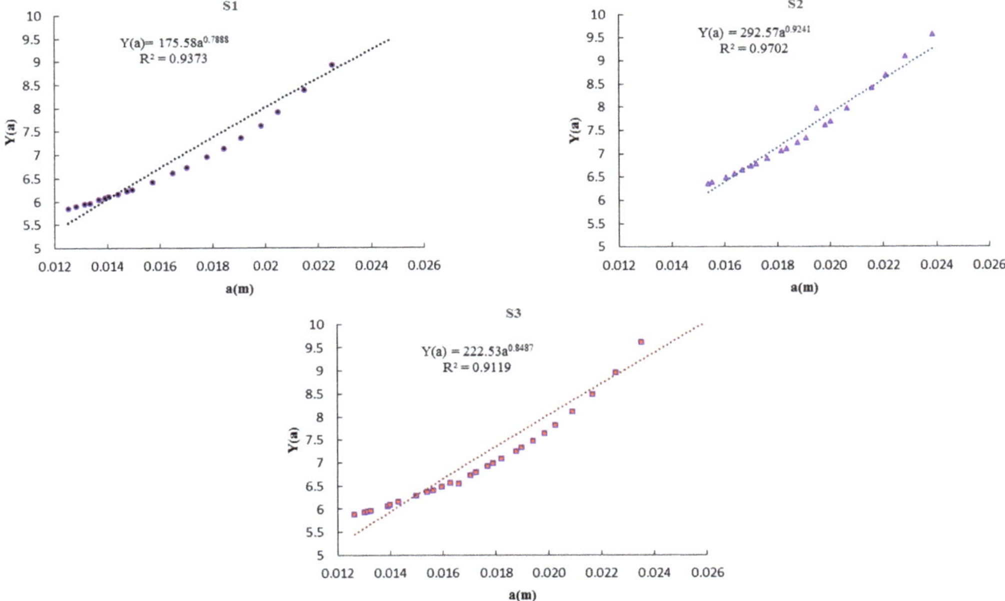

Figure 3. Geometry function and fitted power function of specimens.

4. Probabilistic Modelling of Fatigue Crack Growth

Consider M as a chosen model, which is a function of the vector modelling parameter θ, and D as a quantity of interest to be assessed. The probabilistic description of D can be calculated as [39,40]:

$$D = M(\theta, z) + \varepsilon \tag{6}$$

In Equation (6), ε is the combination of measurement and model parameter uncertainties, and z is the independent model variables where those values are fixed during the analysis. Given the probability distribution function of model parameters (i.e., stress range, material parameters m, C and C_{cr} and geometry function A and B in Equation (4)), the probabilistic result of model output (i.e., crack length at each number of load cycles (N, a_N) in Equation (4)) can be calculated.

However, the real challenge arises when the statistical parameters of the input variables are not available. In this case, to obtain the probabilistic response of a physic-based model such as Equation (4), the probability distribution function of input variables describing the material, geometric, and loads need to be updated based on a given set of observation D [39]. In this study, the experimental results of the pipeline crack growth in a corrosive environment are regarded as the measured data.

The Bayesian approach provides a framework to evaluate the posterior distribution function of model parameters based on prior information and observed experimental data. The main advantage of the Bayesian updating approach lies in its ability to combine different sources of uncertainty based on Bayes' theorem:

$$P(\theta|D) \propto P(D|\theta)P(\theta) \tag{7}$$

where $P(\theta \mid D)$ represents the posterior distribution of the model parameters after being updated. $P(\theta)$ is the probability distribution function of model parameters θ before updating. The prior distribution may be estimated based on various prior information such as expert knowledge, experimental data, and empirical judgment. Estimating the prior distribution depends on the amount of available information [41]. In the practical engineering problem, the normal distribution can be employed as a prior distribution when the mean value of parameters is known, [39]. In addition, uniform prior is assumed when there is no information about the parameters. $P(D \mid \theta)$ represents the likelihood function of the occurrence of the measurement data D given model parameters θ. The likelihood function indicates the agreement between the response of the model and the target measured values. In this study, the normal distribution, which is a common choice of the likelihood function, is regarded as [42]:

$$P(D|\theta) = \frac{1}{\sigma_j \sqrt{2\pi}} \exp\left[-\frac{1}{2\sigma_j^2}\left(D_j - M_j(\theta) - \varepsilon_j\right)^2\right], \quad j = 1, 2, \ldots, k_{data} \tag{8}$$

where k is the number of measured data (experimental crack length for different numbers of cycles), D_j is the jth measured crack length and $M_j(\theta)$ is the jth computed crack length based on the chosen model with the parameters θ. ε_j represents the error between the model output and experimental results of the crack length and σ_j is its standard variation. In fact, ε is the combination of modelling and measurement errors due to bias and noise of the data. Under the assumption that ε is just affected by the measurement noise, it can be modelled by a zero mean normal distribution with the standard deviation of σ [43]. In this study, σ is regarded as an unknown variable and its probability distribution function is calculated based on the updating process. Regarding k observed experimental data $D = \{D_1, D_2, \ldots, D_k\}$, the posterior distribution can be expressed as:

$$P(\theta|D) \propto \left(\prod_{j=1}^{k} \frac{1}{\sigma_j \sqrt{2\pi}}\right) \exp\left[-\sum_{j=1}^{k} \frac{1}{2\sigma_j^2}\left(D_j - M_j(\theta) - \varepsilon_j\right)^2\right] P(\theta) \tag{9}$$

Equation (9) indicates the proportional relation between the posterior distribution with the prior distribution and the likelihood function. A normalizing constant can be multiplied to the right side of Equation (7) to make the posterior distribution integration 1. The posterior distribution in Equation (8) represents the updated distribution of model parameters in the presence of some observed data.

In general, the posterior distribution is a product of complex functions and it is not possible to generate samples directly from it [12,44,45]. In such a case, sampling-based methods such as advanced Monte Carlo sampling methods can be used [12]. In this paper MCMC simulation, specifically the Metropolis-Hasting (MH) algorithm [46], is applied to generate samples based on the posterior distribution of interest. MH algorithm can generate samples from any probability distribution with a given posterior function such as Equation (9). The MH sampler is a random walk algorithm where the next candidate sample θ^{i+1} is generated only based on the current sample θ^i. Each candidate sample is generated based on the so-called proposal distribution $q(\theta^{i+1}, \theta^i)$, which is a symmetrical function, that is $q(\theta^{i+1}, \theta^i) = q(\theta^i, \theta^{i+1})$. The generated candidate samples are accepted or rejected based on an acceptance probability which can be defined as follows:

$$r(\theta^i, \theta^{i+1}) = \min\left[1, \frac{P(\theta^{i+1}|D)}{P(\theta^i|D)}\right] \tag{10}$$

Equation (10) implies that the candidate samples are accepted if they have a higher probability compared to the current samples. In practical problems, this criterion can be achieved by comparing the acceptance probability r with generating a random number s from a uniform distribution between 0 and 1, $s \sim U[0,1]$. The proposed sample is accepted if $r(\theta^i, \theta^{i+1}) \geq s$, otherwise it is rejected.

The main challenges of MCMC are related to the initial samples since they are in general, not distributed according to the stationary distribution. Therefore, the initial samples may be located far away from the posterior distribution and move very slowly to the high-probability regions [42]. In such a case, many samples need to be generated to converge the posterior distribution. To avoid the impact of the inaccurate initial samples, a portion of initial samples $n_{burn-in}$ can be discarded.

For the investigated pipeline, the material parameters m and C, the corrosion-enhancement parameter C_{cr} and the error between model output and observed data ε are considered random variables and their PDF need to be updated. Gao et al. [2] proposed the material parameters of $m = 2.66$ and $lnC = -24.15$ for the submarine pipelines of API X56 steel materials. Also, some studies proposed the C_{cr} of 3 for steel materials under free corrosion conditions [38,47]. In this study, the previous literature results combined with engineering judgment are used to estimate the prior distribution of model parameters. The normal distribution with the mean values of 2.66, −24.15, and 3 and the standard deviation of 0.25, 0.5, and 0.2 are considered for the prior distribution of m, lnC, and C_{cr}. It should be mentioned that for case S_3, which has a larger value of R, $lnC = -22.15$ is considered. In addition, the likelihood function is modelled as a normal distribution with a standard deviation of σ, and its parameters is obtained from the updating process. To avoid the wrong choice of assumed prior distribution, the uniform distribution is also examined and the fatigue crack growth results are compared with the obtained results from the prior normal distribution in Appendix B.

A normal distribution with a mean value of 0.5 mm and a standard deviation of 0.1 mm is regarded for the prior distribution of σ. Half of the experimental data of each specimen (experimental data before 8000 cycles for S_1 and S_2 and also before 11,000 cycles for S_3) have been used to update model parameters. The results of the posterior distribution of model parameters and MCMC chains are presented in Figure 4 and Table 2. It can be observed that the uncertainty of material parameters m and lnC are reduced in comparison with the prior distribution for all three investigated cases. However, the mean and standard deviation values of the corrosion-enhance parameter have changed slightly.

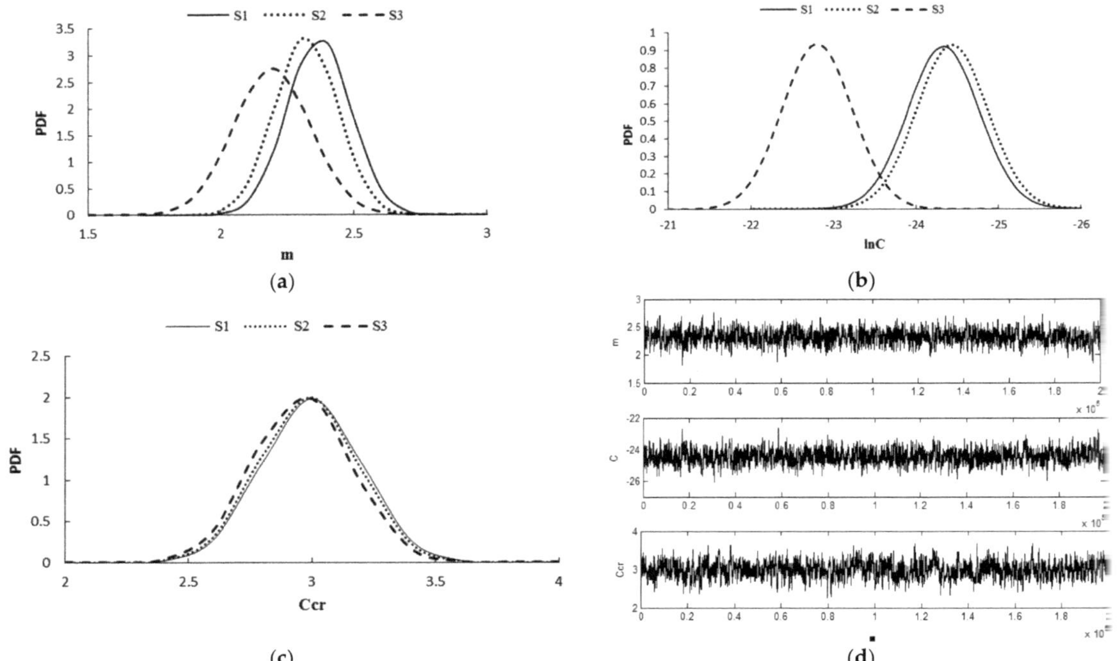

Figure 4. (**a–c**) updated distribution of model parameters m, C and Ccr, respectively. (**d**) chain iteration resulting from MCMC simulation for case S_2.

Table 2. Mean and standard deviation of the updated distribution of model parameters.

	M		**lnC**		**Ccr**	
	Mean	**St.Dev.**	**Mean**	**St.Dev.**	**Mean**	**St.Dev.**
S_1	2.37	0.119	−24.33	0.43	3	0.192
S_2	2.33	0.119	−24.44	0.43	2.98	0.191
S_3	2.18	0.136	−22.77	0.42	2.95	0.184

The probability distribution function of crack length after updating model parameters based on experimental data can be obtained using the fatigue model of Equation (4). The probability of crack length can be expressed as follows:

$$P(a_N) \propto \int_{\Omega(\theta)} P(a_N|\theta, D)P(\theta|D) \tag{11}$$

where $P(a_N \mid \theta, D)$ is the probability of crack length for given model parameters and $P(\theta, D)$ is the posterior distribution of the parameters. 200,000 samples are generated using MCMC simulation to update model parameters and estimate the crack growth curve. The crack length growth of the three investigated cases is presented in Figure 5. Here, the probabilistic results of fatigue crack growth are estimated using a 95% confidence interval and compared with the experimental results. Confidence interval extraction in the crack growth curve is explained in detail in the reference [43]. It can be observed that the estimated crack growth is in acceptable agreement with the experimental results. Although the mean values of the probabilistic model are different from the experimental results in some of the load cycles (particularly in the case of S_1), the 95% confidence interval covers all of the experimental data.

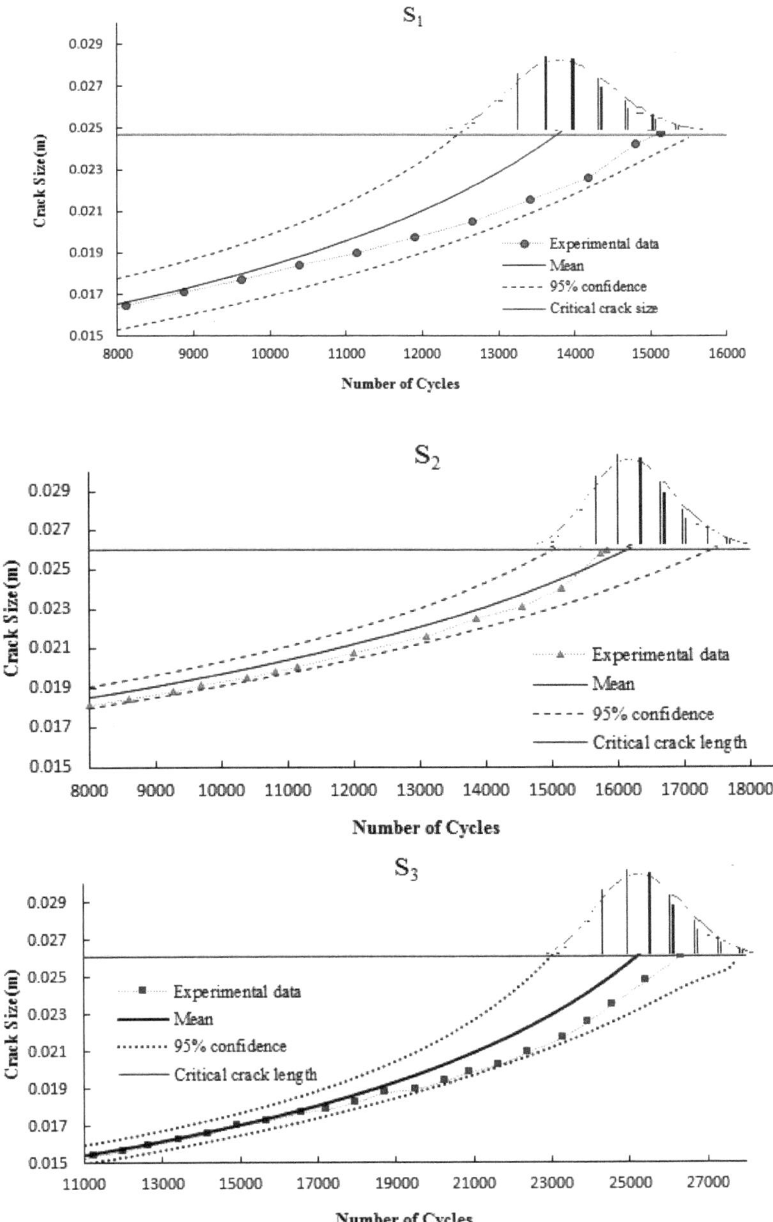

Figure 5. Fatigue crack growth results from Bayesian updating.

5. Reliability Analysis

This section presents the reliability analysis of the investigated pipelines based on the uncertainty modelling of the basic variables which are indicated by the pdf of the posterior distribution of model parameters. The reliability function can be measured by the probability that the crack growth length a_N does not exceed the critical crack size for an arbitrary load cycle within the considered interval, that is:

$$L(N) = P(a_n < a_{cr}) \quad n = 1, 2, \ldots, N \tag{12}$$

The reliability can also be expressed with the help of the hazard function [48]. The "hazard function" $h(N)$ for discrete load cycles is defined as the probability of structural failure at load cycle N, given that the structure has not failed under a previous loading. Regarding N_f as the load cycles to failure, the hazard function can be expressed as [49]:

$$h(N) = P\left[N_f = N \middle| N_f > N - 1\right] = \frac{f_{N_f}(N)}{1 - F_{N_f}(N - 1)} \tag{13}$$

where $f_{Nf}(N)$ and $F_{Nf}(N)$ are the probability distribution function (PDF) and cumulative distribution function (CDF) of the load cycles to failure, respectively. The reliability analysis can be expressed based on the hazard function as follows [50]:

$$L(N) = \exp\left[-\int_0^N h(n)dn\right] \tag{14}$$

Due to a large number of load cycles in fatigue $F_{Nf}(N) = F_{Nf}(N - 1)$. Therefore, Equation (14) leads to well-known relation of $L(N) = 1 - F_{Nf}(N)$. Given the probability distribution of the failure, the structural reliability can be estimated.

The critical crack length for each specimen is considered equal to the values of a_f in Table 1. a_f is the final crack length when the back-face strain gauge fails or the fatigue crack growth rate exceeds 0.01 mm/cycles during the fatigue test process [5]. This criterion corresponds to the end of the stable region of crack growth (region II). In this paper, the values of a_f are considered as the critical values of crack length and the distribution of the number of cycles that lead to $a_N = a_{cr}$ are calculated based on the limit state function obtained from Equation (5) regarding the PDF of updated model parameters as basic variables. The statistical parameters of the distribution of the number of cycles to failure are indicated in Table 3. The probabilistic results are compared with the experimental data and deterministic results obtained from the finite element method (FEM), presented in [5]. It can be observed that the probabilistic results of failure load cycles provide an acceptable estimation of fatigue life. In addition, the PDF of failure load cycles are presented in Figure 6. It should be mentioned that Weibull distribution often has been used to model the failure data [14,51]. However, it is observed that the estimated PDF of failure load cycles in three investigated cases follows lognormal distribution instead of a Weibull distribution. This can be due to the disregarding of the stress range uncertainty, which usually follows the Weibull distribution.

Table 3. Statistical parameters of the estimated failure load cycles.

Specimens	S_1	S_2	S_3
mean	13,908	16,264	25,279
St.Dev.	753	623	1320
Exp.	15,128	15,830	26,313
FEM	13,285	15,287	22,629

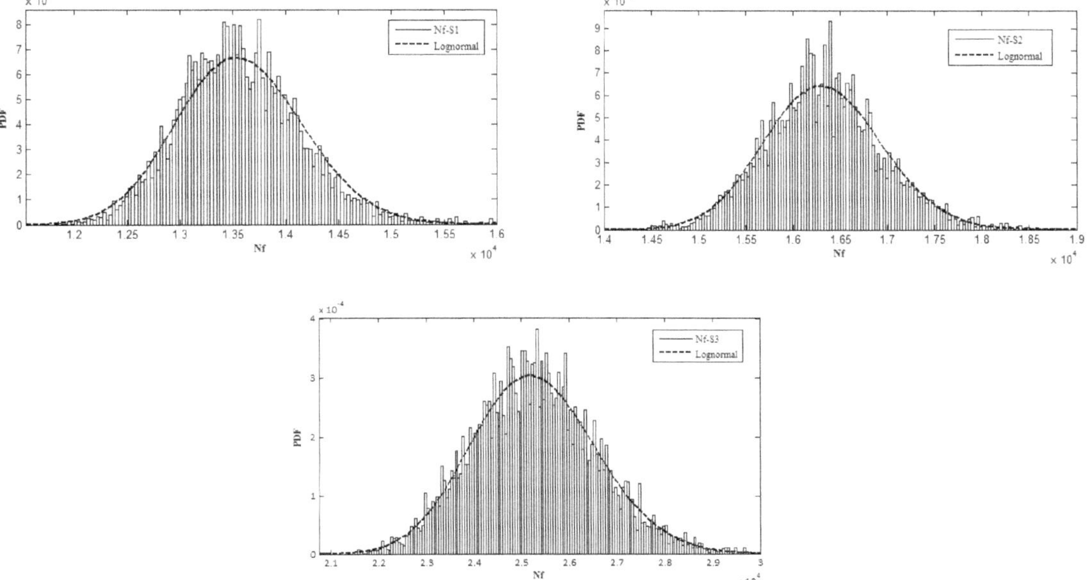

Figure 6. Estimated PDF of failure load cycles.

The reliability analysis of the investigated cases is indicated in Figure 7. It can be observed that the stress ratio R has a significant effect on the reliability level of structures. Increasing the stress ratio from $R = 0.1$ in the case of S_1 to $R = 0.4$ in the case of S_3 leads to a remarkable increase in the reliability function. Also, the reliability function is affected by maximum loads P_{max}. The reliability function of S_2 with $P_{max} = 9$ Kn is higher than the reliability function of S_1 with $P_{max} = 11$ Kn, even if S_2 has a larger initial crack length. In addition, Table 4 indicates the performance assessment of the investigated cases based on the reliability index, which is calculated as:

$$\beta(N) = -\Phi^{-1}(1 - L(N)) \tag{15}$$

where Φ is the standard normal function. In Table 4, $\beta < 2$, $2 \leq \beta \leq 4$ and $\beta > 4$ are regarded as poor performance (red), good performance (yellow) and excellent performance (green) of pipelines, respectively [14], and the associated number of load cycles to each performance levels are indicated.

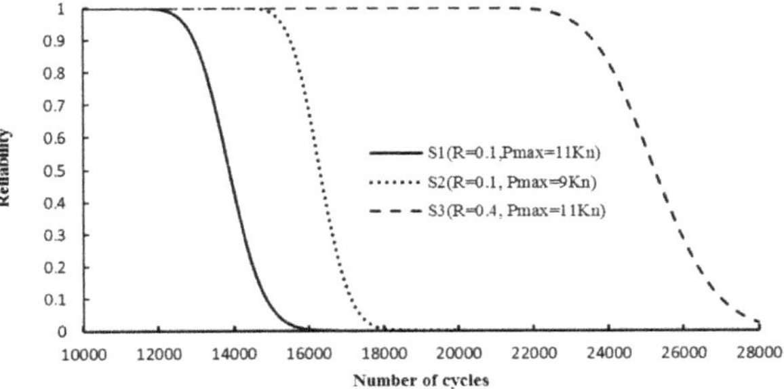

Figure 7. Reliability function of the investigated specimens.

Table 4. Reliability-based performance assessment.

Reliability Index	S_1	S_2	S_3
$\beta < 2$	N > 12,475	N > 15,120	N > 22,755
2	12,475	15,120	22,755
4	11,225	14,025	20,480
$\beta > 4$	N < 11,225	N < 14,025	N < 20,480

It should be noted that the estimated reliability level of the investigated cases is based on the assumption of deterministic stress range and initial crack size. The impact of uncertainties of this parameter is investigated through the reliability assessment of a full-scale API 5L X56 submarine pipeline with a diameter of 219 mm and a thickness of 14 mm. The maximum applied load is 235 Kn and the stress ratio R and load frequency are 0.1 and 0.5, respectively.

The material parameters m and $\ln C$ are only affected by the fatigue test environment and material [2]. Therefore, the updated probabilistic modelling of parameters in case S_2 can be considered in this example. For the sensitivity studies, different values of uncertainties are considered for stress range and initial crack size as described in Table 5. Also, the corrosion-enhancement factor may change from environment to environment. Therefore, different mean values of this parameter are regarded to investigate its impact on the reliability assessment of submarine pipelines. Here the critical value of crack depth is defined as 0.8 of the pipeline wall thickness. It should be mentioned that the finite element results for the stress intensity factor of the investigated submarine pipeline can be found in Gao et.al [2]. The geometry function parameters are calculated based on the finite element results for the stress intensity factor, as explained in Appendix A.

Table 5. Input parameters for sensitivity studies.

Parameter	Mean	Coefficient of Variation	Distribution
m	2.33	0.051	Normal
$\ln C$	-24.44	0.018	Normal
C_{cr}	2, 3 and 4	0.1	Normal
6Δ (Mpa)	23.5	0.1, 0.3 and 0.5	Weibull
a_0 (mm)	1, 2 and 3	1	Exponential
a_{cr} (mm)	11.2	-	Deterministic
A	4.476	-	Deterministic
B	-0.79	-	Deterministic

Figure 8 presents the impact of different parameters on the reliability results of the investigated pipeline. The effect of stress range uncertainty is depicted in Figure 8a, assuming the mean value of a_0 and C_{cr} is 1 mm and 3, respectively. As expected, the increasing uncertainty in the stress range leads to lower values of reliability indices. The slope of the β-N curve increases with the increase of stress range coefficient of variation at the early stage of the fatigue process (lower values of load cycles). However, for the larger number of load cycles, the slope of the β-N curve becomes flatter for different values of coefficient of variation.

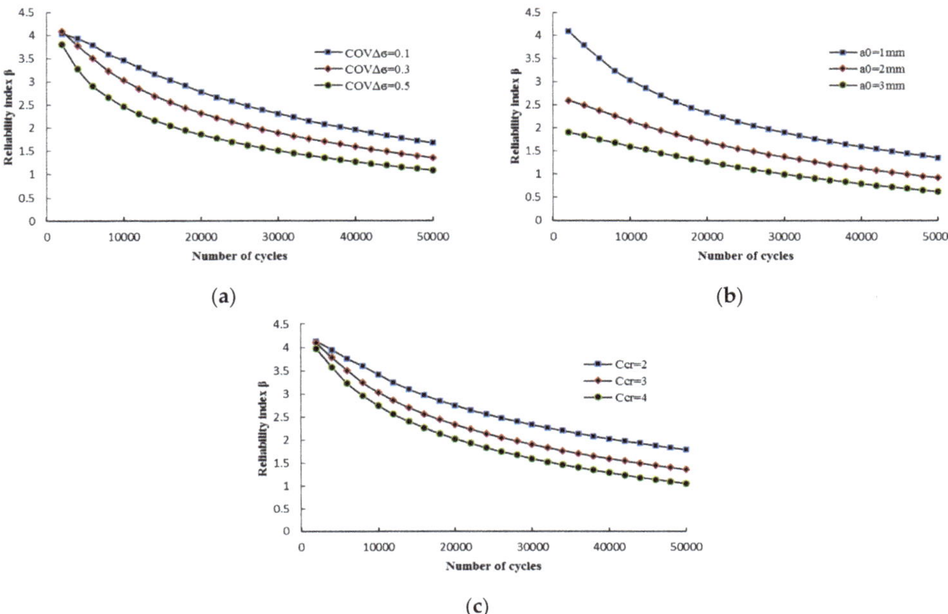

(a)

(b)

(c)

Figure 8. Effect of (**a**) uncertainty of stress range (**b**) initial crack size (**c**) mean value of corrosion enhanced parameter on reliability assessment of investigated pipeline.

Figure 8b shows the impact of initial crack size on the reliability level of the investigated pipeline for $C_{cr} = 3$ and $COV_{\Delta\sigma} = 0.3$. It can be observed that increasing of initial crack size significantly decreases the reliability level, particularly at the early stage of the fatigue process. For the higher number of load cycles, the difference between reliability indices becomes lower for different values of initial crack size. The impact of the corrosion-enhancement factor on reliability assessment is depicted in Figure 8c for $a_0 = 1$ mm and $COV_{\Delta\sigma} = 0.3$. It appears that the corrosion enhancement parameter mostly affects the reliability level in a higher number of load cycles.

Another important parameter that may affect the reliability assessment of submarine pipelines is the correlation between material parameters m and $\ln C$. Some of the previous studies have indicated the relation between material parameters [52,53]. Figure 9 presents the interrelation between the material parameters m and $\ln C$ for investigated case S_2. It can be observed a strong linear relationship with a negative correlation coefficient of 0.985 between the material parameters of the API 5L X56 submarine pipeline.

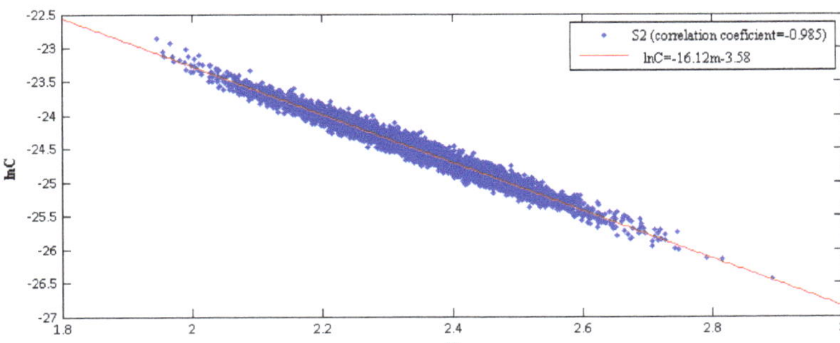

Figure 9. Interrelation between the material parameters m and lnC for investigated case S_2.

To investigate the effect of correlation between material parameters on reliability assessment, a multivariate normal distribution has been used to generate correlated random numbers. The probability density function of the h-dimensional multivariate normal distribution can be expressed as:

$$f(x, \mu, \Sigma) = (2\pi)^{\frac{-h}{2}} |\Sigma|^{\frac{-1}{2}} \exp\left(\frac{-1}{2}(x - \mu)^T \Sigma^{-1}(x - \mu)\right) \tag{16}$$

where x and μ are vectors of random variables and their mean values respectively. Σ is a h-by-h covariance matrix which is a function of the correlation coefficient and standard deviation of random variables. Since two variables (m, lnC) with a normal distribution are considered correlated variables, Equation (16) with $h = 2$ is used to generate correlated random numbers for material parameters. As an alternative method for sampling from correlated random variables, the Nataf transformation method, which assumes a normal Copula function for the joint distribution of random variables, can be also applicable [43].

Figure 10 indicates the impact of different values of the correlation coefficient between material parameters on the reliability assessment of investigated pipeline. It can be observed that increasing negative correlation coefficients leads to higher values of reliability indices. The lowest reliability curve is related to the uncorrelated situation and the highest curve is to a fully negatively correlated situation. This implies that disregarding the negative correlation between material parameters leads to a conservative assessment of the fatigue safety level of submarine pipelines.

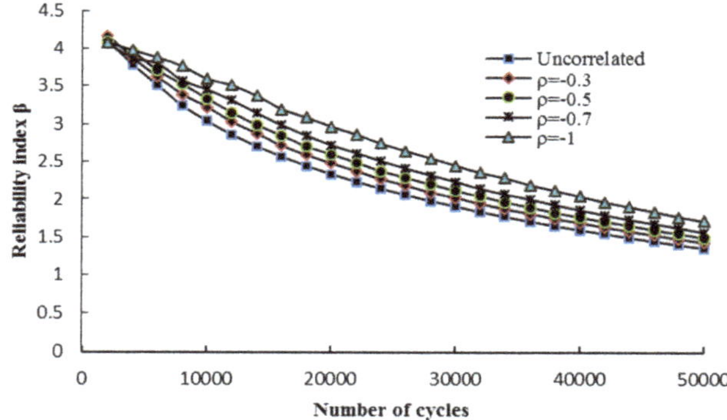

Figure 10. Impact of correlation between material parameters on reliability level of submarine pipeline.

6. Conclusions

In this study, a fracture mechanics-based fatigue reliability analysis of a submarine pipeline in a corrosive environment has been investigated. The main challenge in the fatigue reliability assessment of cracked pipelines is the lack of sufficient experimental or field data. Therefore, the Bayesian updating approach has been used to obtain the updated probability distribution function of the fatigue model parameters in the presence of some available experimental data related to API 5L X56 submarine pipelines. The uncertainties of material parameters m, lnC, and corrosion-enhanced factor C_{cr} have been estimated for the investigated pipeline using MCMC simulation. The predicted fatigue crack growth results have been compared to available experimental and deterministic finite element results to show the ability of the proposed probabilistic model for a submarine pipeline. The probability distribution of failure load cycles and the reliability level of API 5L X56 submarine pipelines for three different cases have been estimated. In addition, the effect of different parameters, including the stress ratio, maximum load, uncertainties of stress range and initial crack size, corrosion-enhance factor, and also the correlation between material parameters on the structural reliability level of submarine pipelines have been investigated. The results indicate the high sensitivity of the reliability level to the stress ratio, maximum applied load, initial crack size, particularly in the early stage of fatigue crack growth, the uncertainty of stress range, and also the mean values of the corrosion-enhanced factor for a higher number of load cycles. In addition, it was found that the increase of negative correlation between material parameters results in higher values for the reliability index of submarine pipelines.

Author Contributions: Conceptualization, A.K., C.G.S., A.K.A. and W.P.; methodology, A.K. and C.G.S.; software, A.K.; validation, A.K.; formal analysis, A.K.; investigation, A.K. and C.G.S.; resources, C.G.S., A.K.A. and W.P.; data curation, A.K.; writing—original draft preparation, A.K.; writing-review and editing, C.G.S.; visualization, A.K.; supervision, C.G.S., A.K.A. and W.P.; project administration, C.G.S.; funding acquisition, C.G.S. All authors have read and agreed to the published version of the manuscript.

Funding: This research was funded by European Project RESET Reliability and Safety Engineering for Large Maritime Engineering Systems (H2020-MSCA-RISE 2016-730888-RESET Research and Innovation Staff Exchange). This work contributes to the Strategic Research Plan of the Centre for Marine Technology and Ocean Engineering (CENTEC), which is financed by the Portuguese Foundation for Science and Technology (Fundação para a Ciência e Tecnologia—FCT) under contract UIDB/UIDP/00134/2020.

Institutional Review Board Statement: Not applicable.

Informed Consent Statement: Not applicable.

Data Availability Statement: Not applicable.

Conflicts of Interest: The authors declare no conflict of interest.

Appendix A. Geometry Function for the Investigated Pipeline

Given the stress range intensity factor ΔK, the geometry function can be calculated as:

$$Y(a) = \frac{\Delta K}{\Delta \sigma \sqrt{\pi a}} \tag{A1}$$

In Table A1, column 2 is the obtained results of the stress range intensity factor obtained from the finite element method based on reference [2]. The geometry function is calculated and depicted in Figure A1. A power function is fitted to the data and parameters A and B are obtained. The error in the estimated geometry function occurred mostly because of limited data.

Table A1. Obtained geometry function for the investigated pipeline based on stress range intensity factor calculated from FEM results adopted from [2].

a (m)	ΔK (Mpa.m$^{0.5}$)	Y (a)
0.0055	20.848	6.761289
0.0060	21.354	6.630558
0.0070	23.006	6.613610
0.0080	24.715	6.646027
0.0090	25.776	6.534927
0.0100	27.180	6.522350
0.0110	27.714	6.355494
0.0120	28.302	6.214025
0.0130	29.561	6.235825
0.0132	30.693	6.425380

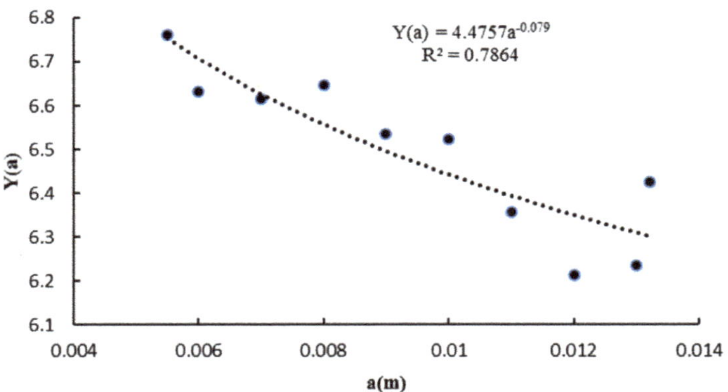

Figure A1. Geometry function and fitted power function for sensitivity study.

Appendix B. Geometry Function for the Investigated Pipeline

The results of the predicted fatigue crack growth regarding uniform and normal prior distribution are presented in Figure A2. the range of the parameters is considered between $\mu 2\sigma$ and $\mu + 2\sigma$ for uniform distribution. μ and σ are mean values and standard deviation of associated normal distribution, respectively. It can be observed that the predicted fatigue crack growth regarding normal prior distribution for cases S1 and S3 are closer to the experimental results compared to the uniform prior distribution. However, normal and uniform prior distribution lead to the approximately same results for fatigue crack growth of case S2. Therefore, the obtained results from normal prior distribution are slightly better than uniform prior distribution in investigated case study.

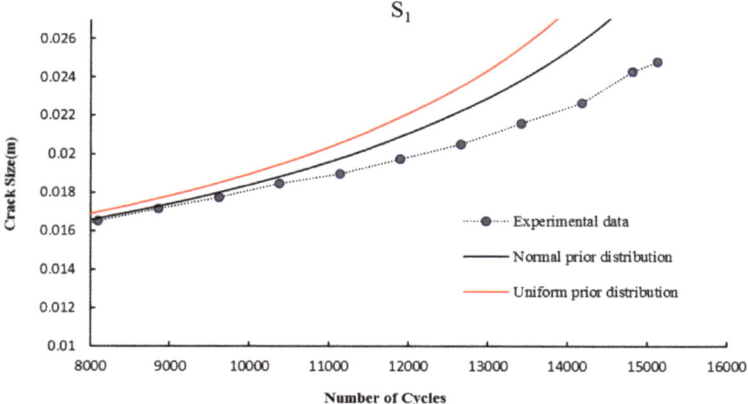

Figure A2. *Cont.*

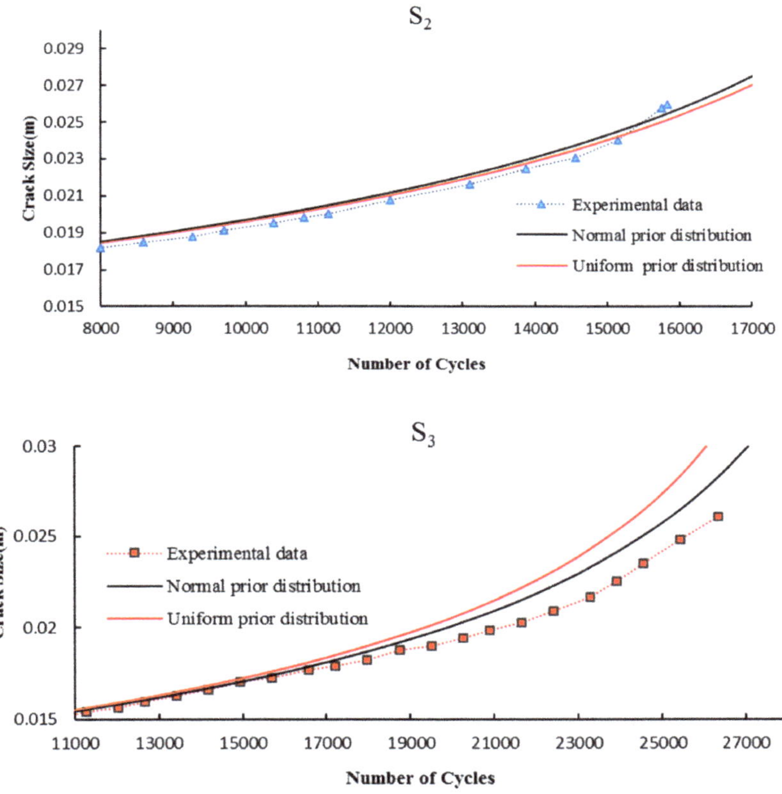

Figure A2. The mean values of fatigue crack growth for normal and uniform prior distribution.

References

1. Silva, L.M.R.; Teixeira, A.P.; Guedes Soares, C. A Methodology to quantify the risk of subsea pipeline systems at the oilfield development selection phase. *Ocean Eng.* **2019**, *179*, 213–225. [CrossRef]
2. Gao, X.; Shao, Y.; Xie, L.; Wang, Y.; Yang, D. Prediction of corrosive fatigue life of submarine pipelines of API 5L X56 steel materials. *Materials* **2019**, *12*, 1031. [CrossRef]
3. Bhardwaj, U.; Teixeira, A.P.; Guedes Soares, C. Uncertainty in the estimation of partial safety factors for different steel-grade corroded pipelines. *J. Mar. Sci. Eng.* **2023**, *11*, 177. [CrossRef]
4. Namazi-saleh, F.; Kurian, V.J.; Mustaffa, Z.; Tahan, M.; Kim, D. Effect of bed vicinity on vortex shedding and force coefficients of fluid flow on an offshore pipeline. *J. Mar. Sci. Appl.* **2017**, *16*, 81–86. [CrossRef]
5. Guo, Y.; Shao, Y.; Gao, X.; Li, T.; Zhong, Y.; Luo, X. Corrosion fatigue crack growth of serviced API 5L X56 submarine pipeline. *Ocean Eng.* **2022**, *256*, 111502. [CrossRef]
6. Dong, Y.; Ji, G.; Fang, L.; Liu, X. Fatigue strength assessment of single-sided girth welds in offshore pipelines subjected to start-up and shut-down cycles. *J. Mar. Sci. Eng.* **2022**, *10*, 1879. [CrossRef]
7. Garbatov, Y.; Guedes Soares, C.; Parunov, J. Fatigue strength experiments of corroded small scale steel specimens. *Int. J. Fatigue* **2014**, *59*, 137–144. [CrossRef]
8. Cheng, A.; Chen, N.-Z. Structural integrity assessment for deep-water subsea pipelines. *Int. J. Press. Vessel. Pip.* **2022**, *199*, 104711. [CrossRef]
9. Xie, M.; Wang, Y.; Xiong, W.; Zhao, J.; Pei, X. A crack propagation method for pipelines with interacting corrosion and crack defects. *Sensors* **2022**, *22*, 986. [CrossRef]
10. Drumond, G.P.; Pasqualino, I.P.; Pinheiro, B.C.; Estefen, S.F. Pipelines, risers and umbilicals failures: A literature review. *Ocean Eng.* **2018**, *148*, 412–425. [CrossRef]
11. Guan, X.; Jha, R.; Liu, Y. Model selection, updating, and averaging for probabilistic fatigue damage prognosis. *Struct. Saf.* **2011**, *33*, 242–249. [CrossRef]

12. Lyathakula, K.R.; Yuan, F.-G. A probabilistic fatigue life prediction for adhesively bonded joints via ANNs-based hybrid model. *Int. J. Fatigue* **2021**, *151*, 106352. [CrossRef]
13. Salemi, M.; Wang, H. Fatigue life prediction of pipeline with equivalent initial flaw size using Bayesian inference method. *J. Infrastruct. Preserv. Resil.* **2020**, *1*, 2. [CrossRef]
14. Garbatov, Y.; Guedes Soares, C. Fatigue reliability of dented pipeline based on limited experimental data. *Int. J. Press. Vessel. Pip.* **2017**, *155*, 15–26. [CrossRef]
15. Feng, G.Q.; Garbatov, Y.; Guedes Soares, C. Fatigue reliability of a stiffened panel subjected to correlated crack growth. *Struct. Saf.* **2012**, *36*, 39–46. [CrossRef]
16. Garbatov, Y.; Guedes Soares, C. Fatigue reliability of maintained welded joints in the side shell of tankers. *J. Offshore Mech. Arct. Eng.* **1998**, *120*, 2–9. [CrossRef]
17. Garbatov, Y.; Guedes Soares, C. Cost and reliability based strategies for fatigue maintenance planning of floating structures. *Reliab. Eng. Syst. Saf.* **2001**, *73*, 293–301. [CrossRef]
18. Bhardwaj, U.; Teixeira, A.P.; Guedes Soares, C. Reliability assessment of a subsea pipe-in-pipe system for major failure modes. *Int. J. Press. Vessel. Pip.* **2020**, *188*, 104177. [CrossRef]
19. Bhardwaj, U.; Teixeira, A.P.; Guedes Soares, C. Uncertainty in reliability of thick high strength pipelines with corrosion defects subjected to internal pressure. *Int. J. Press. Vessel. Pip.* **2020**, *188*, 104170. [CrossRef]
20. Baek, J.; Jang, Y.; Kim, I.; Yoo, J.; Kim, C.; Kim, Y. Structural reliability analysis of in-service API X65 natural gas pipeline using statistical data. *Int. J. Press. Vessel. Pip.* **2022**, *199*, 104699. [CrossRef]
21. Bhardwaj, U.; Teixeira, A.P.; Guedes Soares, C. Probabilistic collapse design and safety assessment of sandwich pipelines. *J. Mar. Sci. Eng.* **2022**, *10*, 1435. [CrossRef]
22. Cai, J.; Jiang, X.; Yang, Y.; Lodewijks, G.; Wang, M. Data-driven methods to predict the burst strength of corroded line pipelines subjected to internal pressure. *J. Mar. Sci. Appl.* **2022**, *21*, 115–132. [CrossRef]
23. Shabani, M.M.; Shabani, H.; Goudarzi, N.; Taravati, R. Probabilistic modelling of free spanning pipelines considering multiple failure modes. *Eng. Fail. Anal.* **2019**, *106*, 104169. [CrossRef]
24. He, Z.; Zhou, W. Fatigue reliability analysis of dented pipelines. *J. Pipeline Sci. Eng.* **2021**, *1*, 290–297. [CrossRef]
25. Pinheiro, B.; Guedes Soares, C.; Pasqualino, I. Generalized expressions for stress concentration factors of pipeline plain dents under cyclic internal pressure. *Int. J. Press. Vessel. Pip.* **2019**, *170*, 82–91. [CrossRef]
26. Dong, Y.; Kong, X.; An, G.; Kang, J. Fatigue reliability of single-sided girth welds in offshore pipelines and risers accounting for non-destructive inspection. *Mar. Struct.* **2022**, *86*, 103268. [CrossRef]
27. Dong, Y.; Garbatov, Y.; Guedes Soares, C. Recent developments in fatigue assessment of ships and offshore structures. *J. Mar. Sci. Appl.* **2022**, *21*, 3–25. [CrossRef]
28. Akpan, U.O.; Koko, T.S.; Ayyub, B.; Dunbar, T.E. Risk assessment of aging ship hull structures in the presence of corrosion and fatigue. *Mar. Struct.* **2002**, *15*, 211–231. [CrossRef]
29. Bai, Y.; Bai, Q. *Subsea Pipeline Integrity and Risk Management*; Gulf Professional Publishing: Houston, TX, USA, 2014; ISBN 0123946484.
30. Chen, N.-Z.; Wang, G.; Guedes Soares, C. Palmgren–Miner's rule and fracture mechanics-based inspection planning. *Eng. Fract. Mech.* **2011**, *78*, 3166–3182. [CrossRef]
31. Garbatov, Y.; Guedes Soares, C. Bayesian updating in the reliability assessment of maintained floating structures. *J. Offshore Mech. Arct. Eng.* **2002**, *124*, 139–145. [CrossRef]
32. Li, X.; Zhang, Y.; Abbassi, R.; Khan, F.; Chen, G. Probabilistic fatigue failure assessment of free spanning subsea pipeline using dynamic Bayesian network. *Ocean Eng.* **2021**, *234*, 109323. [CrossRef]
33. Xie, M.; Zhao, J.; Pei, X. Maintenance strategy optimization of pipeline system with multi-stage corrosion defects based on heuristically genetic algorithm. *Process Saf. Environ. Prot.* **2023**, *170*, 553–572. [CrossRef]
34. Paris, P.; Erdogan, F. A critical analysis of crack propagation laws. *J. Basic Eng.* **1963**, *85*, 528–533. [CrossRef]
35. Guedes Soares, C.; Moan, T. Model uncertainty in the long-term distribution of wave-induced bending moments for fatigue design of ship structures. *Mar. Struct.* **1991**, *4*, 295–315. [CrossRef]
36. Arzaghi, E.; Abaei, M.M.; Abbassi, R.; Garaniya, V.; Chin, C.; Khan, F. Risk-based maintenance planning of subsea pipelines through fatigue crack growth monitoring. *Eng. Fail. Anal.* **2017**, *79*, 928–939. [CrossRef]
37. Garbatov, Y.; Rudan, S.; Guedes Soares, C. Assessment of geometry correction functions of tanker knuckle details based on fatigue tests and finite-element analysis. *J. Offshore Mech. Arct. Eng.* **2004**, *126*, 220–226. [CrossRef]
38. Moan, T.; Ayala-Uraga, E. Reliability-based assessment of deteriorating ship structures operating in multiple sea loading climates. *Reliab. Eng. Syst. Saf.* **2008**, *93*, 433–446. [CrossRef]
39. Lye, A.; Cicirello, A.; Patelli, E. Sampling methods for solving Bayesian model updating problems: A tutorial. *Mech. Syst. Signal Process.* **2021**, *159*, 107760. [CrossRef]
40. Capellari, G.; Chatzi, E.; Mariani, S. Optimal sensor placement through Bayesian experimental design: Effect of measurement noise and number of sensors. *Multidiscip. Digit. Publ. Inst. Proc.* **2016**, *1*, 41.
41. Yuen, K.-V. *Bayesian Methods for Structural Dynamics and Civil Engineering*; John Wiley & Sons: Hoboken, NJ, USA, 2010; ISBN 0470824557.

42. Zárate, B.A.; Caicedo, J.M.; Yu, J.; Ziehl, P. Bayesian model updating and prognosis of fatigue crack growth. *Eng. Struct.* **2012**, *45*, 53–61. [CrossRef]
43. Reddy Lyathakula, K.; Yuan, F.-G. Fatigue damage diagnostics-prognostics framework for remaining life estimation in adhesive joints. *AIAA J.* **2022**, *60*, 4874–4892. [CrossRef]
44. Dong, T.; An, D.; Kim, N.H. Prognostics 102: Efficient Bayesian-based prognostics algorithm in Matlab. In *Fault Detection, Diagnosis and Prognosis*; Books on Demand: Paris, France, 2019; pp. 5–25.
45. Baker, J.W. Probabilistic structural response assessment using vector-valued intensity measures. *Earthq. Eng. Struct. Dyn.* **2007**, *36*, 1861–1883. [CrossRef]
46. Hastings, W.K. Monte Carlo sampling methods using Markov chains and their applications. *Biometrika* **1970**, *57*, 91–109. [CrossRef]
47. Almarnaess, A. *Fatigue Handbook: Offshore Steel Structures*; Tapir: Trondheim, Norway, 1985.
48. Wang, C. *Structural Reliability and Time-Dependent Reliability*; Springer: Berlin/Heidelberg, Germany, 2021; ISBN 3030625052.
49. Melchers, R.E.; Beck, A.T. *Structural Reliability Analysis and Prediction*; John Wiley & Sons: Hoboken, NJ, USA, 2018; ISBN 1119265991.
50. Guedes Soares, C.; Ivanov, L.D. Time-dependent reliability of the primary ship structure. *Reliab. Eng. Syst. Saf.* **1989**, *26*, 59–71. [CrossRef]
51. Weibull, W. A statistical distribution function of wide applicability. *J. Appl. Mech.* **1951**, *18*, 293–297. [CrossRef]
52. Lost, A. The effect of load ratio on the $M-\mathrm{Ln}\,C$ relationship. *Int. J. Fatigue* **1991**, *13*, 25–33. [CrossRef]
53. Li, Y.; Wang, H.; Gong, D. The interrelation of the parameters in the Paris equation of fatigue crack growth. *Eng. Fract. Mech.* **2012**, *96*, 500–509. [CrossRef]

Journal of
Marine Science and Engineering

Article

Bending Deformation and Ultimate Moment Calculation of Screen Pipes in Offshore Sand Control Completion

Yudan Peng [1,2], Guangming Fu [1,2], Baojiang Sun [1,2,*], Xiaohui Sun [1,2], Jiying Chen [3] and Segen F. Estefen [4,*]

[1] School of Petroleum Engineering, China University of Petroleum (East China), Qingdao 266580, China
[2] Key Laboratory of Unconventional Oil & Gas Development (China University of Petroleum (East China)), Ministry of Education, Qingdao 266580, China
[3] Department of Chemical Engineering and Biotechnology, University of Cambridge, Cambridge CB3 0AS, UK
[4] Ocean Engineering Department, Federal University of Rio de Janeiro, Rio de Janeiro 21945-970, Brazil
[*] Correspondence: sunbj1128@vip.126.com (B.S.); segen@lts.coppe.ufrj.br (S.F.E.)

Abstract: Horizontal wells, extended-reach wells, and multi-branch wells were often used to exploit subsea oil and gas efficiently. However, during the sand control screen completion of those wells, the sand control screen pipe was easily deformed. Failure occurred when passing through the bending section due to the large bending section in the wellbore trajectory. A parametric analysis model of the screen pipe was established based on ABAQUS and Python software under pure bending load first. Then, deformation patterns and mechanisms were identified and discussed. The effects of parameters on the screen pipe bending deformation patterns and the ultimate moment were analyzed. Finally, an empirical formula for calculating the ultimate moment of the screen pipe was established. The results showed that the deformation of the screen pipe was complex, and three deformation patterns were related to the hole parameters. Due to an increase in the diameter and number of circumferential and axial holes, the ultimate moment of the screen pipe gradually decreased, and the circumferential holes had a more significant effect on the ultimate moment than the axial holes. The established empirical formula could accurately calculate the ultimate moment of the screen pipe, and the average difference between the formula and numerical simulation results was 3.25%.

Keywords: sand control screen pipes; pipe bending; deformation patterns; screen pipe empirical formula

Citation: Peng, Y.; Fu, G.; Sun, B.; Sun, X.; Chen, J.; Estefen, S.F. Bending Deformation and Ultimate Moment Calculation of Screen Pipes in Offshore Sand Control Completion. *J. Mar. Sci. Eng.* **2023**, *11*, 754. https://doi.org/10.3390/jmse11040754

Academic Editor: José António Correia

Received: 14 March 2023
Revised: 25 March 2023
Accepted: 28 March 2023
Published: 31 March 2023

1. Introduction

Considering onshore oil and gas resource depletion, attention had been gradually focused on offshore oil exploration and development. However, sand production problems were often encountered during offshore and onshore unconsolidated sandstone reservoir development [1–3]. Excessive sand production in reservoir formations during exploitation reduces the production and damages the production equipment [4–6]. Therefore, it was necessary to adopt reasonable sand control measures to prevent excess sand from entering the wellbore. Currently, sand control screen pipes were the leading sand control equipment in fields. During the development of onshore and offshore oilfields, it is necessary to run sand control screens pipe to prevent massive formation sand production. For offshore screen pipes completion, the screen pipes enter the production formation of oil and gas through the riser and wellbore from the well head of the offshore platform. The screen enters the formation through the curved well section of the horizontal well is one of the processes of offshore screen completion. The screen pipe is made by perforating the pipe according to certain rules, and its shape is similar to the complete pipe. The geometry and hole parameters of screen pipes are designed according to operating conditions and oil and gas production. Commonly used materials for screen pipes are K55, J55, N80 steel, and so on. The base pipe made of perforated pipes was the main component of the sand control screen pipes to bear the loading.

With the increase in oil and gas production and consequent economic benefits, horizontal wells, extended reach wells, and multi-branch wells were often used on-site. Those well types could expand the exploitation scope of oil and gas wells without offshore platform restrictions. However, the well trajectories of those well types had a large, curved section [7,8], as shown in Figure 1. During the sand control screen completion of those wells, the screen pipes encounter a large bending load when entering the formation through the curved section [9,10], as shown in Figure 1. Because a complete pipe perforates the base of the screen pipe, the holes reduce the bearing capacity of the screen pipe, as shown in Figure 1. Moreover, the screen pipes may be bent and damaged due to the formation collapse or formation settlement during the production process. Since the acidic component exists in oil and gas in formation, the screen pipes were corroded. The longer the production time, the more serious the corrosion of the screen pipe was, and the bending strength performance of the screen pipes decreased. Plastic deformation and fracture occurred around the screen pipe body and hole when the pipe was subjected to a large bending load. This results in the failure of the sand control screen, which considerably threatened the average production of oil and gas wells. Therefore, it was necessary to analyze the deformation behavior of a screen pipe under a bending load and calculate the ultimate bending strength.

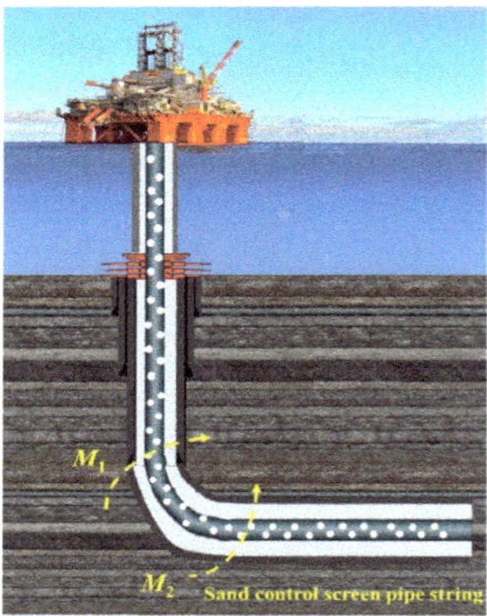

Figure 1. Loads acting on the curved section of the screen pipe.

The research activities on the ultimate bending moment of intact and corroded pipes under bending loads were relatively mature. The main research approaches include the finite element method, experimental tests, and analytical treatment. Compared with the previous research object, the structure of the screen pipe was more complex. The bending deformation mechanism of the screen pipe under bending load needed to be clarified, and the lack of a corresponding formula for calculating the ultimate bending moment of the screen pipe caused difficulties in evaluating the ultimate strength.

Chen et al. [11] proposed a simplified method for predicting the ultimate bending strength of a pipeline based on Hencky's total strain theory. The results showed that the simplified model was consistent with the experimental results in predicting the ultimate bending capacity of steel pipes, and the strain-hardening effect significantly influenced the ultimate bending capacity of steel pipes. Adam et al. [12] established a nonlinear finite element model (FEM) of the plastic buckling of cylindrical metal tubes and shells under a bending load using ABAQUS software. They compared the applicability and economy of continuous and shell finite elements in simulating the pipe-bending deformation. Chen et al. [13] established a theoretical model of the residual bending strength of a corroded screen pipe under combined internal pressure and axial load. They analyzed the influence of different corrosion shapes on the residual strength of the pipe. Using ABAQUS, Kumar et al. [14] established a FEM of the buckling of cylindrical shells under a pure bending load. They analyzed the influence, sensitivity, and impact of steel strain hardening models on the bending behavior of cylindrical shells. Moreover, using ABAQUS, Erling et al. [15,16] also established a FEM for a pipeline with a peripheral surface crack under a pure bending load and combined bending load and internal pressure. They analyzed the evolution law of the crack tip opening of the cracked pipeline and the effects of crack depth, length, diameter–thickness ratio, and material hardening on the fracture response. Behrouz [17] analyzed the influence of corrosion depth and shape on the strength of corroded pipes under combined internal pressure and bending load based on experiments and FEMs. Based on numerical simulation data, Hieu et al. [18] established an empirical formula for the ultimate bending capacity of pipes with corrosion defects under axial loads. Sjors [19,20] analyzed the ultimate bending capacity of spiral-welded steel tubes based on experiments and FEMs. Limam A et al. [21,22] analyzed the inelastic wrinkling of intact tubes and the effect of local dents on the collapse curvature of pipes under combined bending and internal pressure with experiment and FEM method. Fu Guangming et al. [23] and Peng Yudan [24] established finite element model of the screen pipes under external pressure and combined external pressure and bending load with the finite element method, and the influence of different parameters on the collapse strength of the screen pipes was analyzed, and the corresponding calculation formula for the collapse strength of the screen pipes were established. Kyriakides and Corona [25] and Karampour and Albermani [26] studied the plastic bending strength of pipes under bending load based on experiment and finite element method, and the results of simulation and experiment were compared. Karampour [27] studied the lateral buckling of pipelines with nonlinear soil pipe interaction based on the finite element analysis method, and an analytical solution for lateral buckling of pipes with a single defect was proposed. Taheri et al. [28] analyzed the linear eigenvalue bucking and nonlinear post bucking of sandwich pipes under bending loads based on the finite element method, and the influence of structural parameters of sandwich pipes on pre-bucking, bucking, and post bucking responses was analyzed. Binazir et al. [29] analyzed the linear bifurcation and geometrically nonlinear behavior of pipe in pipe under bending loads based on experiments and finite element methods and proposed a formula for calculating the ultimate bending moment of pipe in pipe.

In the present work, a FEM of the bending deformation of a screen pipe under pure bending was established based on ABAQUS and Python script. The sand control screen pipe's bending deformation patterns and mechanism were discussed in detail, and the influence of the screen pipe and hole arrangement parameters on the bending deformation and ultimate moment of the screen pipe were analyzed. Furthermore, an empirical formula for calculating the ultimate moment of the screen pipe under a bending load was established.

2. Numerical Simulation Model of the Sand Control Screen Pipe under Bending

2.1. Finite Element Model

This study adopted the parallel hole arrangement in a perforated screen pipe. Because a screen pipe with a parallel hole arrangement was symmetrical, a 1/2 symmetrical model was used to simplify the calculation, as shown in Figure 2. The screen pipe was severely deformed under a bending load, and C3D8R element, which had the advantages of high accuracy and no shear self-locking even when the mesh was severely deformed [30], was employed to simulate the screen pipe under bending. The mesh sensitivity analysis was carried out, as shown in Figure 3. When the mesh number is around 210,570, the ultimate moment of the screen pipes tends to stabilize, and as the mesh number continues to increase, the ultimate moment of the screen pipes changes slightly. Therefore, the mesh number of 210,570 was employed in the present work. The length of the screen was ten times greater than the diameter of the pipe to reduce the influence of the ending effect on the bending deformation of the screen pipe.

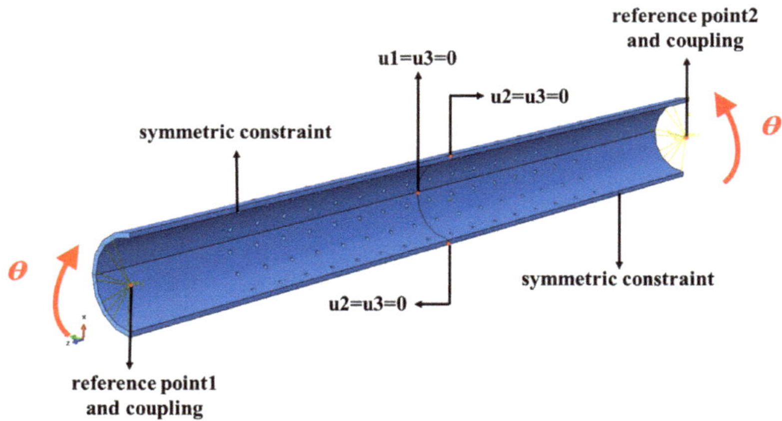

Figure 2. The FEM of screen pipe and boundary condition.

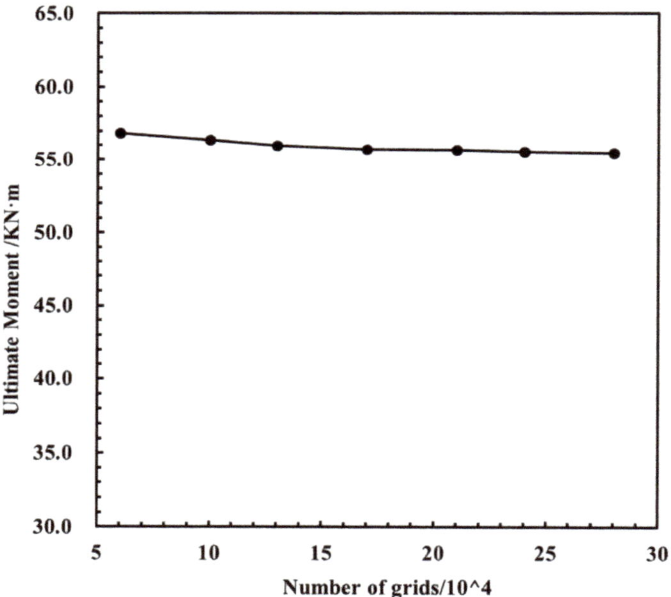

Figure 3. Mesh sensitivity analysis.

Figure 2 shows the load and boundary conditions of the model. The reference points were set at the center of the two end faces of the screen pipe and were coupled with the faces. Symmetry constraints were applied to the symmetry plane, and a corresponding displacement constraint was applied to the one half screen pipe length nodes to eliminate the rigid body displacement of the pipe in different directions during loading [31]. The same rotation angle was applied to the coupling points at both ends of the pipe to apply a bending load. Additionally, general analysis steps were adopted, and the screen pipe's ultimate moment and rotation angles were obtained in a rotation angle-moment curve. The elastic modulus, plastic strength, and Poisson's ratio of this material were 203 GPa, 464.36 MPa, and 0.3, respectively [23]. The isotropic hardening behavior of screen pipe material was assumed, and the stress–strain curve of the screen pipe material is shown in Figure 4.

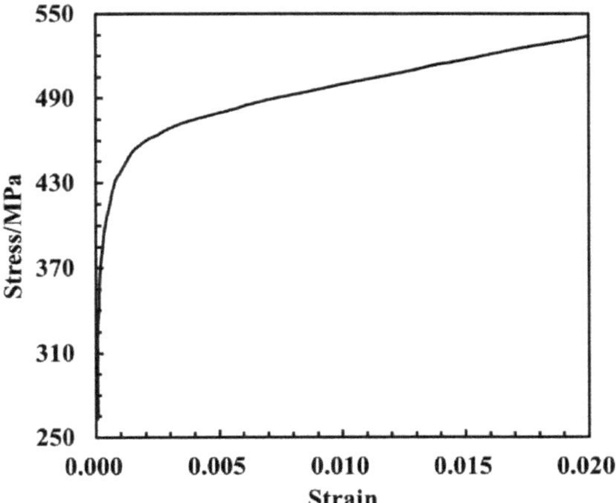

Figure 4. Material property of the sand control screen.

2.2. Secondary Development of FEM

ABAQUS provided a secondary development interface [32,33]. A parametric analysis model of the ultimate moment of screen pipe under bending load was established based on the secondary development of ABAQUS by Python script. The secondary development process is shown in Figure 5, and its main difficulty was realizing a regular arrangement of hole parameters and parameterization. Secondary development of FEMs significantly simplified the tedious modeling process and laid the data foundation for the study of the deformation pattern of screen pipe as well as the establishment of the empirical formula of the ultimate moment.

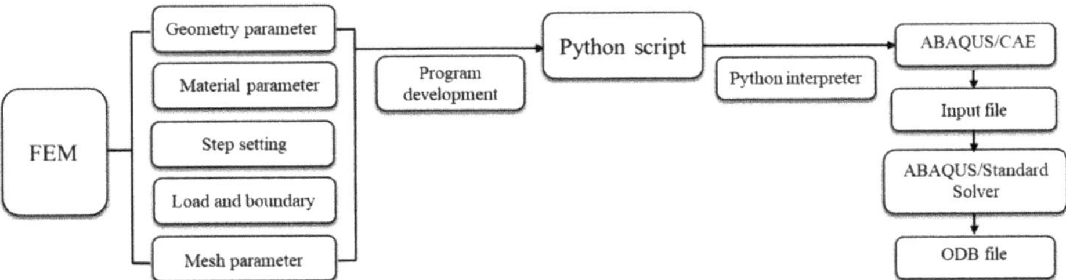

Figure 5. Secondary development process of FEM.

2.3. Numerical Results Discussion

The screen pipe's bending deformation patterns and mechanism under bending load were currently unclear. Therefore, the bending deformation of the screen pipes with different screen parameters, as listed in Table 1, was analyzed based on FEM.

Table 1. Screen pipes parameter commonly used in the field.

D/t	D/mm	t/mm	d/mm	N_1	N_2	L/mm
			6.35	4	12	
17	127.0	7.52	9.5	8	16	
19	177.8	9.19	12.7	12	20	1600.0
23	152.6	6.46	14.0	16	24	
			16.0	20	28	

D is the diameter of the screen pipes, mm; t is wall thickness of screen pipes, mm; d is the diameter of the holes, mm; N_1 is the number of circumferential holes; N_2 is the number of axial holes; L is the length of screen pipe, mm.

Numerous results of screen pipe deformation indicated that the deformation could be divided into three patterns: In pattern-1, the screen pipe buckled from the middle position, and the hole area did not fracture. In pattern-2, the screen pipe first buckled from the middle position, and plastic deformation occurred around the hole area, which eventually fractured as the bending load increased. In pattern-3, plastic deformation occurred around the fractured hole area. However, the deformation around the middle of the screen pipe was small.

The three bending deformation patterns of screen pipes were discussed in detail as follows from different aspects, such as the relationship between the rotation angle and moment, ovality change laws at the middle section of the screen pipe and fracture section around the holes, and stress and strain laws at the middle position of the screen pipes and fracture position around the holes.

2.3.1. Pattern-1

As the bending load increased, the screen pipe buckled from the middle position. However, the hole area did not fracture. Figure 6 shows the screen pipe's moment-rotation angle relationship in the pattern-1, which was divided into four stages. In stage-1, as the bending load increased, the bending moment sharply increased, and the stress and bending deformation were small, as shown in #1. In stage-2, as the bending load increased, the bending moment gradually reduced, whereas the deformation and stress gradually increased, as shown in #2. In stage-3, as the bending load gradually increased, the increasing speed of the bending moment tended to stabilize, and the bending deformation gradually increased, as shown in #3. The deformation and stress of the screen pipe when the ultimate bending moment was reached are shown in #4. In stage-4, when the bending load was increased after reaching the ultimate bending moment, the bending moment of the screen sharply decreased. However, the bending deformation continued to increase, whereas the stress at both ends of the screen gradually decreases, causing the screen pipe to buckle from the middle position, as shown in #5. Meanwhile, there was no deformation around the hole area.

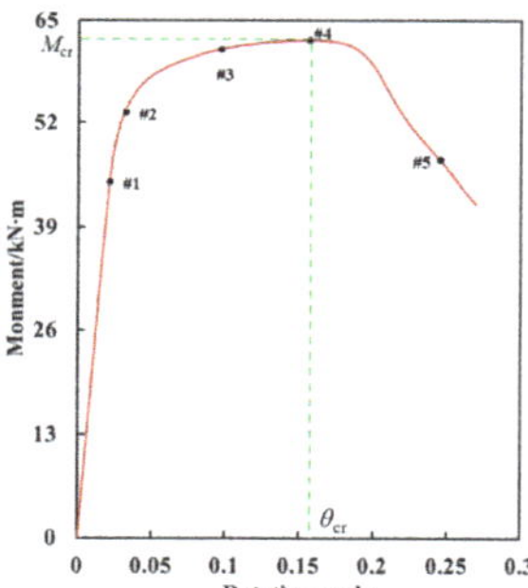
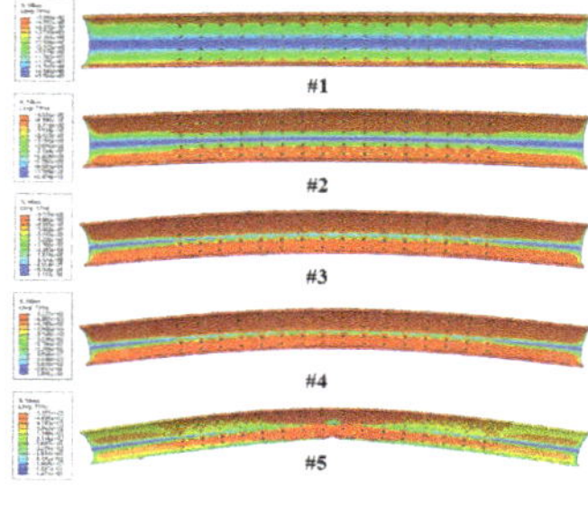

Figure 6. Moment versus rotation angle and stress contour of screen pipes.

The deformation was most significant at the middle position of the screen pipe, which is the central area of concern. Therefore, the ovality, stress, and strain in the middle of the screen tube are discussed in detail here. Figure 7a,b show the middle section of the screen pipe and the ovality curve with the rotation angle of the middle section, respectively. The curve could be roughly divided into three stages: In the first stage, when the rotation angle was less than the critical rotation angle, the ovality of the section gradually increased as the rotation angle increased, and the deformation of the screen pipe was small. In the second stage, the ovality and deformation sharply increased with an increase in the rotation angle after the angle reached the critical angle. In the third stage, as the rotation angle increased, the increase in ovality tended to stabilize, and the deformation of the screen pipe continued to increase.

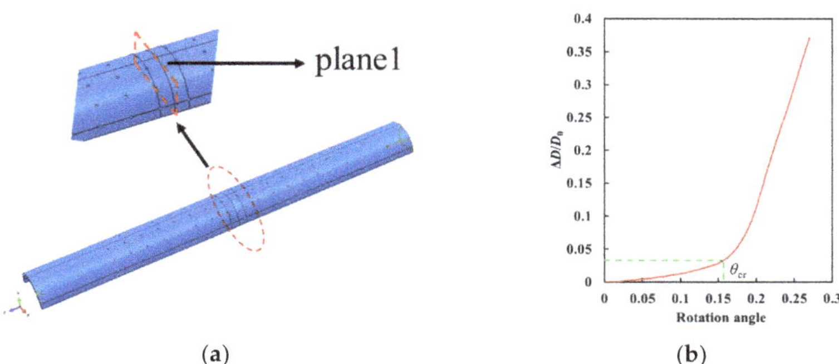

(a)   (b)

Figure 7. (**a**) Middle section of screen pipe; (**b**) Curve of ovality versus rotation angle.

Figure 8 shows the selection of normalized elements and nodes in the middle of the screen pipe. Figure 9 shows the stress and strain contours of the selected nodes and elements on the compression and tension sides of the screen pipe during the bending process. In Figure 9a,b, the stress and strain contours about the middle position were symmetrical. When $\theta < 0.016$, the stress difference of the nodes at different positions was small. As the rotation angle increased, the nodes on both sides reached the yield stress first compared to the middle nodes at the same θ. The stress in the middle position and on both sides fluctuated after $\theta > \theta_{cr}$. In contrast, when $\theta < 0.0195$, the strain of the elements on both sides was more significant than that of the central elements. It was because the stress at both sides reached the yield state first. The strain in the middle position was gradually greater than those on both sides, and the closer the position was to the middle, the greater the strain, causing the screen pipe to buckle from the middle position on the compression side.

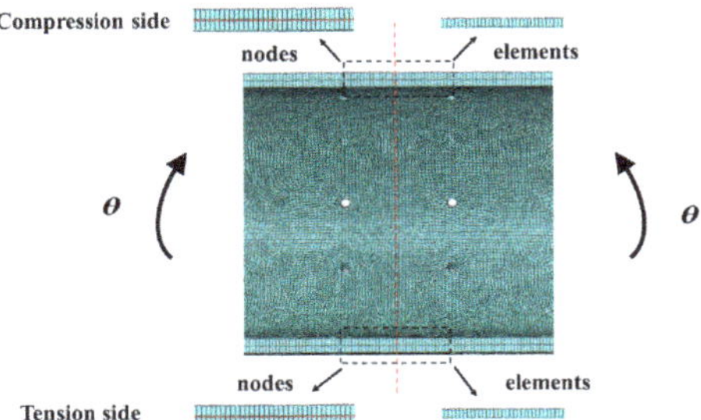

Figure 8. The selected elements and nodes of screen pipe.

Figure 9c,d shows the stress and strain contours on the tensile side, which were symmetrical about the middle position. Although the stress variation law on the tensile side was similar to that on the compression side, the rotation angle to yield stress at the middle position was more significant on the tensile side. The stress of the nodes at different positions did not change after reaching the yield stress. However, the strain on the tensile side of the screen pipe was opposite to that on the compression side: the closer the element was to the middle position, the smaller the strain. In addition, the strain on the tensile side was less than on the compression side at the same angle rotation and corresponding elements.

2.3.2. Pattern-2

As the bending load increased, the screen pipe first exhibited overall bending deformation, and buckling deformation occurred at the middle position. In contrast, plastic deformation and fracture occurred around the hole area. The moment–rotation angle curve was divided into five stages, as shown in Figure 10. The first four stages were similar to those of pattern-1 and will not be repeated here. In stage-5, as the bending load increased, the deformation of the middle position of the screen pipe increased, and plastic deformation appeared around the hole area, which gradually evolved into a fracture. The deformation and stress of the screen pipe are shown in #6.

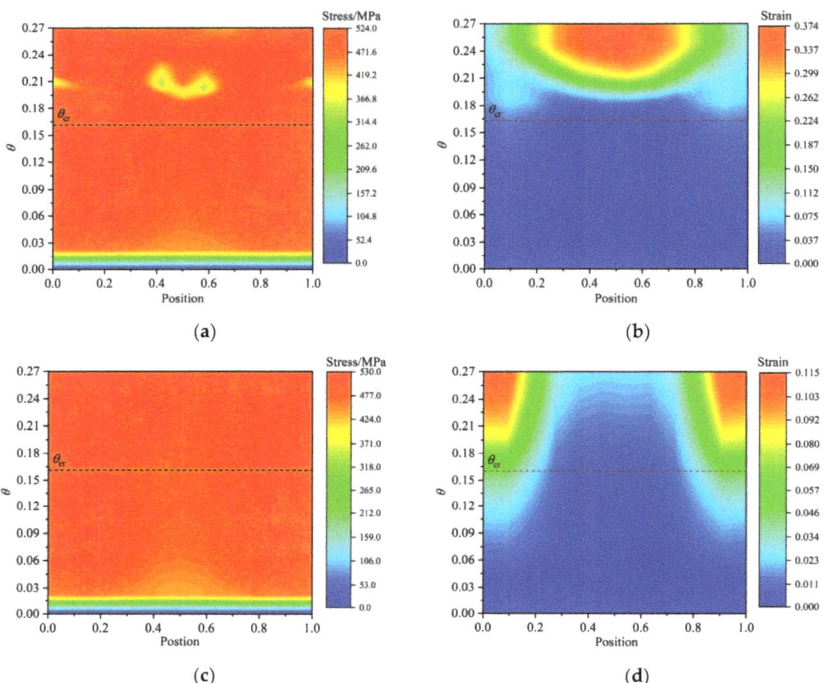

Figure 9. (**a**,**b**) stress and strain on compression side; (**c**,**d**) stress and strain on the tensile side.

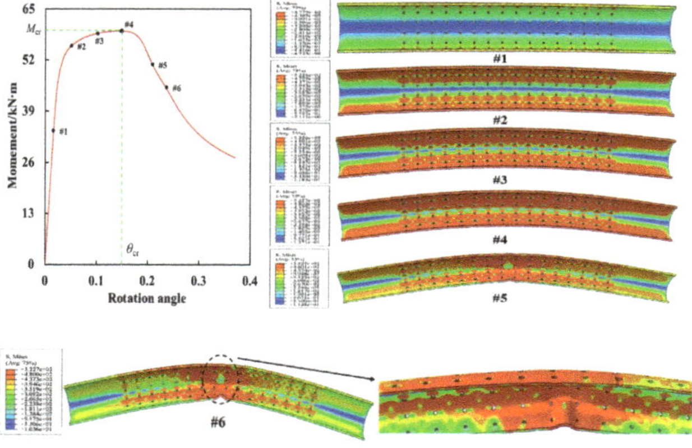

Figure 10. Moment versus rotation angle, and stress contour of screen pipes.

Meanwhile, the deformations at the middle position and hole area were significant, the main areas of concern. Therefore, the ovality, stress, and strain at the middle position and the area around the hole are discussed in detail here. Figure 11a shows the middle and plastic deformation hole sections (planes 1 and 2, respectively), and Figure 11b shows the ovality curve with the rotation angles of planes 1 and 2. The curve could be divided into three regions: In the first region ($\theta < \theta_{cr}$), as the rotation angle increased, the ovality of planes 1 and 2 gradually increased, and the deformation of the screen pipe was small. Additionally, the ovality of planes 1 and 2 was approximately equal, indicating that the bending deformations of the two positions were the same. In the second region ($\theta > \theta_{cr}$), the ovality of plane 1 was greater than that of plane 2. It increased approximately in a straight line, indicating that the deformation of the hole area was more significant than that of the middle position of the screen pipe. In the third region, as the rotation angle increased, the ovality of plane 1 gradually decreased and tended to be constant, whereas that of plane 2 sharply increased, indicating that the radial deformation of the screen tube mainly occurred at the hole position.

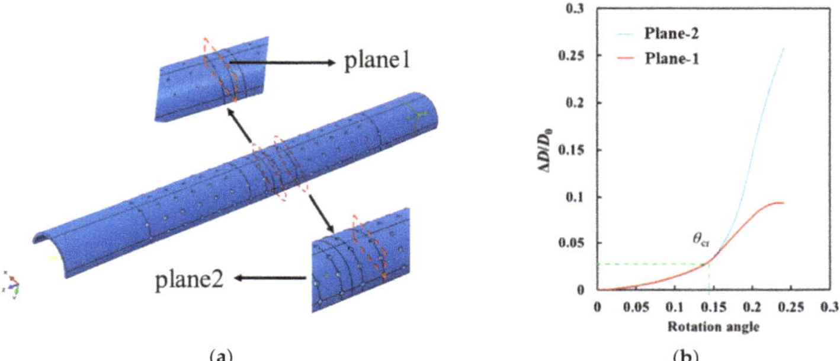

(a)

(b)

Figure 11. (**a**) Middle section and hole section of screen pipe; (**b**) Ovality versus rotation angle.

The node stress and element strain at the middle position and the hole areas on the compression and tensile sides of the screen were discussed in detail here. Figure 12 shows the selected elements and nodes. Figure 13 shows the stress and strain of the nodes and elements in the middle position of the tensile and compression sides, respectively. The stress and strain on the compression side were similar to those of pattern-1, as shown in Figure 13a,b, which was not repeated here. The stress on the tensile side was comparable to that in the pattern-1, as shown in Figure 13c,d. Yield stress occurred only in the nodes at the middle position on the tensile side after the rotation angle reached a significant value. The strain in the nodes at the middle position gradually approached and exceeded those at both sides when the rotation angle increased to a more significant value. This difference may be due to the appearance of the plastic deformation of the hole area at the tensile side of the screen pipe, which significantly influenced the stress and strain distributions at the tensile side but slightly influenced those at the compression side.

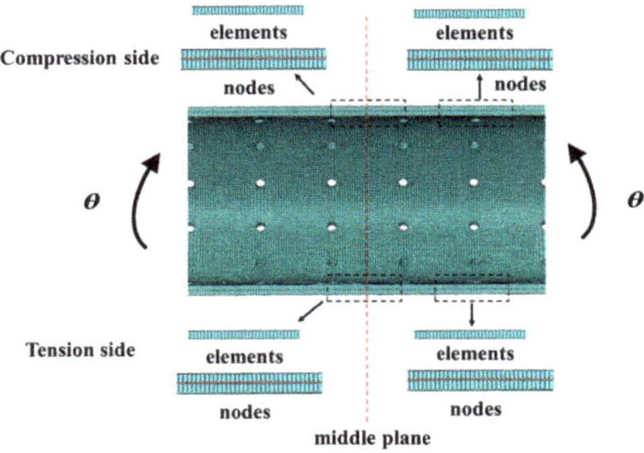

Figure 12. Selected elements and nodes.

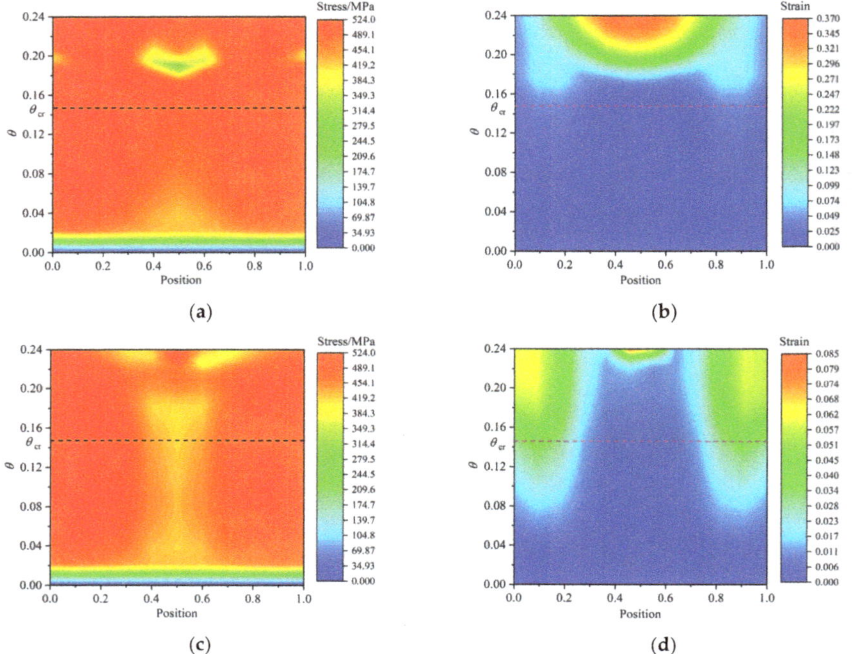

Figure 13. (**a**,**b**) Stress and strain on the compression side of screen pipe in the middle position; (**c**,**d**) Stress and strain on the tensile side of screen pipe in the middle position.

Figure 14 shows the stress and strain of the nodes and elements around the holes on the compression and tension side. As shown in Figure 14a, when $\theta < 0.02$, the node stresses at different positions were equal under the same rotation angle. When $0.02 < \theta < 0.08$, the stress increased and then decreased as the position was closer to the middle. When $0.08 < \theta < 0.2$, the node stress values at different positions were consistent under the same rotation angle. When $0.2 < \theta$, the node stress gradually decreased at the same node position, which differed from the stress variation in the middle position of the screen pipe. As shown in Figure 13b, the strain in the middle position of the hole was the smallest. When $\theta < 0.04$, the strain difference of each element was small. When $0.04 < \theta$, the strain of the screen pipe first increased and then decreased as the unit position approached the middle position of the hole.

Figure 14c,d shows the stress and strain at the nodes and elements on the tensile side. As shown in Figure 14c, when $\theta > 0.08$, the stress in the middle position of the hole gradually reached the maximum stress and was more significant than those on both sides of the pipe. When $\theta < 0.16$, the strain difference of the elements around the hole was small at the same rotation angle. When $\theta > 0.16$, the strain increased as the element position was closer to the middle of the hole. Moreover, the strain of the area around the hole on the tension side was much more significant than on the compressive side.

Comparing the stress and strain variations at the middle and hole positions of the tension and compression sides of the screen pipe, when $\theta < \theta_{cr}$, the strain in the middle position of the pipe on the compression side was greater than that in the hole position on the tensile side. However, when $\theta > \theta_{cr}$, the situation was reversed, and the strain at the hole position was considerable. Therefore, the screen tube first buckled at the middle position on the compression side. Then, plastic fracture occurred at the hole position on the tension side.

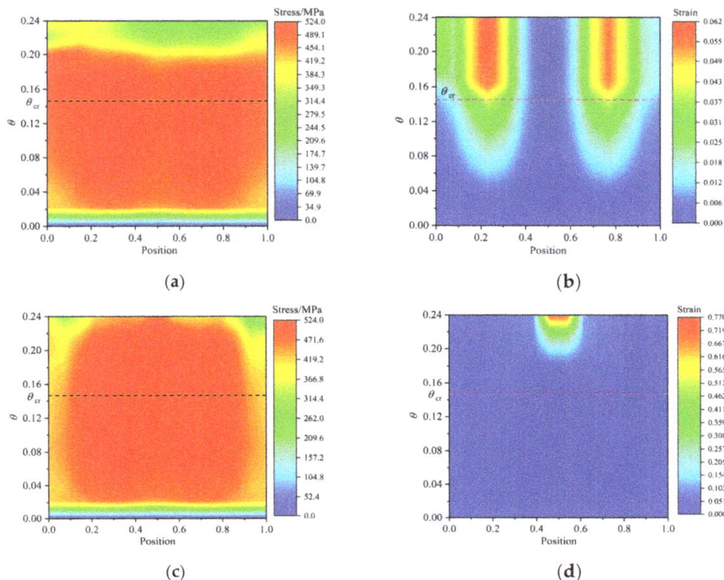

Figure 14. (**a**,**b**) stress and strain on the compression side around the holes; (**c**,**d**) stress and strain on the tension side around the holes.

2.3.3. Pattern-3

In this pattern, as the bending load increased, the screen pipe deformed and fractured at the hole position, whereas the middle position of the screen pipe did not buckle. Figure 15 shows the moment–rotation angle curve divided into five stages. As shown in Figure 15, the first three stages were similar to those in pattern-1 and not repeated here. In stage-4, as the bending load further increased after reaching the ultimate moment, the bending moment of the screen tube sharply decreased, and the bending deformation of the screen pipe continued to increase. However, there was no buckling deformation in the middle position of the screen pipe. The deformation and stress of the screen tube are shown in #5. In stage-5, with a continuous increase in the bending load, the overall bending deformation of the screen pipe further increased, and plastic deformation appeared in the hole position, which gradually evolved into a fracture. The deformation and stress of the screen tube are shown in #6.

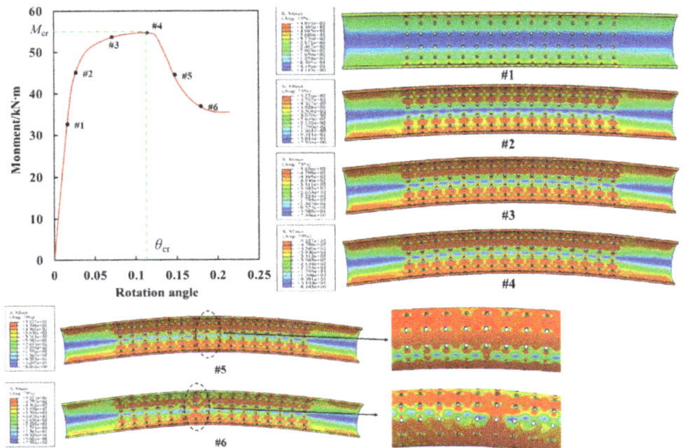

Figure 15. Moment versus rotation angle and stress contour of screen pipes.

Meanwhile, the deformation at the middle position and the area around the hole was significant, which were the main areas of concern. Therefore, the ovality, stress, and strain at the middle position and area around the hole are discussed in detail. Figure 11a shows the middle and hole sections, and Figure 16 shows the ovality curve with the rotation angles of planes 1 and 2. As shown in Figure 16, the change in ovality with the rotation angle could be divided into three stages. In the first stage, when $\theta < \theta_{cr}$, the ovality difference between planes 1 and 2 was small as the rotation angle increased, indicating that the radial deformations of planes 1 and 2 were consistent. In the second stage, when $\theta > \theta_{cr}$, the ovality of plane 1 was gradually greater than that of plane 2 with an increase in θ. In the third stage, as θ increased, the ovality of planes 1 and 2 gradually increased, reaching a maximum and decreasing sharply. Meanwhile, the ovality of plane 2 was much smaller and reached the maximum value earlier than that of plane 1.

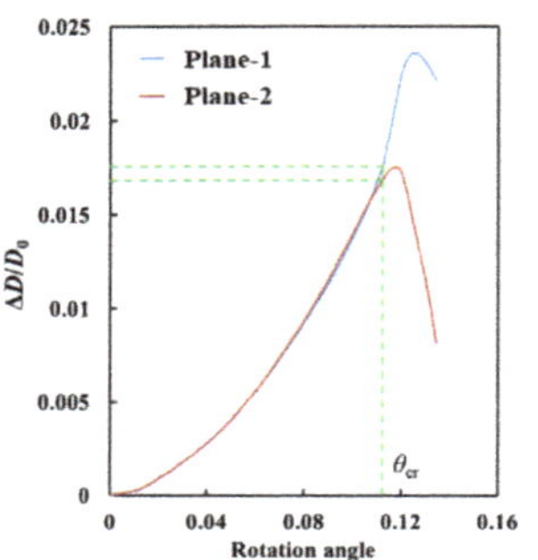

Figure 16. Ovality versus rotation angle at middle position and hole position of screen pipes.

The node stress and element strain at the middle position and the area around the hole on the compression and tension sides of the screen are discussed in detail here. Figure 12 shows the selected elements and nodes, and Figure 17 shows the element strain and node stress at the middle position on the compression and tensile side. As shown in Figure 17a, when $\theta < 0.022$, the stress difference of different nodes was small under the same θ. When $0.022 < \theta < \theta_{cr}$, the stress first increased and then decreased as the node position was closer to the middle position of the screen pipe, and the stresses on both sides were more significant than that in the middle position under the same θ. When $\theta > \theta_{cr}$, the stress values at different locations fluctuated. Concerning the element strain, as shown in Figure 17b, the strain difference of the elements at different positions was small under the same θ. When $0.045 < \theta < 0.127$, the strains of the elements on both sides were more significant than that of the middle elements under the same θ. However, when $\theta > 0.127$, the situation was reversed.

Figure 17c,d shows the node stress and element strain in the middle of the screen pipe at the tension side. The observations were similar to those on the compression side and will not be repeated here. Compared to the first two screen pipe deformation patterns, the strain values on the compression and tension sides of the middle position of pattern-3 were small. The difference between them was also small, indicating that the deformation at the middle position was small during the screen pipe bending process.

Figure 18a,b shows the node stress and element strain on the tension and compression sides of the hole. Figure 18a shows the node stress around the hole on the compression side. When $\theta < 0.022$, the stress difference of different nodes was small under the same θ. When $0.022 < \theta < 0.06$, the stress first increased and then decreased as the node position was closer to the middle position of the hole. When $\theta > 0.06$, the stress of the middle nodes of the hole was greater than that of the nodes on both sides. Figure 18b shows the element strain around the hole on the compression side. When $\theta < 0.03$, the stress difference of different nodes was small under the same θ. When $\theta > 0.03$, the element strains on both sides of the hole were more significant than that in the middle position. As the element position was closer to the middle of the hole, the element strain first increased and then decreased.

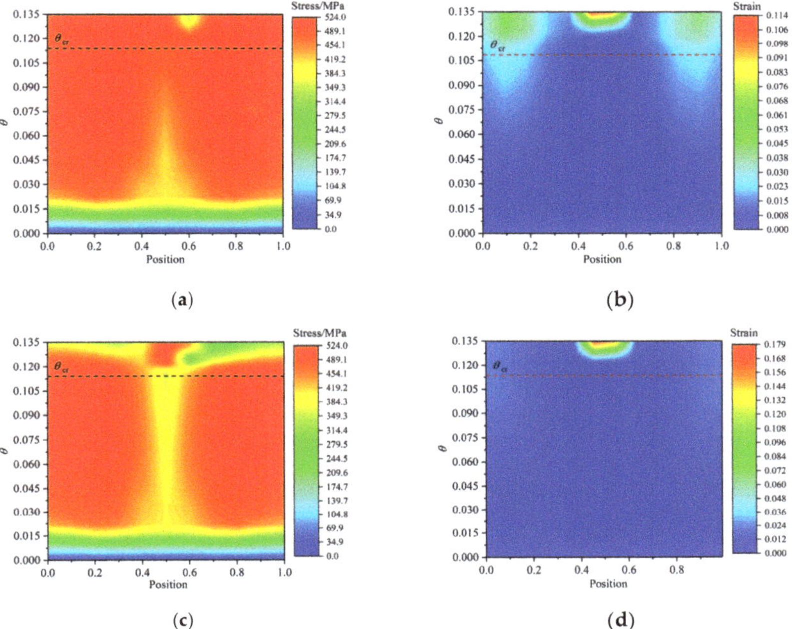

Figure 17. (**a**,**b**) stress and strain on the compression side of screen pipe at middle position; (**c**,**d**) stress and strain on the tension side of screen pipe at middle position.

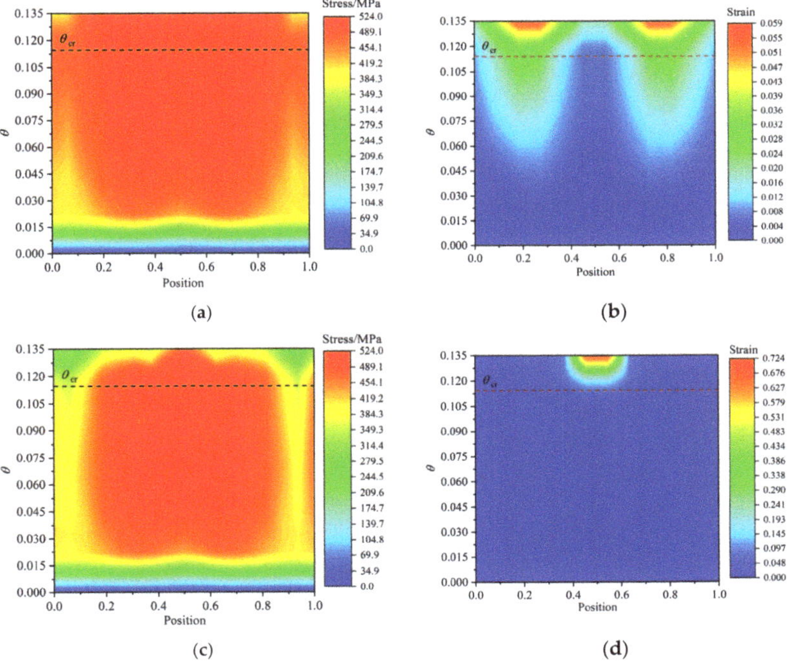

Figure 18. (**a**,**b**) stress and strain on the compression side of screen pipe at holes position; (**c**,**d**) stress and strain on the tension side of screen pipe at holes position.

Figure 18c,d shows the node stress and element strain on the tension side of the hole position. The node stress at the hole position on the tension side was similar to that on the compression side and will not be repeated here. Concerning the element strain on the tension side, when $\theta < \theta_{cr}$, the strain difference of the different elements was small under the same rotation angle. When $\theta > \theta_{cr}$, the strain of the central elements of the screen hole was much larger than those of the elements on both sides. Comparing the strain at the hole position on the tension and compression sides, the maximum strain on the tension side was 0.724, which was much greater than the maximum strain of 0.059 on the compression side, indicating that the deformation at the hole position of the tension side was significant compared to that of the compression side, resulting in the plastic deformation of the screen pipe at the tension side.

Comparing the element strain results at the middle position of the screen pipe and hole position, when $\theta > \theta_{cr}$, the strain and, consequently, the deformation at the hole position on the tension side were much larger than those at the middle position of the screen pipe. As a result, deformation and yielding occurred from the hole position on the tension side of the screen pipe, which evolved into a fracture, whereas the deformation in the middle position of the screen pipe was small.

3. Influence of Different Parameters on the Ultimate Moment of Screen Pipes

The deformation patterns of screen pipes were different under different hole distribution parameters. However, when the moment was less than the ultimate moment, the change law of the rotation angle-moment is similar under different deformation patterns according to the above analysis. The screen tube's ultimate moment was the field's main parameter of interest. Therefore, the influence of different hole and screen pipe parameters on the ultimate moment of the screen pipe was analyzed. The selection of the parameters is shown in Table 1.

3.1. Effect of Diameter and Diameter-to-Thickness Ratio on the Ultimate Moment of Screen Pipes

The parameters N_1 and N_2 were 12 and 20. The other parameters were shown in Table 2. The effects of different hole diameters and D/t values on the ultimate bending moment of the screen pipe were analyzed, and the results were shown in Table 2. As shown in Table 2, as the hole diameter increased, the ultimate moment of the screen pipes with different D/t values gradually decreased, and the ultimate moment of the screen pipe with $D/t = 19$ was the largest with the same hole diameter. This is because the outer diameter of the screen pipe is inconsistent. According to the formula for calculating the ultimate bending moment of the complete pipe [34], the ultimate bending moment of the complete pipe is related to D and t. D significantly impacts the ultimate bending moment of the complete pipe. In addition, with an increase in the hole diameter, the ultimate moment of the screen pipe with $D/t = 17$, $D/t = 19$, and $D/t = 23$ decreased by 36.8%, 21.2%, and 26.3%, respectively, compared to the ultimate moment of the screen tube with a hole diameter of 6.35 mm.

Table 2. Effect of D/t and hole diameter on ultimate moment of screen pipes.

D/t	D/mm	t/mm	d/mm	$M_{\text{cr-}d/t\,=\,17}$/kNm	$M_{\text{cr-}d/t\,=\,19}$/kNm	$M_{\text{cr-}d/t\,=\,23}$/kNm
			6.35	50.36	125.30	64.62
17	127.0	7.52	9.5	45.94	119.20	60.78
19	177.8	9.19	12.7	39.42	110.32	54.90
23	152.6	6.46	14.0	36.28	105.70	52.10
			16.0	31.80	98.76	47.58

3.2. Effect of Diameter-to-Thickness Ratio and Number of Axial Holes on the Ultimate Moment of Screen Pipes

The parament N_1 and d were 12 and 9.5 mm. The parameters were shown in Table 3. The effects of the different number of axial holes and D/t values on the ultimate moment of the screen pipes were analyzed, and the results are shown in Table 3. As shown in Table 3, as the number of axial holes increased, the ultimate moment of the screen tubes with different D/t values gradually decreased. The ultimate moment of the screen pipe with $D/t = 19$ was the largest with the same number of axial holes. In addition, the number of axial holes slightly affected the screen pipes' moment. Compared to the ultimate moment of the screen pipes with eight axial holes, the reductions of the $D/t = 17$, $D/t = 19$, $D/t = 23$ values were 2.2%, 6.4%, and 4.7%, respectively.

Table 3. Effect of D/t and axial holes number on ultimate moment of screen pipes.

D/t	D/mm	t/mm	N_2	$M_{\text{cr-d/t = 17}}$/kNm	$M_{\text{cr-d/t = 19}}$/kNm	$M_{\text{cr-d/t = 23}}$/kNm
			8	46.44	121.72	61.68
17	127.0	7.52	12	46.22	120.76	61.16
19	177.8	9.19	16	46.18	120.1	60.96
23	152.6	6.46	20	45.94	119.2	60.78
			24	45.84	116.58	60.2
			28	45.44	113.96	58.78

3.3. Effect of Diameter-to-Thickness Ratio and Number of Circumferential Holes on the Ultimate Moment of Screen Pipes

The parament N_2 and d were 12 and 9.5 mm. The parameters were shown in Table 4. The effects of the different number of circumferential holes and D/t values on the ultimate moment of the screen pipes were analyzed, and the results are shown in Table 4. As shown in Table 4, the ultimate moment of the screen tubes with different D/t values gradually decreased as the number of circumferential holes increased. The ultimate moment of the screen pipes with $D/t = 19$ was the largest with the same number of circumferential holes. Compared to the axial holes' effect on the screen pipes' ultimate moment, the number of circumferential holes had a significant influence on the ultimate moment of the screen pipe. The reductions of the $D/t = 17$, $D/t = 19$, and $D/t = 23$ values were 28.4%, 15.4%, and 20.5%, respectively, compared to the ultimate moment of the screen pipes with four circumferential holes.

Table 4. Effect of D/t and circumferential holes on ultimate moment of screen pipes.

D/t	D/mm	t/mm	N_1	$M_{\text{cr-d/t = 17}}$/kNm	$M_{\text{cr-d/t = 19}}$/kNm	$M_{\text{cr-d/t = 23}}$/kNm
			4	52.04	127.26	66.26
17	127.0	7.52	8	49.78	124.64	64.14
19	177.8	9.19	12	46.22	120.76	61.16
23	152.6	6.46	16	41.84	114.88	56.10
			20	37.24	107.62	52.64

3.4. Effect of Diameter and Number of Axial Holes on the Ultimate Moment of Screen Pipes

The number of circumferential holes and D/t of the screen pipes were 12 and 19, respectively, and the hole diameters were 6.35, 9.5, 12.7, 14.0, and 16.0 mm. The numbers of axial holes were 16, 20, and 24. The effects of the different number of axial holes and diameters on the ultimate moment of the screen pipes were analyzed, and the results are shown in Table 5. As shown in Table 5, the ultimate moment of the screen pipes with different numbers of axial holes gradually decreased as the diameter increased. Under the same diameter, as the number of axial holes increased, the ultimate moment of the screen pipes decreased. With an increase in the hole diameter, the ultimate moment of the screen pipe with 16, 20, and 24 axial holes decreased by 21.2%, 21.2%, and 21.9%, respectively, compared to the ultimate moment of the screen tube with a hole diameter of 6.35 mm.

Table 5. Effect of diameter and axail holes number on ultimate moment of screen pipes.

N_2	d/mm	$M_{\text{cr-N2}=16}$/kNm	$M_{\text{cr-N2}=20}$/kNm	$M_{\text{cr N2}=24}$/kNm
	6.35	125.86	125.30	123.90
16	9.5	120.10	119.20	116.58
20	12.7	111.16	110.32	107.20
24	14.0	106.24	105.70	102.92
	16.0	99.24	98.76	96.78

3.5. Effect of Diameter and Number of Circumferential Holes on the Ultimate Moment of Screen Pipes

The number of axial holes and D/t of the screen pipes were 16 and 23, respectively, and the hole diameters were 6.35, 9.5, 12.7, 14.0, and 16.0 mm, respectively. The numbers of circumferential holes were 8, 12, and 16. The effects of the different number of circumferential holes and diameters on the ultimate moment of the screen pipes were analyzed, and the results are shown in Table 6. As the diameter increased, the screen pipes' ultimate moment under different circumferential holes gradually decreased. For the same diameter, the greater the number of circumferential holes, the smaller the ultimate moment of the screen and the more significant the reduction in the ultimate moment. Compared to the ultimate moment of the screen pipe with a diameter of 6.35 mm, the ultimate moment values of the screen pipe with eight, 12, and 16 circumferential holes were reduced by 15.1%, 26.5%, and 42.9%, respectively. Compared to the effects of the axial holes on the ultimate moment, the effects of the circumferential holes were more significant.

Table 6. Effect of diameter and circumferential holes number on ultimate moment of screen pipes.

N_1	d/mm	$M_{\text{cr-N1}=8}$/kNm	$M_{\text{cr-N1}=12}$/kNm	$M_{\text{cr-N1}=16}$/kNm
	6.35	66.30	64.78	64.78
8	9.5	63.80	60.96	57.12
12	12.7	60.40	54.94	47.96
16	14.0	58.78	52.16	43.38
	16.0	56.28	47.60	36.96

4. The Ultimate Moment Formula of Screen Pipes

A simplified calculation formula was established to simplify the calculation of the ultimate bending moment of the screen pipe under pure bending and facilitate its field application. It was assumed that the ultimate moment of the screen pipe was related to the ultimate moment of the complete casing before perforation and the hole arrangement parameters (d, l, c), which could be expressed as the following relationship:

$$U = \frac{M_b}{M_i} = f(d, l, c) \tag{1}$$

$$M_i = (1.05 - 0.0015D/t) \times \gamma \times D^2 \times t \tag{2}$$

$$c = \frac{\pi d}{N_1} \tag{3}$$

where U is the ultimate moment coefficient of the screen pipe, dimensionless, M_b the plastic ultimate bending moment the screen pipe, N·m, M_i the plastic ultimate bending moment of the complete casing, which can be calculated using the formula in the literature [34], N·m; d the diameter, mm; l the axial hole spacing, mm; c the circumferential hole spacing, mm; D the outer diameter of the screen pipe, mm; t the wall thickness, mm; and N the number of circumferential holes. γ was specified minimum yield strength.

Through fitting with the numerical simulation data, the relationship between U, d, c, and l was obtained, as shown in Figure 19. Thus, the dimensionless parameters d/c and d/l were introduced, and Equation (1) is expressed as:

$$U = \frac{M_b}{M_i} = F\left(\frac{d}{l}, \frac{d}{c}\right) \tag{4}$$

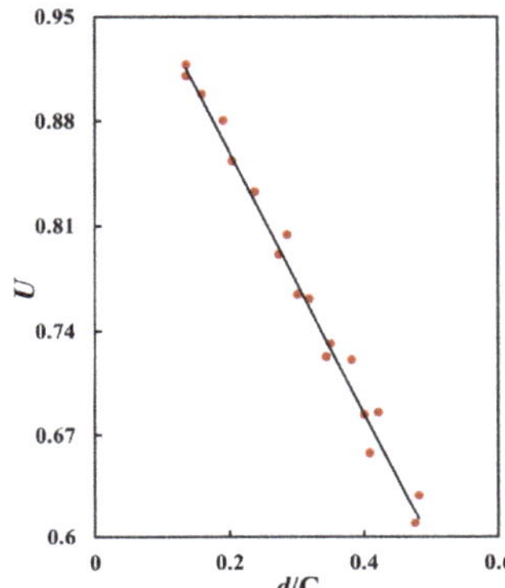
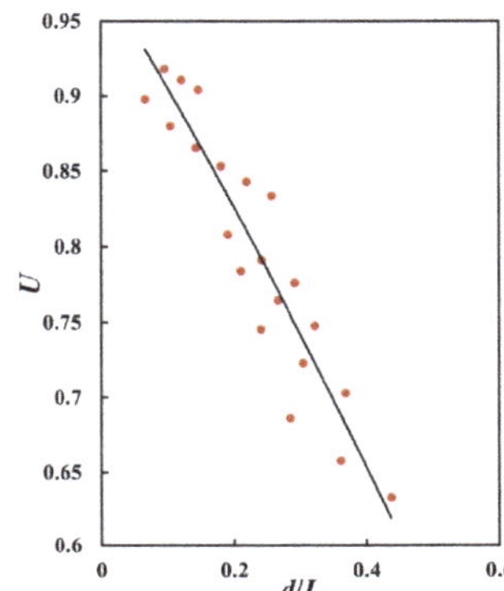

Figure 19. The relationship between U and d, l, c.

Assuming that the F function in Equation (4) could be expanded into a power series, Equation (4) could be expressed as:

$$U = \frac{M_b}{M_i} = \sum_{i=0}^{\infty} B_i \left[\left(\frac{d}{l}\right)^{n1} \left(\frac{d}{c}\right)^{n2}\right]^i \tag{5}$$

Considering that the ultimate moment of the screen pipe was less than that of the casing before perforation, $U \le 1$ when $n = 0$ and $B_0 = 1$. Ignoring the influence of higher-order terms in Equation (5) [35–37], it was approximately expressed as:

$$U = \frac{M_b}{M_i} \approx 1 - B_1\left(\frac{d}{l}\right)^{n1} \left(\frac{d}{c}\right)^{n2} \tag{6}$$

The numerical simulation results were used to fit Equation (6) and determine the parameter values in the formula. It was given by:

$$U = \frac{M_b}{M_i} = 1 - 1.32\left(\frac{d}{l}\right)^{0.175} \left(\frac{d}{c}\right)^{1.25} \tag{7}$$

Figure 20 shows the comparison between the numerical simulation results and the calculation results obtained using Equation (7). As shown in Figure 20, the difference between the FEM results and formula calculation results had a range of -0.04–4.79%, and the average error is 1.26%. Therefore, the established formulas can accurately reflect the numerical results.

Due to the lack of literature results on the bending test of screen pipe, the eight groups of screen pipe finite element results shown in Table 7 are used to verify the accuracy of the established formulas. Those results were outside the fitting data of the formulas. As shown in Table 7, the minimum difference between the numerical simulation and the formula results was 0.02%, and the maximum error was 6.16%. The average error for the FEM results of the five groups of screen pipes was 3.25%. Therefore, the established empirical formula can accurately calculate the ultimate moment of screen pipe under bending load.

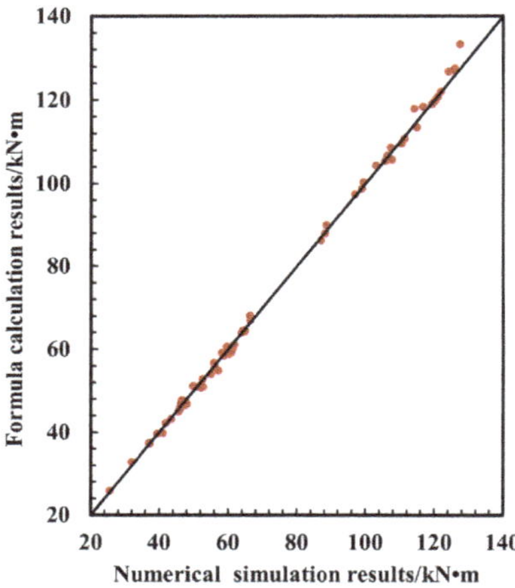

Figure 20. Comparison of numerical simulation and formula calculation results.

Table 7. Ultimate moment comparison between numerical and formula results.

CASE	D/mm	t/mm	N_1	N_2	L/mm	d/mm	$M_{\text{cr-FEM}}$/kNm	$M_{\text{cr-calculate}}$/kNm	Error
CASE-1	127.0	7.52	12	20	52.63	14.00	36.28	37.17	2.46%
CASE-2	177.8	9.19	8	12	90.91	9.50	124.64	127.50	2.30%
CASE-3	152.6	6.46	12	20	52.63	12.70	54.90	53.40	−2.73%
CASE-4	127.0	7.52	4	12	90.91	9.50	52.04	54.95	5.58%
CASE-5	177.8	9.19	12	20	52.63	16.00	98.76	98.78	0.02%
CASE-6	152.6	6.46	16	16	66.67	6.35	64.78	61.90	−4.45%
CASE-7	152.6	6.46	16	16	66.67	19.00	26.68	28.32	6.16%
CASE-8	127.0	7.52	12	20	52.63	19.00	25.42	26.00	2.28%

5. Conclusions

The finite element model of a screen pipe with a parallel hole arrangement under pure bending was defined based on ABAQUS and Python script. The deformation patterns and ultimate moment of screen pipes with different parameters were analyzed. Three deformation patterns were identified and discussed in detail. An empirical formula of the ultimate moment of screen pipes was proposed. The main conclusions are summarized as follows:

(1) The deformation patterns of screen pipes could be divided into three categories under a bending load: In Pattern-1, the screen pipes buckled from the middle position, and the area around the holes did not fracture. In Pattern-2, the screen pipes first buckled from the middle position. As the bending load increased, plastic deformation appeared around the holes, and fracture occurred. In Pattern-3, plastic deformation appeared around the hole area, and fracture occurred. However, the deformation around the middle of the screen pipes was small.

(2) The proposed empirical formula could accurately calculate the ultimate moment of screen pipe under bending load, and the average difference between the empirical formula and numerical simulation results was 3.25%.

(3) With an increase in the diameter and number of circumferential and axial holes, the ultimate moment of the screen pipe gradually decreased, and the circumferential holes had a more significant effect on the ultimate moment than the axial holes.

Author Contributions: Numerical simulation model, Y.P. and G.F.; Python Script, Y.P.; Formal analysis, B.S. and S.F.E.; investigation, X.S.; data curation, J.C.; writing, Y.P.; review and editing, G.F.; supervision, B.S. and S.F.E.; funding acquisition, B.S. and S.F.E. All authors have read and agreed to the published version of the manuscript.

Funding: This work was supported by the National Natural Science Foundation of China [51890914, 51709269, 52104056], the National Ministry of Industry and Information Technology Innovation Special Project-Engineering Demonstration Application of Subsea Oil and Gas Production System-Subject 4: Research on Subsea Christmas Tree and Wellhead Offshore Testing Technology [MC-201901-S01-04]. The last author, Estefen, acknowledges the support from the Brazilian Research Council—CNPQ [303182/2022-9].

Institutional Review Board Statement: Not applicable.

Informed Consent Statement: Not applicable.

Data Availability Statement: Not applicable.

Conflicts of Interest: The authors declare no conflict of interest.

References

1. Ma, C.; Deng, J.; Dong, X. A new laboratory protocol to study the plugging and sand control performance of sand control screens. *J. Petrol. Sci. Eng.* **2020**, *184*, 106548. [CrossRef]
2. Dong, C.; Gao, K.; Dong, S. A new integrated method for comprehensive performance of mechanical sand control screens testing and evaluation. *J. Petrol. Sci. Eng.* **2017**, *158*, 775–783. [CrossRef]
3. Galkin, V.; Martyushev, D.; Ponomareva, I. Developing features of the near-bottom hole zones in productive formations at fields with high gas saturation of formation oil. *J. Min. Inst.* **2021**, *249*, 386–392. [CrossRef]
4. Ranjith, P.; Perera, M.; Perera, W. Effective parameters for sand production in unconsolidated formations: An experimental study. *J. Petrol. Sci. Eng.* **2013**, *105*, 34–42. [CrossRef]
5. Mahmud, H.; Leong, H.; Lestariono, Y. Sand production: A smart control framework for risk mitigation. *Petroleum* **2020**, *6*, 1–13. [CrossRef]
6. Wang, D.-B.; Zhou, F.-J.; Li, Y.-P. Numerical simulation of fracture propagation in Russia carbonate reservoirs during refracturing. *Petrol. Sci.* **2022**, *19*, 2781–2795. [CrossRef]
7. Kuriyama, Y.; Tsukano, Y.; Mimaki, T. Effect of wear and bending on casing collapse strength. In Proceedings of the SPE Annual Technical Conference and Exhibition, Washington, DC, USA, 4–7 October 1992; OnePetro: Richardson, TX, USA, 1992.
8. Zhang, J.; Yin, G.; Fan, Y. The helical buckling and extended reach limit of coiled tubing with initial bending curvature in horizontal wellbores. *J. Petrol. Sci. Eng.* **2021**, *200*, 108398. [CrossRef]
9. Wang, J.D.; Dou, Y.; Cao, Y. Collapsing Strength Analysis of Bending Part Casing in Horizontal Wells under Three-point Bending. In *Applied Mechanics and Materials*; Trans Tech Publications Ltd.: Stafa-Zurich, Switzerland, 2015; Volume 713, pp. 57–60.
10. Wu, J. Drill-pipe bending and fatigue in rotary drilling of horizontal wells. In Proceedings of the SPE Eastern Regional Meeting, Columbus, OH, USA, 23–25 October 1996.
11. Chen, Y.; Zhang, J.; Zhang, H. Ultimate bending capacity of strain hardening steel pipes. *China Ocean Eng.* **2016**, *30*, 231–241. [CrossRef]
12. Sadowski, A.; Rotter, J. Solid or shell finite elements to model thick cylindrical tubes and shells under global bending. *Int. J. Mech. Sci.* **2013**, *74*, 143–153. [CrossRef]
13. Chen, Y.; Zhang, H.; Zhang, J. Residual bending capacity for pipelines with corrosion defects. *J. Loss. Prevent. Proc.* **2014**, *32*, 70–77. [CrossRef]
14. Kshitij, K.; Gerasimidis, S. Instability of thin steel cylindrical shells under bending. *Thin-Walled Struct.* **2019**, *137*, 151–166.
15. Erling, Ø.; Jayadevan, K.; Thaulow, C. Fracture response of pipelines subject to large plastic deformation under bending. *Int. J. Press. Vessel. Pip.* **2005**, *82*, 201–215.
16. Jayadevan, K.; Erling, Ø.; Thaulow, C. Fracture response of pipelines subjected to large plastic deformation under tension. *Int. J. Press. Vessel. Pip.* **2004**, *81*, 771–783. [CrossRef]
17. Chegeni, B.; Sachith, J.; Sreekanta, D. Effect of corrosion on thin-walled pipes under combined internal pressure and bending. *Thin-Walled Struct.* **2019**, *143*, 106211–106218. [CrossRef]
18. Hieu, C.; Le, T.; Bui, N. An empirical model for bending capacity of defected pipe combined with axial load. *Int. J. Press. Vessel. Pip.* **2021**, *191*, 104368.
19. Es, S.; Gresnigt, A.; Vasilikis, D. Ultimate bending capacity of spiral-welded steel tubes—Part I: Experiments. *Thin-Walled Struct.* **2016**, *102*, 286–304.
20. Es, S.; Gresnigt, A.; Vasilikis, D. Ultimate bending capacity of spiral-welded steel tubes—Part II: Predictions. *Thin-Walled Struct.* **2016**, *102*, 305–319.
21. Limam, A.; Lee, L.; Corona, E. Inelastic wrinkling and collapse of tubes under combined bending and internal pressure. *Int. J. Mech. Sci.* **2010**, *52*, 637–647. [CrossRef]
22. Limam, A.; Lee, L.; Kyriakides, S. On the collapse of dented tubes under combined bending and internal pressure. *Int. J. Mech. Sci.* **2012**, *55*, 1–12. [CrossRef]
23. Fu, G.; Peng, Y.; Sun, B. Collapse pressure calculation of sand control screen tube under combined external pressure and bending. *J. China Univ. Pet. (Ed. Nat. Sci.)* **2021**, *45*, 78–86. (In Chinese)

24. Peng, Y. Research on the Influence of Hole Distribution Parameters on the Strength Performance of Sand Control Screen. Master's Thesis, China University of Petroleum, Beijing, China, 2020. (In Chinese).
25. Kyriakides, S.; Corona, E. Localization and propagation of curvature under pure bending in steel tubes with Lüders bands. *Int. J. Solids Struct.* **2008**, *45*, 3074–3087. [CrossRef]
26. Karampour, H.; Albermani, F. Experimental and numerical investigations of buckle interaction in subsea pipelines. *Eng. Struct.* **2014**, *66*, 81–88. [CrossRef]
27. Karampour, H. Effect of proximity of imperfections on buckle interaction in deep subsea pipelines. *Mar. Struct.* **2018**, *59*, 444–457. [CrossRef]
28. Arjomandi, K.; Taheri, F. Bending capacity of sandwich pipes. *Ocean Eng.* **2012**, *48*, 17–31. [CrossRef]
29. Binazir, A.; Karampour, H.; Sadowski, A.J.; Gilbert, B.P. Pure bending of pipe-in-pipe systems. *Thin-Walled Struct.* **2019**, *145*, 106381. [CrossRef]
30. Lyu, Y.; Li, G.Q.; Cao, K. Bending behavior of splice connection for corner-supported steel modular buildings. *Eng. Struct.* **2022**, *250*, 113460. [CrossRef]
31. Yong, B.; Tang, J.; Xu, W. Collapse of reinforced thermoplastic pipe (RTP) under combined external pressure and bending moment. *Ocean Eng.* **2015**, *94*, 10–18.
32. Luo, L.; Zhao, M. The Secondary Development of ABAQUS by using Python and the Application of the Advanced GA. *Phys. Procedia* **2011**, *22*, 68–73. [CrossRef]
33. He, T.; Duan, M.; An, C. Prediction of the collapse pressure for thick-walled pipes under external pressure. *Appl. Ocean Res.* **2014**, *47*, 199–203. [CrossRef]
34. SØren, R.; Bai, Y. Bending Moment Capacity of Pipes. *J. Offshore Mech. Arct. Eng.* **2000**, *4*, 243–252.
35. Kyriakides, S.; Lee, L. On the arresting efficiency of slip-on buckle arrestors for offshore pipelines. *Int. J. Mech. Sci.* **2005**, *7*, 1035–1055.
36. Fu, G.; Li, M.; Yang, J. A simplified equation for the collapse pressure of sandwich pipes with different core materials. *Ocean Eng.* **2022**, *254*, 111292. [CrossRef]
37. Li, G.; Gong, S. Buckle propagation of pipe-in-pipe systems under external pressure. *Eng. Struct.* **2015**, *84*, 207–222.

Journal of
Marine Science and Engineering

Article

CFD Investigation on Secondary Flow Characteristics in Double-Curved Subsea Pipelines with Different Spatial Structures

Fenghui Han [1,2], Yuxiang Liu [1], Qingyuan Lan [1], Wenhua Li [1,2] and Zhe Wang [1,2,*]

1   Marine Engineering College, Dalian Maritime University, Dalian 116026, China
2   National Center for International Research of Subsea Engineering Technology and Equipment, Dalian Maritime University, Dalian 116026, China
*   Correspondence: wang.zhe@dlmu.edu.cn

Citation: Han, F.; Liu, Y.; Lan, Q.; Li, W.; Wang, Z. CFD Investigation on Secondary Flow Characteristics in Double-Curved Subsea Pipelines with Different Spatial Structures. *J. Mar. Sci. Eng.* **2022**, *10*, 1264. https://doi.org/10.3390/jmse10091264

Academic Editors: Baiqiao Chen and Carlos Guedes Soares

Received: 7 August 2022
Accepted: 2 September 2022
Published: 7 September 2022

Publisher's Note: MDPI stays neutral with regard to jurisdictional claims in published maps and institutional affiliations.

Abstract: Double-curved pipes are widely employed as essential components of subsea pipeline systems. Considering the layout flexibility and application diversity, there are various spatial structures for the double-curved combinations. However, few studies have compared the flow characteristics in different double-curved pipes. The dissipations of the corresponding downstream flow have not been thoroughly investigated, which are crucial for the measurement accuracy and flow assurance. In this paper, the turbulent flow in double-curved pipes with different spatial structures (i.e., Z-, U-, and spatial Z- type) was numerically studied by employing the ω-Reynolds stress model. The major purpose was to develop an in-depth knowledge on the secondary flow characteristics in different double-curved pipes and quantify the dissipations of the downstream flow. The effects of the spatial angle and interval distance of the two curves on the flow fields are taken into consideration, and the swirl intensity S_i is introduced to evaluate the secondary flow dissipation. It is found that the secondary flows in the Z- and U-type structures are in opposite directions when the interval distance is short ($3D$), and the secondary flow in the spatial Z-type exhibits an oblique symmetric form. Only in the Z-type pipe with a short interval distance the secondary flow exhibits an exponential dissipation, and the fully developed flow is easier to achieve than the other cases. However, as the interval distance increases, the directions of the secondary flow in the U- and Z-type structures are the same, and the flow dissipations in all the structures return to the exponential types. The obtained dissipation rates for the secondary flow downstream of Z-, U-, and spatial Z-pipes with the $9D$ interval distance were 0.40, 0.25, and 0.20, respectively. The results are expected to guide the design of pipeline layouts and provide a reference for the arrangements of flowmeters in a complex subsea pipeline system.

Keywords: subsea pipeline; double-curved pipe; secondary flow; swirl intensity

1. Introduction

The subsea pipeline is the main part of the transportation of oil and gas products in the underwater production system and gained much attention [1,2]. To ensure the safety of energy transportation, many researchers have focused on corrosion [3], residual stress [4], buckling [5], and blocking [6] of the straight subsea pipelines. Moreover, the pipelines are not always straight in the underwater production system. To meet the flexibility of the spatial layout, they are designed as multicurved structures in which the flow fields are more complex than those in straight pipes.

In all curved structures, the 90° bends are the most basic components and widely employed in the subsea pipeline system. Different from the straight pipe, the curvatures of the bends cause the generation of secondary flow superimposed on the main flow. When passing through a 90° bend, the fluid with a high speed near the outer corner turns to the inner corner along the pipe wall, while the fluid with a low speed near the inner corner flows toward the outer corner as a result of the combined effect of the wall pressure gradient and centrifugal force. This fluid motion is firstly investigated in curved pipes by Dean and Chapman [7]. Therefore, the induced vortices are called as Dean vortices. In the decades since, the Dean motion in a curved pipe has been widely studied. The bifurcation of Dean vortices from a single pair to multipairs induced by the flow conditions and pipe geometries has been numerically discussed by Nandakumar and Masliyah [8], Yang et al. [9], and Yanase et al. [10]. Bovendeerd et al. [11] carried out an experiment to measure the secondary flow field in a bend at a Reynolds number of 700. Moreover, Sudo et al. [12,13] experimentally investigated the turbulent flow through the bent channels with square and circle sections. Jurga et al. [14] employed the Explicit Algebraic Reynolds Stress Model (EARSM) to investigate the turbulent flow in a bent pipe. Li et al. [15] adopted the computational fluid dynamics with discrete element method (CFD-DEM) to simulate the solid and liquid phase flow in a bent pipe and evaluated the degree of wall wear. Ning et al. [16] also used the CFD-DEM to study the solid–liquid flow through the channels with different curvature ratios. In recent years, the swirl-switching of the turbulent flow in a bent pipe has attracted increasing interest since this unsteady flow motion may cause fatigue damage to the pipelines. Hellström et al. [17] obtained that the two characteristic Strouhal numbers of the swirl-switching are 0.16 and 0.33 through a proper orthogonal decomposition (POD) method, which are in good agreement with those of Rütten et al. [18] and Kalpakli et al. [19]. Hufnagel et al. [20] conducted a direct numerical simulation (DNS) on the bent flow and concluded that the switching phenomenon is intrinsic to the bent geometry and independent of the upstream flow conditions. In addition, the complex flow physics in some other types of bending structures such as T-, plugged T-, and Y-junctions have been investigated. Sakowitz et al. [21] employed a large eddy simulation (LES) to investigate the turbulent flow mechanisms in a T-type junction. Ong et al. [22] and Han et al. [23,24] numerically investigated the laminar flow characteristics in the plugged T-junctions and reported the effect of structural parameters of the structures. Hu et al. [25] conducted a numerical and experimental study on the motion of particles in the Y-type bend and revealed that the particle transport is strongly affected by the secondary flow.

Since the pipeline system consists of straight, curved, and multicurved pipes, many researchers conducted studies on the flow in some kinds of bend combinations. Fiedler [26] conducted an experiment on the flow in double-curved pipes where the second bend is perpendicular to the first one and explained the asymmetries of the velocity profiles in the second bend. Mazhar et al. [27] carried out an experiment on the turbulent flow in S-shape 90° bends and found the higher turbulence kinetic energy near the downstream bend. In terms of the U-type bend, Sudo et al. [28] conducted an experiment to measure the flow field in the 180° bend. Moreover, some recent efforts have focused on the flow behaviors in double- [29] and triple-curved [30] pipes when the Reynolds number exceeds 10^7; the researchers attributed the flow-accelerated corrosion to the unsteady motions of the secondary flows in the bends. In terms of the multicurved structures, Liu et al. [31] numerically studied the flow characteristics and mixing conditions along the M-type jumper tubes with plugged T-junctions. Kim and Srinil [32] numerically studied the slug flows in the subsea M-type jumpers and evaluated the deformation and stress of the pipe.

In terms of the flow in multicurved pipes, the researchers generally focused on the specific curved structures such as Z-, S-, and U-type bend combinations. However, the effects of different spatial structures of the double-curved pipes on the flow behaviors and secondary flow characteristics have not been thoroughly investigated. Moreover, the secondary flow dissipation at the downstream of the double-curved pipes with various spatial structures has not been taken into consideration so far. Mattingly and Yeh [33] pointed out that the elbow-produced secondary flow will influence the flowmeter measurement accuracy. Research based on the elbow-produced secondary flow characteristics will guide the location of the flowmeter in the subsea pipeline system. Hence, in the present study, the turbulent flow in double-curved pipes with different spatial structures (i.e., Z-, U-, and spatial Z- type) was numerically studied using the Reynolds stress model based on the

ω-equation (ω-RSM) to consider the anisotropy of turbulence. In addition, the RSM based on the ω-equation can provide a more accurate near-wall treatment, which was proved by Di Piazza and Ciofalo [34] to predict the satisfactory flow field in a coiled tube. The effects of the spatial angle and interval distance on secondary flow fields were thoroughly analyzed. The development of the velocity distributions, generations of the Dean vortices, and the dissipation of the swirl intensity were discussed in detail. The present study was intended to reveal the turbulent flow behaviors in the double-curved subsea pipelines and clarify the influence of spatial structures on secondary flow fields. The results are expected to guide the design of the subsea pipelines and provide a reference for the location of flowmeters.

2. Methodology

2.1. Governing Equations

In this study, the steady Reynolds–average Navier–Stokes (RANS) equations were solved, which can be described as

$$\partial U_i / \partial x_i = 0 \tag{1}$$

$$\partial(U_i U_j) / \partial x_j = \frac{1}{\rho}\partial p / \partial x_i + \frac{\partial}{\partial x_j}\left(v \cdot \frac{\partial U_i}{\partial x_j} - \overline{u_i' u_j'}\right) \tag{2}$$

where x_i and x_j (i, j = 1, 2, and 3) represent the three directions of the coordinate system, respectively; U_i and U_j represent the corresponding time average velocity component; p represents the time average pressure; and ρ and v represent the density and kinematic viscosity of the fluid, respectively. $\overline{u_i' u_j'}$ is the Reynolds stress tensor, which is the time average value of the product of the fluctuations of the velocity component. The ω-Reynolds stress model (ω-RSM) is employed to solve the RANS equations, which avoids using the Boussinesq assumption employed in the eddy-viscosity model. It has been reported that the RSM is expected to capture more exact flow details when the secondary flows are induced by the curvatures [35]. The RSM directly resolves the transport equations of the Reynolds stress, which can be described as (ignoring the buoyancy)

$$\frac{\partial}{\partial t}\left(\rho\overline{u_i' u_j'}\right) + \frac{\partial}{\partial x_k}\left(\rho u_k \overline{u_i' u_j'}\right)$$
$$= -\frac{\partial}{\partial x_k}\left(\rho\overline{u_i' u_j' u_k'} + \overline{p'\left(\delta_{kj}u_i' + \delta_{ik}u_j'\right)}\right) + \overline{p'\left(\frac{\partial u_i'}{\partial x_j} + \frac{\partial u_j'}{\partial x_i}\right)} - 2\mu\overline{\frac{\partial u_i'}{\partial x_k}\frac{\partial u_j'}{\partial x_k}} + \frac{\partial}{\partial x_k}\left(\mu\frac{\partial}{\partial x_k}\left(\overline{u_i' u_j'}\right)\right) \tag{3}$$
$$-\rho\left(\overline{u_i' u_k'}\frac{\partial u_j}{\partial x_k} + \overline{u_j' u_k'}\frac{\partial u_i}{\partial x_k}\right)$$

where the right-hand side of the equation includes the terms of turbulent diffusion, pressure strain correlation, dissipation, molecular diffusion, and stress production, respectively.

The ω-equation is defined as

$$\frac{\partial(\rho\omega)}{\partial t} + \frac{\partial(U_k \rho\omega)}{\partial x_k} = \alpha\rho\frac{\omega}{k}P_k - \beta\rho\omega^2 + \frac{\partial}{\partial x_k}\left(\left(\mu + \frac{\mu_t}{\sigma_\omega}\right)\frac{\partial\omega}{\partial x_k}\right) \tag{4}$$

where the coefficients $\sigma_\omega = 2$, $\alpha = 5/9$ $\beta = 0.075$, $\mu_t = \rho k/\omega$, and P_k is the production rate of turbulence.

The turbulent diffusion term $D_{T,ij}$, pressure strain correlation term ϕ_{ij}, and dissipative term ε_{ij} should be modelled to close the equations. The model equations (ω-based) are as follows:

$$D_{T,ij} = \frac{\partial}{\partial x_k}\left(\frac{\rho k}{\omega\sigma_k}\frac{\partial\overline{u_i' u_j'}}{\partial x_k}\right) \tag{5}$$

$$\phi_{ij} = -C_1\rho\beta'\omega\left(\overline{u_i' u_j'} - \frac{2}{3}\delta_{ij}k\right) - \hat{a}_0\left(-\rho\overline{u_i' u_k'}\frac{\partial U_j}{\partial x_k} - \overline{u_j' u_k'}\frac{\partial U_i}{\partial x_k} - \frac{2}{3}\delta_{ij}p\right)$$
$$-\hat{\beta}_0\left(-\rho\overline{u_i' u_k'}\frac{\partial U_k}{\partial x_j} - \rho\overline{u_j' u_k'}\frac{\partial U_k}{\partial x_i} - \frac{2}{3}\delta_{ij}p\right) - \hat{\gamma}_0\rho k\left(S_{ij} - \frac{1}{3}S_{kk}\delta_{ij}\right) \tag{6}$$

$$\varepsilon_{ij} = \frac{2}{3}\delta_{ij}\rho\beta'k\omega \tag{7}$$

where the coefficients are $\beta' = 0.09$, $\hat{a}_0 = \frac{8+C}{11}$, $\hat{\beta}_0 = \frac{8C-2}{11}$, $\hat{\gamma}_0 = \frac{60C-4}{55}$, $C = 0.52$, and $C_1 = 1.8$.

ANSYS CFX is employed to solve the RANS equations by an element-based finite-volume method. The total-variation-diminishing (TVD) scheme is employed for spatial discretization, which is of second-order accuracy. Since the ω-RSM does not use wall functions, the near-wall grids were densified in this study. The details of the corresponding meshing strategy will be discussed in Section 2.3.

2.2. Computational Models

The calculation domains of the present study are displayed in Figure 1. The spatial angles between the upstream and downstream bends in three domains are $0°$, $90°$, and $180°$, which represent the Z-, spatial Z-, and U-type pipes, respectively. The pipe diameter is defined as $D = 1$ m. The upstream length is $8D$, and the downstream length is $15D$. The curvature radius of the bend is $2D$, and the interval distances between the two bends varied from $3D$ to $9D$. For simplified description, different geometries analyzed in this paper are named with the double-bend angle and interval distance. As an example, the double-curved pipe with a spatial angle of $0°$ (Z-type) and an interval distance of $3D$ (i.e., the entity in Figure 1) is named as *Case-0°-3D*.

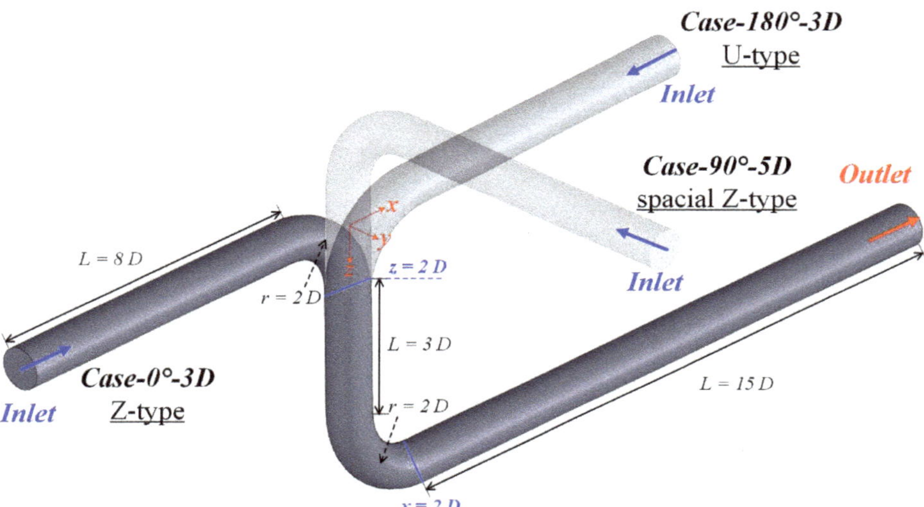

Figure 1. Computational domain of double-curved pipe.

The zero normal gradient is specified for the inlet pressure condition, and the inlet velocity condition employs a modified power law profile from Salama [36]:

$$U = \frac{U_{bulk}}{\beta(1,1+n)}\left(1-\left(\frac{r}{R}\right)^2\right)^{\frac{1}{n}} \tag{8}$$

where β is the Euler integral of the first kind, $n = 0.77\ln(Re)-3.47$, $U_{bulk} = 10$ m/s is the inlet bulk velocity, r is the distance from the location to the center of the cross section, and R is the pipe radius.

The reference pressure is zero at the outlet, where the velocity is defined as the zero normal gradient. The zero normal gradient is specified for the wall surface, and the flow velocity employs a nonslip boundary condition.

2.3. Verification and Validation Study

In this section, the mesh convergence study was firstly carried out, and then the numerical method employed in this study was validated by comparing with the published results. The verification study was conducted to obtain a suitable mesh for the current study. The flow in the Z-type double-curved pipe with the interval distance of $3D$ (i.e., the entity in Figure 1) at a Reynolds number of 10,000 is provided as an example. Three sets of structured grids for the computational domain have been generated, and the distributions of the velocity at the outlets of the upstream and downstream bends (i.e., $z = 2D$ and $x = 2D$) are shown in Figure 2. Obvious divergences can be observed between the results in mesh 1 with 757,307 elements and mesh 2 with 1,417,843 elements, while good consistency can be found between mesh 2 with 1,417,843 elements and mesh 3 with 2,316,733 elements. It is found that the mean deviations between mesh 1 and mesh 2 at the outlet of the first and second bends are 2.46% and 2.27%, respectively. However, the mean deviations between mesh 2 and mesh 3 are 0.11% and 0.23%, respectively. In addition, the max deviations between mesh 1 and mesh 2 at the outlet of the first and second bends are 11.59% and 17.34%, while the corresponding deviations between mesh 2 and mesh 3 are only 0.498% and 0.723%, respectively. Hence, mesh 2 (1,417,843 elements), which can provide sufficient numerical accuracy, was employed in the present study. The contour of y^+ and details of mesh 2 are shown in Figure 3. The average y^+ value is calculated as 0.798 ($y^+ = \Delta y \cdot u_* / v$, where u_* is the friction velocity, and Δy is the distance from the first grid to the wall).

Then, a validation study was carried out to confirm the reliability of the numerical method. The turbulent flow through a 90° bend with the curvature ratio of 2 at a Reynolds number of 60,000 has been widely investigated with experiments [12] and numerical simulations [37,38]. For validation purposes, the numerical study was carried out on the same bend structure under the same flow condition using the obtained meshing strategy and ω-RSM. Then, the outlet velocity distribution of the bend obtained by the current simulation was compared with the published results to validate the numerical method in the present study. Figure 4 shows the velocity distributions at the elbow outlet obtained by the experiment and different numerical methods. Except for the region near the inner-side wall, the numerical results are close to the experiment data. However, the velocity distribution obtained with the ω-RSM is closer to the results of LES prediction [37] than that of the RNG k-ε model [38]. It indicates that the ω-RSM can obtain a more accurate result than the RNG k-ε model in predicting the curved flow, which is consistent with the conclusion of Di Piazza and Ciofalo [34]. Furthermore, the ω-RSM with a lower computational cost can provide a similar near-wall prediction as the LES method. Therefore, comprehensively considering the advantages on the numerical accuracy and computational cost, the ω-RSM was validated and employed in the present study. It also shares the same view as Wallin and Johansson [35] and Pruvost et al. [39].

2.4. Definition of Swirl Intensity

In order to quantitatively evaluate the developments of the secondary flow and swirling strength, the swirl intensity S_i is employed in the following studies, which is introduced by Sudo et al. [12]. The swirl intensity S_i can be described as

$$S_i = \int \left[\vec{U} - \left(\vec{U} \cdot \hat{n} \right) \hat{n} \right]^2 dA / \left(U_{bulk}^2 \int dA \right) \tag{9}$$

where $\hat{n}$ represents the unit vector parallel to the flow direction, and $\vec{U}$ represents the vector of the flow velocity.

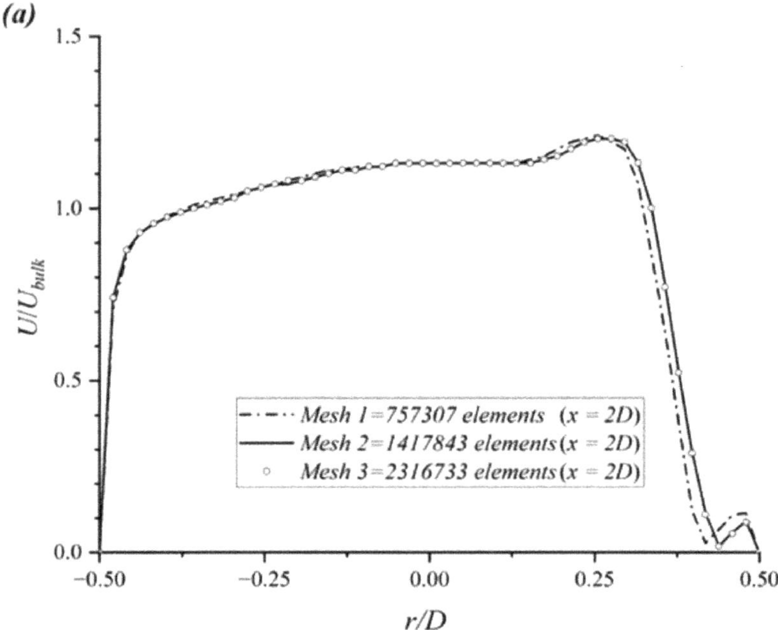

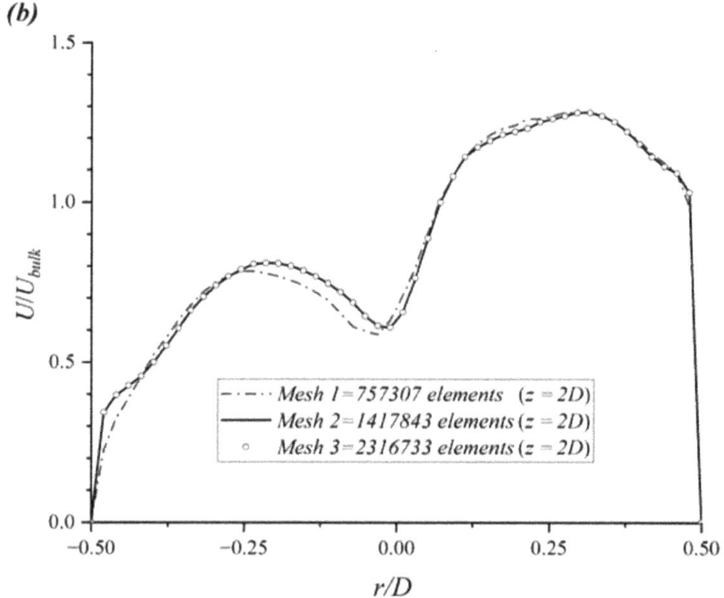

Figure 2. Velocity profiles at (**a**) outlet of upstream bend ($z = 2D$) and (**b**) outlet of downstream bend ($x = 2D$) for verification study.

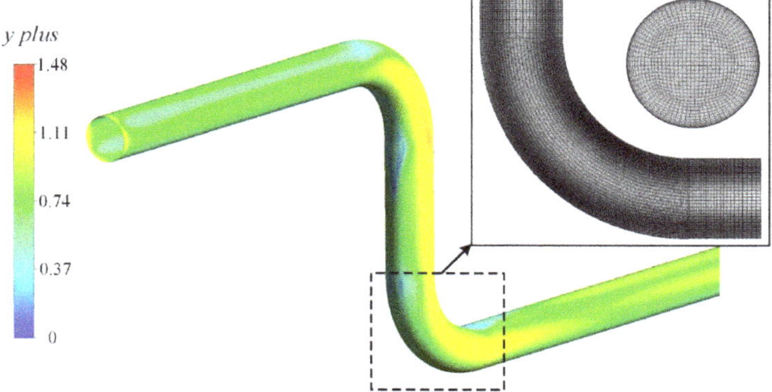

Figure 3. Contour of y^+ and details of mesh 2.

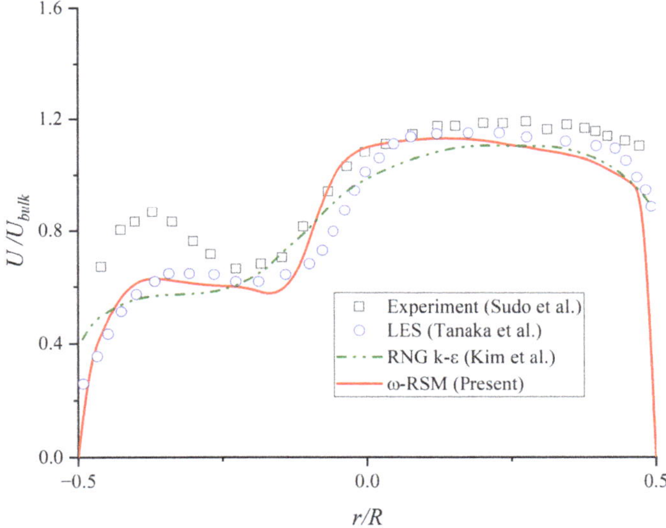

Figure 4. Velocity profiles at the outlet of elbow for validation study at Reynolds number of 60,000.

3. Results and Discussions

Since the numerical method has been validated in Section 2.3 with the studies of Sudo et al. [12], Tanaka et al. [37], and Kim et al. [38], the ω-RSM and verified meshing strategy are employed in the following sections to investigate the turbulent flow in the double-curved pipes with different spatial structures at a Reynolds number of 10,000. The effects of the spatial angle and interval distance between two bends have been discussed.

3.1. Effect of Spatial Angle

To provide an intuitive understanding on how the upstream bends at different spatial angles affect the downstream flow fields, the cross-sectional velocity fields at the outlets of the first bends, the global streamlines, and the velocity vectors downstream of the first bends in *Case-0°-3D*, *Case-90°-3D*, and *Case-180°-3D* are shown in Figure 5. The vector fields are displayed in the symmetric plane of the second bend (i.e., the *x-z* plane) in all the structures to compare the effect of the spatial angle on the flow fields in the downstream bends. For the first bend, the flow conditions in different structures are the same as shown in Figure 5a. However, the vector field in the *x-z* plane changes with the variation of the spatial angle from 0° to 180°, leading to different inlet conditions of the downstream bends.

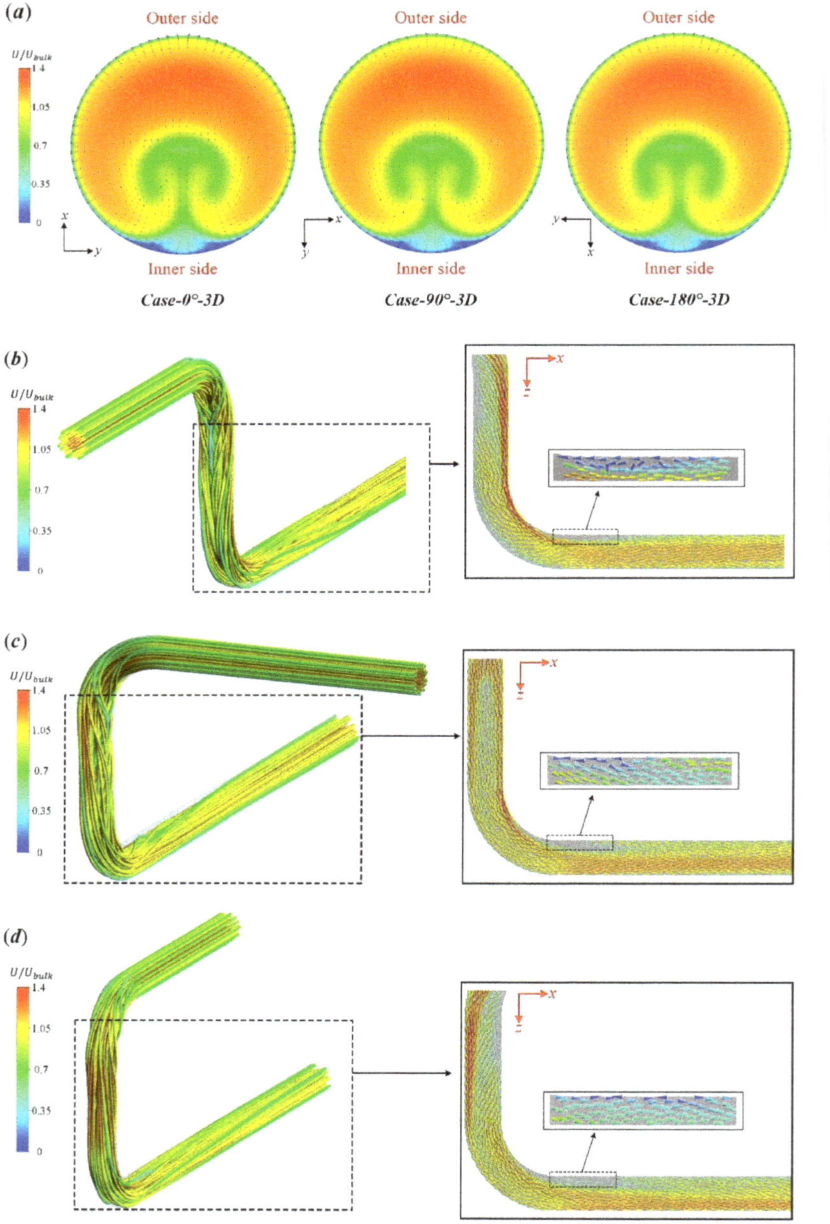

Figure 5. (**a**) Cross-sectional flow fields at the outlet of first bends. Global streamlines and local velocity vectors in (**b**) *Case-0°-3D*, (**c**) *Case-90°-3D*, and (**d**) *Case-180°-3D*.

In addition, the velocity vectors downstream of the first bends show significant differences with the spatial angles between two bends varying from 0° to 180°. A high-velocity region appears near the +*x* direction in the intermediate pipe of *Case-0°-3D*, while the ve-

gion appears near the $-x$ direction in *Case-180°-3D*. In addition, the velocity vectors in the intermediate pipe of *Case-90°-3D* display an asymmetrical distribution. The high-velocity region appears near the pipe wall, while the low-velocity region appears at the center of the cross-section. At the downstream of the second bend, the flow becomes quite uniform in *Case-0°-3D*. However, as the spatial angle increases, the velocity near the outer corner of the downstream bend increases. The result indicates that it is easier for the flow to achieve full development when the spatial angle is 0° as compared with the other cases. Since the low-velocity areas near the inner corner of the second bend (black-dotted rectangles) are hard to be observed in the whole vector fields, the velocity vectors in these areas are magnified alongside. It can be found that the velocity gradient in *Case-0°-3D* is higher than the other cases. In addition, the flow separation can be clearly observed in *Case-0°-3D*, while it greatly reduces in *Case-180°-3D*.

Figure 6 illustrates the developments of the flow distribution in the (a) first bends, (b) intermediate pipe between two bends, (c) second bends, and (d) downstream of second bends in *Case-0°-3D*, *Case-90°-3D*, and *Case-180°-3D*. The locations of the selected profiles are shown in Figure 6e. It is found that the velocity distributions in the first bends are very similar to each other, which is physically sound. In the intermediate pipe, the velocity profiles of *Case-90°-3D* display symmetric bimodal distributions since the upstream bend is perpendicular to the *x-z* plane. The velocity profiles in *Case-0°-3D* and *Case-90°-3D* show the contrary unimodal distributions due to the opposite directions of the upstream bends. It is worth noting that the divergences in Figure 6b are due to the different spatial angles and same selected coordinates. When the fluid enters the second bend, as shown in Figure 6c, the symmetries of velocity distributions are broken, and the velocity near the inner corner increases in *Case-90°-3D*. In addition, the peak velocity gradually moves toward the center of the pipe as the bending angle θ increases. A similar phenomenon can be observed in *Case-0°-3D*; however, the velocity near the inner corner is lower than that in *Case-90°-3D*. In *Case-180°-3D*, the velocity peaks appear in both inner and outer corners. More specifically, the outer-corner peak is due to the upstream flow distribution, and the inner-corner peak is induced by the bend curvature. At the outlet of the second bend, velocity fluctuations are found near the inner corner at $x/D = 2$ and 2.5 in *Case-0°-3D*, implying the flow separation area. However, the separation area can hardly be seen in *Case-90°-3D* and *Case-180°-3D*, which is consistent with the qualitative analysis in Figure 5. As the downstream distance increases, the flow in *Case-0°-3D* more rapidly restores uniformity than the other cases. High-velocity gradients can still be observed near the inner corner at $x/D = 4.5$ in *Case-0°-3D*, *Case-90°-3D*, and *Case-180°-3D*. It is concluded that, for structures with a short interval distance, the flow downstream the first bend is not fully developed before the second bend. Therefore, the downstream flow is deeply influenced by the upstream bend with different spatial structures, and the flow fields downstream of the second bend are more complex than those of the first bend.

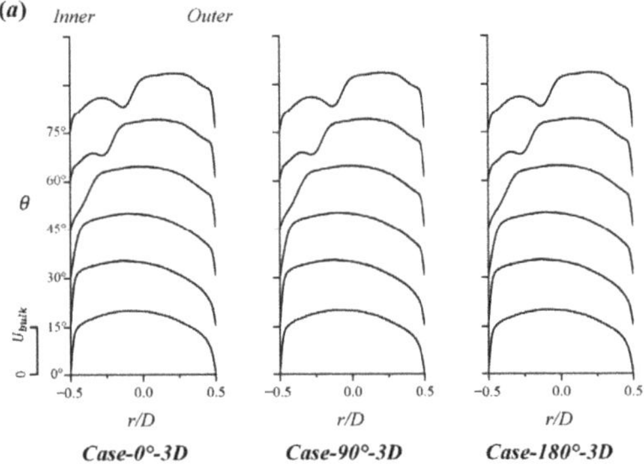

Figure 6. *Cont.*

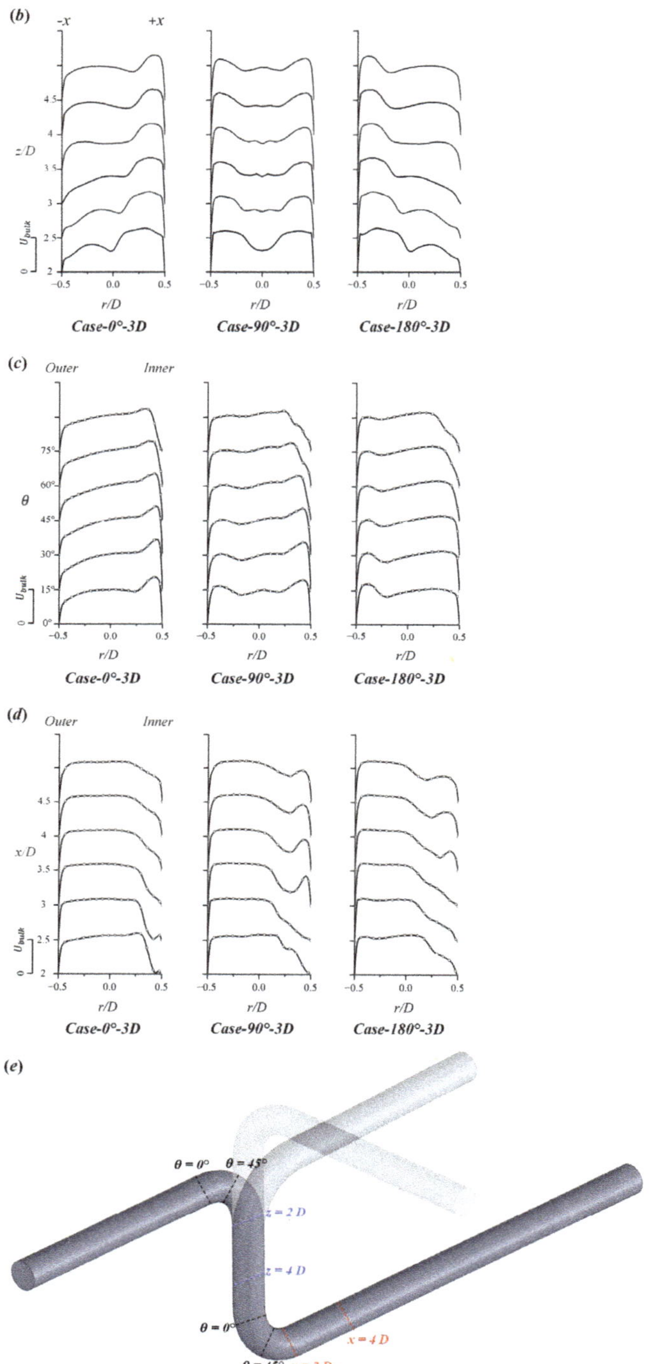

Figure 6. Developments of velocity profiles in (**a**) first bends, (**b**) intermediate pipe, (**c**) second bends, and (**d**) downstream of second bends in *Case-0°-3D*, *Case-90°-3D*, and *Case-180°-3D*, and (**e**) locations of selected profiles.

Furthermore, the effect of different spatial structures on the secondary flow characteristics downstream of the double-curved pipes was investigated. Figures 7–9 display the streamlines on the selected positions (i.e., x/D = 2, 2.5, 3, 3.5, 4, 4.5, and 5) at the downstream of *Case-0°-3D*, *Case-180°-3D*, and *Case-90°-3D*, respectively. The streamlines are colored with dimensionless tangential velocity $U_{t_x}/U_{bulk} = \sqrt{U_z^2 + U_y^2}/U_{bulk}$, where U_z and U_y represent the velocity components in the z and y directions, respectively. As shown by *Case-0°-3D* in Figure 7, the fluid near the inner corner rushes to the outer corner at a high tangential velocity owing to the influence of centrifugal force at $x = 2D$. Then, the outer-corner fluid turns back through the center line and finally forms a pair of vortices, which is called the Base vortices defined by Bhunia and Chen [40]. In addition, an extra pair of vortices can be observed near the inner corner and finally disappears in 1.5D downstream (i.e., $x = 3.5D$). This pair of vortices characterizes the same motion as the inner-corner Dean vortex described by Bhunia and Chen [40] and Dutta and Nandi [41]. However, a branching generates near the inner corner and leads to another pair of vortices near the inner-corner Dean vortex at $x = 2.5D$, which has not been reported in the single bend so far. Hence, this branching is considered to be caused by the upstream bend.

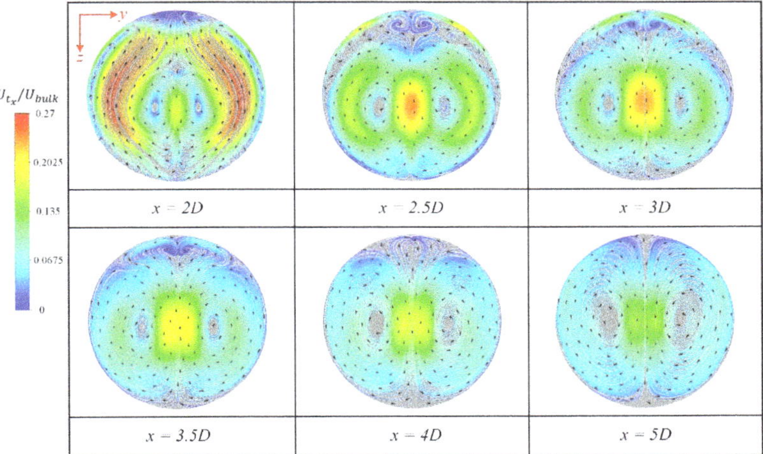

Figure 7. Streamlines and tangential velocity vectors downstream the second bend in *Case-0°-3D*.

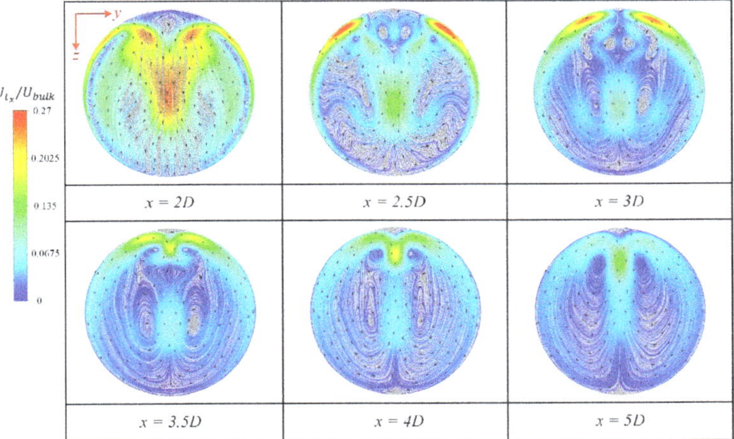

Figure 8. Streamlines and tangential velocity vectors downstream the second bend in *Case-180°-3D*.

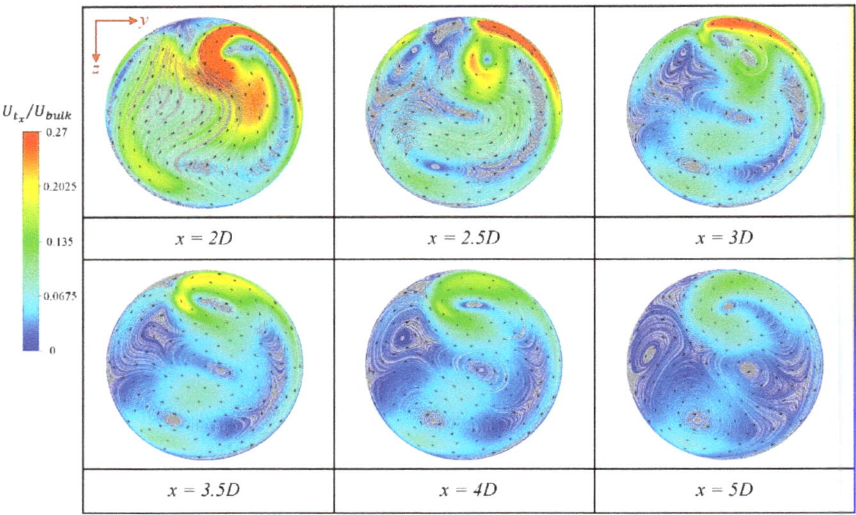

Figure 9. Streamlines and tangential velocity vectors downstream the second bend in *Case-90°-3D*.

Figure 8 displays the secondary flow fields in *Case-180°-3D*. The secondary flow motion in *Case-180°-3D* is contrary to that in *Case-0°-3D*. The fluid near the inner corner rushes to the outer corner through the center line, while the outer-corner fluid turns to the inner corner along the pipe wall on both sides. It is found that there is no vortex generation at $x = 2D$. However, as the downstream distance increases, two pairs of vortices first generate near the inner corner at $x = 2.5D$. Subsequently, the vortices near the inner corner split, and an extra pair of vortices generates near the pipe center at $x = 3D$. As the downstream distance increases from 1.5D (i.e., $x = 3.5D$) to 3D (i.e., $x = 5D$), the split vortices merge by degrees and eventually form one pair of large vortices at $x = 5D$. In terms of the velocity distribution, the high tangential velocity areas appear near the inner corner in *Case-180°-3D*, whereas they appear near the pipe center in *Case-0°-3D*. Then, as shown by the streamlines in Figure 9, the inclined secondary flow motion in *Case-90°-3D* can be found, which is induced by the vertical upstream bend. A high tangential velocity area occurs at one side of the inner corner and forms a main vortex. As the downstream distance increases, the other three vortices generate and eventually form a tilt symmetric four-vortex structure. It is found that the directions of the adjacent vortices are opposite, and the main vortex near the inner corner occupies the most tangential momentum. According to the above discussion, it is revealed that changing the spatial angle between the upstream and downstream bends in the double-curved pipe will redistribute the velocity field, reverse the flow direction, and break the central symmetry of the secondary flow at the downstream.

Subsequently, to quantify the intensities of the secondary flow in *Case-0°-3D*, *Case-90°-3D*, and *Case-180°-3D*, the swirl intensity S_i introduced in Section 2.4 is employed. Figure 10 shows the developments of the swirl strengths along *Case-0°-3D*, *Case-90°-3D*, and *Case-180°-3D*. It is worth noting that the swirl strengths are selected every 0.5D length in the straight pipe sections and every 15° from 0° to 90° in the bends. The shaded sections represent the upstream and downstream bends. It is found that S_i has slightly increased before the first bend and then increases by a wide margin in the first bend. Subsequently, S_i shows a sharp decline at the intermediate pipe after outflowing from the first bend. The developments of S_i in *Case-0°-3D*, *Case-90°-3D*, and *Case-180°-3D* are quite similar in the above process. However, significant differences can be observed inside and downstream of the second bends. S_i increases in the second bend of *Case-0°-3D* and *Case-90°-3D*, whereas it decreases in the second bend of *Case-180°-3D*. The phenomenon indicates that the effect of the downstream bend on S_i is strongly related to the spatial structure of the bend. At the outlet sections of the double-curved pipes, the dissipation rate of S_i in *Case-90°-3D* is larger than that in *Case-0°-3D*, although their initial values are almost the same. In addition, the dissipation rate of S_i in *Case-180°-3D* is the highest.

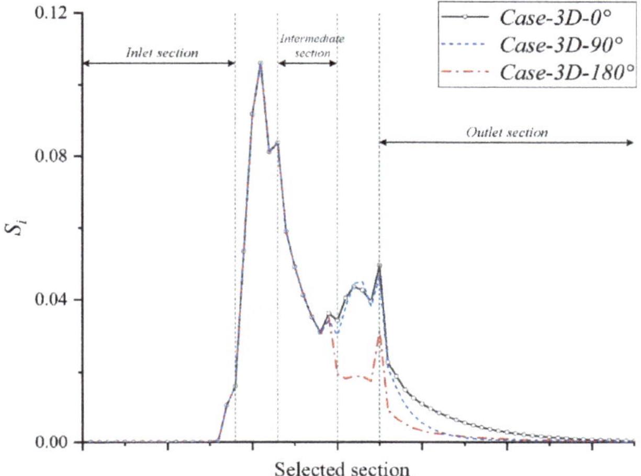

Figure 10. Swirl strengths S_i along the pipes in *Case-0°-3D*, *Case-90°-3D*, and *Case-180°-3D*.

In order to further analyze the dissipation trends of the swirl intensities, the values of *Case-0°-3D*, *Case-90°-3D*, and *Case-180°-3D* obtained with Equation (9) are displayed in Figure 11 with a logarithmic ordinate and compared with the result of a traditional single bend reported by Kim et al. [38]. The dissipation of swirl intensity S_i in the single bend is an exponentially decreasing function given as follows [38]:

$$S_i = S_{i_0} \cdot e^{-\beta_s L_s / D} \tag{10}$$

where S_{i_0} represents the swirl intensity at the bend outlet, β_s represents the dissipation rate, L_s represents the downstream distance from the outlet of the second bend, and D represents the diameter of the pipe.

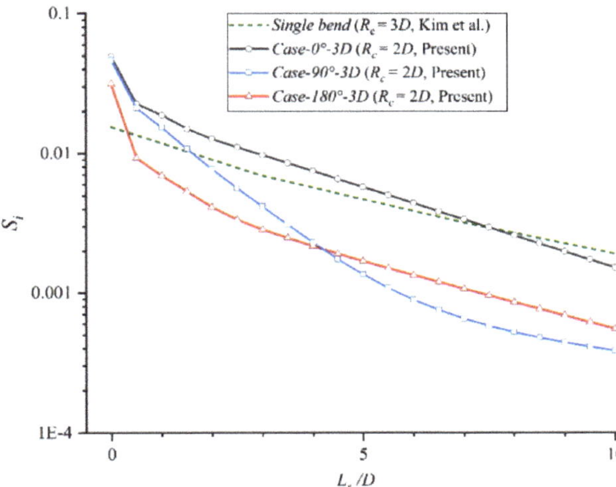

Figure 11. Dissipations of swirl intensity downstream a single bend and second bends in *Case-0°-3D*, *Case-90°-3D*, and *Case-180°-3D*.

The value of β_s is reported as 0.21, and the dissipation is weakly related to the Reynolds number [38,42]. However, as shown in Figure 11, the dissipations of swirl intensities are not exactly exponential in double-curved pipes. Only the swirl intensity in *Case-0°-3D* can be considered to express as exponential dissipation (exclude the initial value), and the β_s is fitted as 0.30. In addition, it is found that the swirl intensity of *Case-90°-3D* is stronger than that of *Case-180°-3D* when $L_s/D \leq 4$ and becomes weaker when $L_s/D > 4$. Hence, the different spatial structures of the double-curved pipes can greatly influence the secondary-flow strength and break the exponential dissipation at the downstream of the pipes. Only the downstream flow in *Case-0°-3D* approximately conforms the exponential dissipation.

3.2. Effect of Interval Distance

In this section, the interval distance is increased to 9D, and the corresponding velocity profiles, secondary flow characteristics, and dissipations of the swirl intensity have been investigated and compared with the results in *Case-3D*. Figure 12 shows the comparisons of velocity profiles at the inlets and outlets of the downstream bends in (a) *Case-0°-3D* and *Case-0°-9D*; (b) *Case-90°-3D* and *Case-90°-9D*; and (c) *Case-180°-3D* and *Case-180°-9D*. As the interval distance increases, the velocity profiles at the inlets of the downstream bends become gentler, which is physically sound. The high-velocity areas near the outer corner become smaller, and the velocities decrease near the center lines in all configurations when the interval distances increase to 9D. The velocity profiles at the bend outlet in the double-curved pipe approach to the single-bend distributions when the interval distance increases. In addition, the separation area in *Case-0°-3D* disappears when the distance between the two bends increases to 9D. It implies that the flow before the second bend starts to be developed with the increase in the interval distance, resulting in a more stable flow condition near the inner corner of the second bend.

To further compare the secondary flow characteristics in the double-curved pipe with interval distances of 3D and 9D, the streamlines at the outlets of the downstream bends (i.e., $x = 2D$) in *Case-0°-9D*, *Case-90°-9D*, and *Case-180°-9D* are displayed in Figure 13 and compared with the streamlines of *Case-3D* in Figure 9. The downstream tangential velocity in *Case-9D* is higher than that in *Case-3D*, indicating that the too short interval distance (3D) will limit the tangential momentum exchange in the double-curved pipe. Moreover, the high-velocity area appears near the pipe center in *Case-0°-3D* and forms a pair of vortices, while the area is near the inner corner of the pipe in *Case-0°-9D*. A similar phenomenon can be observed in *Case-180°-3D* and *Case-180°-9D*. It has been mentioned that the secondary flow motions are greatly influenced by the spatial angle of the upstream bend when the interval distance is short (3D). However, when the interval distance increases to 9D, the fluid rushes to the outer corner from the center line and turns back from the pipe wall in all configurations, showing similar secondary flow motions. In addition, the vortex direction of the secondary flow in *Case-0°-9D* is opposite to that in *Case-0°-3D*. Hence, it can be concluded that increasing the interval distance of the two bends will weaken the effect of the upstream bend and lead to contrary secondary flow motions.

Figure 14 shows the dimensionless vortices iso-surfaces by the Q-criterion [43] in *Case-0°-3D*, *Case-90°-3D*, *Case-180°-3D*, *Case-0°-9D*, *Case-90°-9D*, and *Case18-0°-9D* to discuss the effect of the interval distance on the vortex structures in double-curved pipes. The Q-criterion can be defined as $Q = \left(\Omega^2 - S^2\right)/2$, where Ω represents the rotation tensor, and S represents the strain tensor. The dimensionless value is $Q^* = QD^2/U_{bulk}^2 = 0.5$, which is colored by the velocity. In addition, the criterion contours of the x-components Q_x^* at $x = 4D$ are printed alongside. It can be clearly observed that the pair of vortex cores moves toward the inner corner at the downstream of the pipe in *Case-0°* when the interval distance increases. In *Case-90°*, the vortices are inclined since the angle between the two bends are perpendicular. In terms of the vortices in *Case-180°*, additional sweeping structures generate when the interval distance increases. It is found that the distributions of downstream vortices are various in different structures when the interval distance is 3D. However, the vortices are close to the inner sides in all structures when the interval distance increases to 9*D*. In addition, as the interval distance increases, more vortices generate at the downstream of the pipes. The above phenomenon implies that a short interval distance limits the generation of the vortex at the downstream of the pipe. Increasing the interval distance will lead to similar vortex structures in the double-curved pipe.

(*a*)

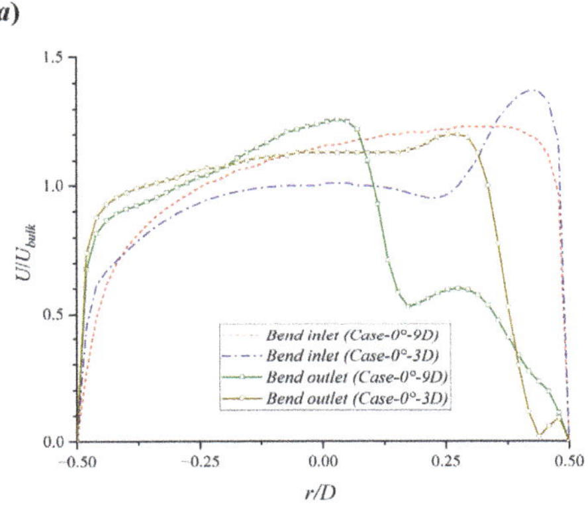

(*b*)

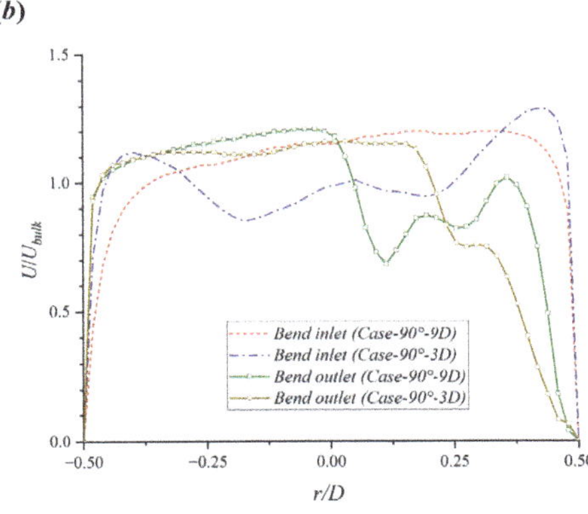

(*c*)

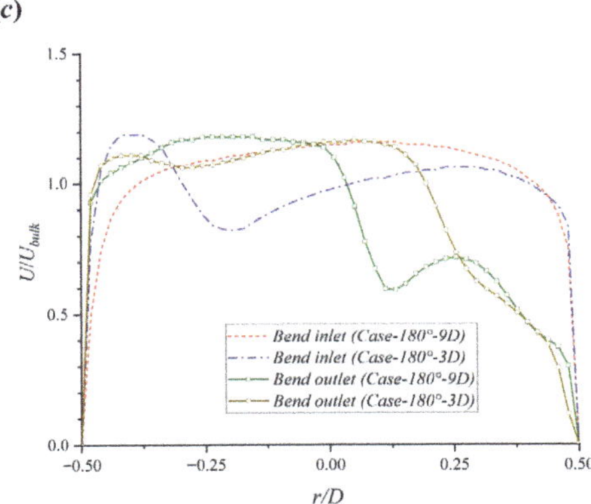

Figure 12. Velocity profiles at inlets and outlets of downstream bends in (**a**) *Case-0°-3D* and *Case-0°-9D*; (**b**) *Case-90°-3D* and *Case-90°-9D*; and (**c**) *Case-180°-3D* and *Case-180°-9D*.

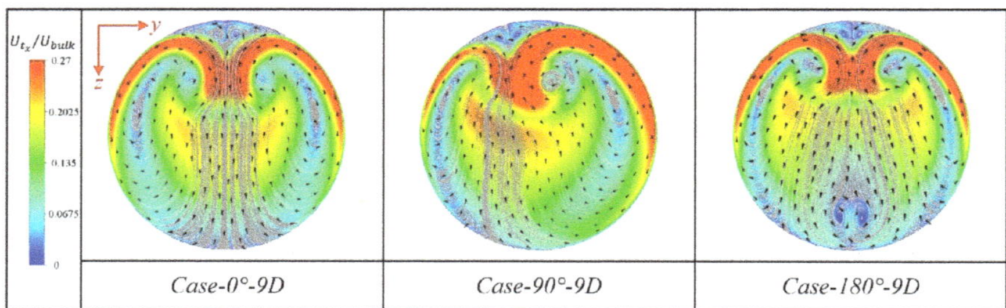

Figure 13. Streamlines and tangential velocity vectors at outlets of downstream bends in *Case-0°-9D*, *Case-90°-9D*, and *Case-180°-9D*.

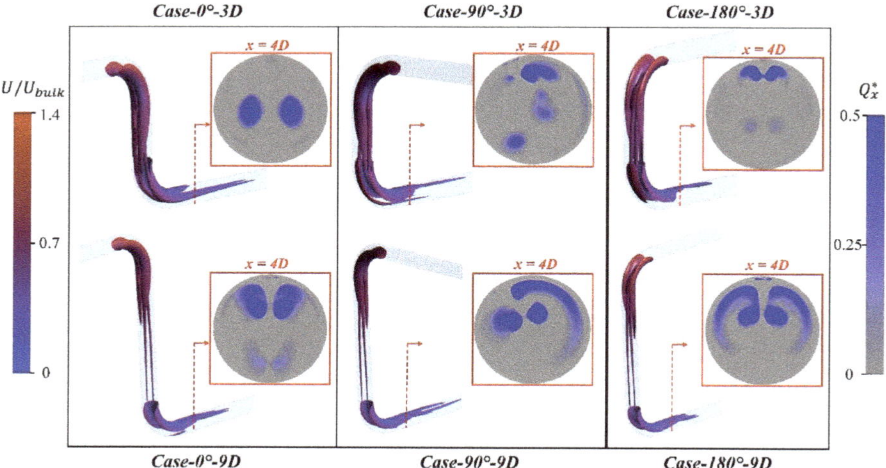

Figure 14. Vortex structures in *Case-0°-3D*, *Case-90°-3D*, *Case-180°-3D*, *Case-0°-9D*, *Case-90°-9D*, and *Case18-0°-9D*.

Subsequently, a comparison of swirl intensities downstream of *Cases-9D* (i.e., *Case-0°-9D*, *Case-90°-9D*, and *Case-180°-9D*) and *Cases-3D* (i.e., *Case-0°-3D*, *Case-90°-3D*, and *Case-180°-3D*) is shown in Figure 15a. The initial values of S_i in *Cases-9D* are higher than those in *Cases-3D* since the higher tangential velocity in *Cases-9D* enhances the strength of the downstream swirls (see Figures 9 and 13). In $1D$ downstream, sharp decreases in swirl intensities occur in *Cases-3D*. However, the swirl intensities exhibit the exponential dissipations in *Cases-9D*. To reveal the dissipation rates of the swirl intensities in *Cases-9D*, Figure 15 (b) displays the dissipation of S_i in a logarithmic coordinate. The dissipation rates β_s for *Case-0°-9D*, *Case-90°-9D*, and *Case-180°-9D* (exclude the initial values) can be fitted as 0.40, 0.20, and 0.25, respectively, indicating the highest dissipation rate in *Case-0°-9D*. To sum up, the short interval distance ($3D$) will limit the swirl intensity downstream of the double-curved pipes, which coincides with the result from the analysis of vortex structures. With the increase in the interval distance, the flow before the second bend starts to be developed. As a result, the effect of the spatial angle is weakened, and the dissipation of the swirl intensity downstream the second bend gradually conforms to an exponential form, which is similar to a single-bend case.

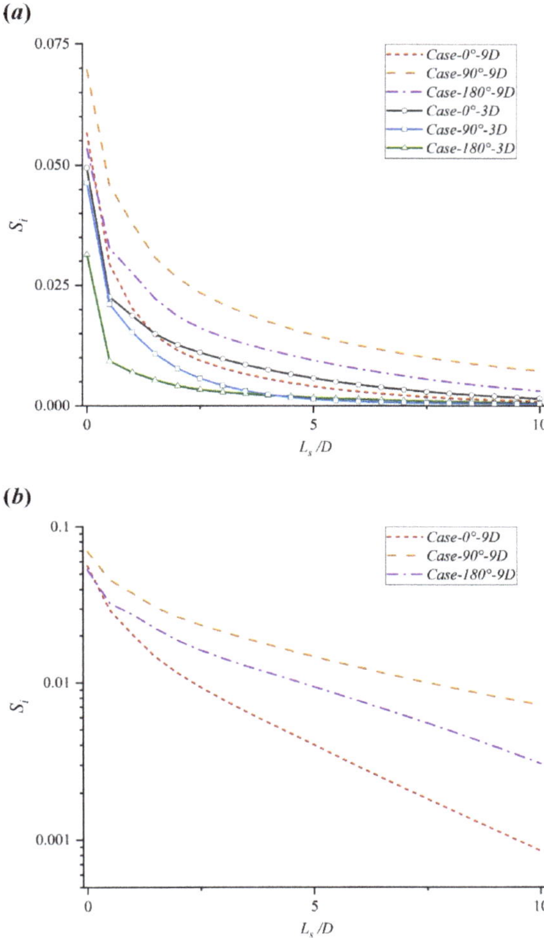

Figure 15. Dissipations of swirl intensities downstream the second bend: (**a**) comparisons of *Cases-9D* and *Cases-3D*; (**b**) results of *Cases-9D* in a logarithmic coordinate.

4. Conclusions

In this paper, a numerical study was conducted on the turbulent flow in double-curved pipes with different spatial structures. The effects of the spatial angle and interval distance between the two bends on the secondary flow were thoroughly investigated with the vector fields, velocity distributions, vortex developments, and dissipations of swirl intensity. Major conclusions are listed as follows:

1. The directions of the secondary flows in the Z- and U-type pipes are opposite when the interval distance between the two bends is short (3D). In addition, the secondary flow in the spatial Z-type structure is biased by the upstream bend and exhibits an oblique symmetric type.
2. The vortex generations downstream of different double-curved pipes are limited when the interval distance is short. However, increasing the interval distance of the two bends will lead to similar secondary flow motions and vortex structures even if their spatial structures are different.

3. When the interval distance is short, only in the Z-type pipe, the downstream flow dissipates in an exponential form, and it is easier to achieve a fully developed flow than the other cases. However, the downstream flow recovers the exponential dissipations in all the structures when the interval distance increases to $9D$. The corresponding dissipation rates of the downstream swirl intensities in the Z-, U-, and spatial Z-pipes are 0.40, 0.25, and 0.20, respectively.

The present study provided an in-depth knowledge on the secondary flow characteristics in double-curved subsea pipelines with different spatial structures. The results can provide guidance for the layout design of subsea pipelines and the arrangement of flowmeters. In terms of the pipeline design, a short interval distance between two bends will limit the swirl strength to obtain a more accurate flow measurement, and the U-type double-curved pipe inducing the weakest swirl is the most beneficial. In addition, increasing the interval distance between the two bends will weaken the effect of the spatial structure and strengthen the swirls. For the locations of flowmeters, it is not recommended to be arranged within $1D$ downstream the bend outlet to avoid the sharp variation of the swirl strength. For further research, the effect of the bend curvature and inlet flow condition should be taken into consideration.

Author Contributions: Conceptualization, F.H. and Z.W.; methodology, F.H. and Y.L.; software, Y.L.; validation, Y.L. and Q.L.; investigation, F.H., Y.L., Q.L. and Z.W.; writing—original draft preparation, F.H. and Y.L; writing—review and editing, F.H., W.L. and Z.W.; supervision, Z.W.; project administration, W.L.; funding acquisition, F.H. and Z.W. All authors have read and agreed to the published version of the manuscript.

Funding: This work was supported by the National Natural Science Foundation of China (No. 52006022), Liaoning Provincial Natural Science Foundation of China (No. 2022-MS-154), Dalian High-Level Talent Innovation Support Project (No. 2021RQ040), Scientific Research Funding Project of the Education Department of Liaoning Province (No. LJKZ0058), China Postdoctoral Science Foundation (No. 2020M670726), Fundamental Research Funds for the Central Universities (No. 3132022350), and 111 Project (No. B18009).

Data Availability Statement: Not applicable.

Conflicts of Interest: No potential conflict of interest was reported by the authors.

References

1. Chen, B.-Q.; Videiro, P.M.; Guedes Soares, C. Opportunities and Challenges to Develop Digital Twins for Subsea Pipelines. *J. Mar. Sci. Eng.* **2022**, *10*, 739. [CrossRef]
2. Chen, B.-Q.; Zhang, X.; Guedes Soares, C. The effect of general and localized corrosions on the collapse pressure of subsea pipelines. *Ocean Eng.* **2022**, *247*, 110719. [CrossRef]
3. Yu, J.; Xu, W.; Yu, Y.; Fu, F.; Wang, H.; Xu, S.; Wu, S. CFRP Strengthening and Rehabilitation of Inner Corroded Steel Pipelines under External Pressure. *J. Mar. Sci. Eng.* **2022**, *10*, 589. [CrossRef]
4. Ma, W.; Bai, T.; Li, Y.; Zhang, H.; Zhu, W. Research on Improving the Accuracy of Welding Residual Stress of Deep-Sea Pipeline Steel by Blind Hole Method. *J. Mar. Sci. Eng.* **2022**, *10*, 791. [CrossRef]
5. Seth, D.; Manna, B.; Shahu, J.T.; Fazeres-Ferradosa, T.; Pinto, F.T.; Rosa-Santos, P.J. Buckling Mechanism of Offshore Pipelines: A State of the Art. *J. Mar. Sci. Eng.* **2021**, *9*, 1074. [CrossRef]
6. Yin, G.; Ong, M.C. On the wake flow behind a sphere in a pipe flow at low Reynolds numbers. *Phys. Fluids* **2020**, *32*, 103605. [CrossRef]
7. Dean, W.R.; Chapman, S. Fluid motion in a curved channel. *Proc. R. Soc. Lond. Ser. A Contain. Pap. A Math. Phys. Charact.* **1928**, *121*, 402–420. [CrossRef]
8. Nandakumar, K.; Masliyah, J.H. Bifurcation in steady laminar flow through curved tubes. *J. Fluid Mech.* **1982**, *119*, 475–490. [CrossRef]
9. Yang, Z.-h.; Keller, H.B. Multiple laminar flows through curved pipes. *Appl. Numer. Math.* **1986**, *2*, 257–271. [CrossRef]
10. Yanase, S.; Yamamoto, K.; Yoshida, T. Effect of curvature on dual solutions of flow through a curved circular tube. *Fluid Dyn. Res.* **1994**, *13*, 217–228. [CrossRef]
11. Bovendeerd, P.; Steenhoven, A.; Vosse, F.; Vossers, G. Steady entry flow in a curved pipe. *J. Fluid Mech.* **1987**, *177*, 233–246. [CrossRef]

12. Sudo, K.; Sumida, M.; Hibara, H. Experimental investigation on turbulent flow in a circular-sectioned 90-degree bend. *Exp. Fluid* **1998**, *25*, 42–49. [CrossRef]
13. Sudo, K.; Sumida, M.; Hibara, H. Experimental investigation on turbulent flow in a square-sectioned 90-degree bend. *Exp. Fluid* **2001**, *30*, 246–252. [CrossRef]
14. Jurga, A.P.; Janocha, M.; Yin, G.; Ong, M.C. Numerical simulations of turbulent flow through a 90-degree pipe bend. *J. Offshore Mech. Arct. Eng.* **2022**, *144*, 1–17. [CrossRef]
15. Li, Y.; Cao, J.; Xie, C. Research on the Wear Characteristics of a Bend Pipe with a Bump Based on the Coupled CFD-DEM. *J. Mar. Sci. Eng.* **2021**, *9*, 672. [CrossRef]
16. Ning, C.; Li, Y.; Huang, P.; Shi, H.; Sun, H. Numerical Analysis of Single-Particle Motion Using CFD-DEM in Varying-Curvature Elbows. *J. Mar. Sci. Eng.* **2022**, *10*, 62. [CrossRef]
17. Hellström, L.H.O.; Zlatinov, M.B.; Cao, G.; Smits, A.J. Turbulent pipe flow downstream of a bend. *J. Fluid Mech.* **2013**, *735*, R7. [CrossRef]
18. Rütten, F.; Schröder, W.; Meinke, M. Large-eddy simulation of low frequency oscillations of the Dean vortices in turbulent pipe bend flows. *Phys. Fluids* **2005**, *17*, 035107. [CrossRef]
19. Kalpakli Vester, A.; Örlü, R.; Alfredsson, P.H. POD analysis of the turbulent flow downstream a mild and sharp bend. *Exp. Fluid* **2015**, *56*, 57. [CrossRef]
20. Hufnagel, L.; Canton, J.; Örlü, R.; Marin, O.; Merzari, E.; Schlatter, P. The three-dimensional structure of swirl-switching in bent pipe flow. *J. Fluid Mech.* **2017**, *835*, 86–101. [CrossRef]
21. Sakowitz, A.; Mihaescu, M.; Fuchs, L. Turbulent flow mechanisms in mixing T-junctions by Large Eddy Simulations. *Int. J. Heat Fluid Flow* **2014**, *45*, 135–146. [CrossRef]
22. Ong, M.C.; Liu, S.; Liestyarini, U.C.; Xing, Y. Three dimensional numerical simulation of flow in blind-tee pipes. In Proceedings of the 9th National Conference on Computational Mechanics, Trondheim, Norway, 11–12 May 2017.
23. Han, F.; Ong, M.C.; Xing, Y.; Li, W. Three-dimensional numerical investigation of laminar flow in blind-tee pipes. *Ocean Eng.* **2020**, *217*, 107962. [CrossRef]
24. Han, F.; Liu, Y.; Ong, M.C.; Yin, G.; Li, W.; Wang, Z. CFD investigation of blind-tee effects on flow mixing mechanism in subsea pipelines. *Eng. Appl. Comput. Fluid Mech.* **2022**, *16*, 1395–1419. [CrossRef]
25. Hu, Q.; Zou, L.; Lv, T.; Guan, Y.; Sun, T. Experimental and Numerical Investigation on the Transport Characteristics of Particle-Fluid Mixture in Y-Shaped Elbow. *J. Mar. Sci. Eng.* **2020**, *8*, 675. [CrossRef]
26. Fiedler, H.E. A note on secondary flow in bends and bend combinations. *Exp. Fluid* **1997**, *23*, 262–264. [CrossRef]
27. Mazhar, H.; Ewing, D.; Cotton, J.S.; Ching, C.Y. Measurement of the flow field characteristics in single and dual S-shape 90° bends using matched refractive index PIV. *Exp. Therm. Fluid Sci.* **2016**, *79*, 65–73. [CrossRef]
28. Sudo, K.; Sumida, M.; Hibara, H. Experimental investigation on turbulent flow through a circular-sectioned 180° bend. *Exp. Fluid* **2000**, *28*, 51–57. [CrossRef]
29. Yuki, K.; Hasegawa, S.; Sato, T.; Hashizume, H.; Aizawa, K.; Yamano, H. Matched refractive-index PIV visualization of complex flow structure in a three-dimentionally connected dual elbow. *Nucl. Eng. Des.* **2011**, *241*, 4544–4550. [CrossRef]
30. Ebara, S.; Takamura, H.; Hashizume, H.; Yamano, H. Characteristics of flow field and pressure fluctuation in complex turbulent flow in the third elbow of a triple elbow piping with small curvature radius in three-dimensional layout. *Int. J. Hydrogen Energy* **2016**, *41*, 7139–7145. [CrossRef]
31. Liu, Y.; Han, F.; Zhang, H.; Wang, D.; Wang, Z.; Li, W. Numerical simulation of internal flow in jumper tube with blind tee. In Proceedings of the 2021 IEEE 16th Conference on Industrial Electronics and Applications (ICIEA), Chengdu, China, 1–4 August 2021; pp. 363–368. [CrossRef]
32. Kim, J.; Srinil, N. 3-D Numerical Simulations of Subsea Jumper Transporting Intermittent Slug Flows. In Proceedings of the International Conference on Ocean, Offshore and Arctic Engineering, Madrid, Spain, 17–22 June 2018. [CrossRef]
33. Mattingly, G.E.; Yeh, T.T. Effects of pipe elbows and tube bundles on selected types of flowmeters. *Flow Meas. Instrum.* **1991**, *2*, 4–13. [CrossRef]
34. Di Piazza, I.; Ciofalo, M. Numerical prediction of turbulent flow and heat transfer in helically coiled pipes. *Int. J. Therm. Sci.* **2010**, *49*, 653–663. [CrossRef]
35. Wallin, S.; Johansson, A.V. Modelling streamline curvature effects in explicit algebraic Reynolds stress turbulence models. *Int. J. Heat Fluid Flow* **2002**, *23*, 721–730. [CrossRef]
36. Salama, A. Velocity Profile Representation for Fully Developed Turbulent Flows in Pipes: A Modified Power Law. *Fluids* **2021**, *6*, 369. [CrossRef]
37. Tanaka, M.-A.; Ohshima, H.; Monji, H. Numerical Investigation of Flow Structure in Pipe Elbow With Large Eddy Simulation Approach. In Proceedings of the ASME 2009 Pressure Vessels and Piping Conference, Prague, Czech Republic, 26–30 July 2009; pp. 449–458. [CrossRef]
38. Kim, J.; Yadav, M.; Kim, S. Characteristics of secondary flow induced by 90-degree elbow in turbulent pipe flow. *Eng. Appl. Comput. Fluid Mech.* **2014**, *8*, 229–239. [CrossRef]
39. Pruvost, J.; Legrand, J.; Legentilhomme, P. Numerical investigation of bend and torus flows, part I: Effect of swirl motion on flow structure in U-bend. *Chem. Eng. Sci.* **2004**, *59*, 3345–3357. [CrossRef]

40. Bhunia, A.; Chen, C.L. Flow Characteristics in a Curved Rectangular Channel With Variable Cross-Sectional Area. *J. Fluids Eng.* **2009**, *131*, 091102. [CrossRef]
41. Dutta, P.; Nandi, N. Numerical analysis on the development of vortex structure in 90° pipe bend. *Prog. Comput. Fluid Dyn. Int. J.* **2021**, *21*, 261–273. [CrossRef]
42. Qiao, S.; Zhong, W.; Wang, S.; Sun, L.; Tan, S. Numerical simulation of single and two-phase flow across 90° vertical elbows. *Chem. Eng. Sci.* **2021**, *230*, 116185. [CrossRef]
43. Hunt, J.C.R.; Wray, A.A.; Moin, P. Eddies, Streams, and Convergence Zones in Turbulent Flows. Center For Turbulence Research, Report CTR-S88. 1988. Available online: https://ntrs.nasa.gov/citations/19890015184 (accessed on 1 September 2022).

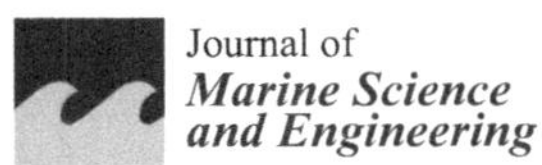
Journal of
Marine Science and Engineering

Article

Variation in Vortex-Induced Vibration Phenomenon Due to Surface Roughness on Low- and High-Mass-Ratio Circular Cylinders: A Numerical Study

Muhammad Usman Anwar [1], Niaz Bahadur Khan [1,*], Muhammad Arshad [2], Adnan Munir [1], Muhammad Nasir Bashir [1], Mohammed Jameel [3], Muhammad Faisal Rehman [4] and Sayed M. Eldin [5]

1   School of Mechanical & Manufacturing Engineering, National University of Sciences and Technology (NUST), Islamabad 44000, Pakistan
2   Department of Chemical Engineering, College of Engineering, King Khalid University, Abha 61421, Saudi Arabia
3   Department of Civil Engineering, College of Engineering, King Khalid University, Asir, P.O. Box 960, Abha 61421, Saudi Arabia
4   The Department of Architecture, University of Engineering and Technology Peshawar, Abbottabad Campus, Peshawar 25000, Pakistan
5   Center of Research, Faculty of Engineering, Future University in Egypt, New Cairo 11835, Egypt
*   Correspondence: niaz.bahadur@smme.nust.edu.pk

Citation: Anwar, M.U.; Khan, N.B.; Arshad, M.; Munir, A.; Bashir, M.N.; Jameel, M.; Rehman, M.F.; Eldin, S.M. Variation in Vortex-Induced Vibration Phenomenon Due to Surface Roughness on Low- and High-Mass-Ratio Circular Cylinders: A Numerical Study. *J. Mar. Sci. Eng.* 2022, 10, 1465. https://doi.org/10.3390/jmse10101465

Academic Editors: Bai-qiao Chen and Carlos Guedes Soares

Received: 26 September 2022
Accepted: 5 October 2022
Published: 10 October 2022

Publisher's Note: MDPI stays neutral with regard to jurisdictional claims in published maps and institutional affiliations.

Abstract: Fluid–structure interaction has been widely studied in the last few decades due to its wide range of applications in engineering fields. This phenomenon plays an important design role, for example, in offshore risers, high slender buildings, chimney stacks and heat exchangers. The vortex shedding generated from a bluff body can induce high-amplitude oscillations, known as vortex-induced vibrations (VIVs). This study presents a numerical analysis to investigate the impact of surface roughness on VIV in the crossflow direction of a circular cylinder. The study also investigates the impact of surface roughness with variation in mass ratio from 2.4 to 11 at a high Reynolds number (Re) = 10^4 using Reynolds-averaged Navier–Stokes (RANS) equations. The study concludes that roughness on a cylinder results in a reduction in amplitude response. Furthermore, the lock-in region is narrower compared to that of a smooth cylinder, irrespective of the mass ratio. However, it is observed that the impact of surface roughness is more significant in high-mass-ratio cylinders where the lock-in region is more squeezed and shifted toward lower reduced velocities. Furthermore, the vortex mode beyond reduced velocities Ur = 5.84 and 7.52 was observed to be 2S for high and low mass ratios, respectively.

Keywords: CFD; vortex-induced vibration; mass ratio; circular cylinder; surface roughness; flow

1. Introduction

Flow over rigid bodies is such a common phenomenon that it can be encountered in almost every aspect of our daily experience. This fluid–structure interaction is of great importance, as the fluid flowing past the structure can induce the vortex-induced vibration (VIV) phenomenon. Vortex-induced vibrations can be extremely destructive in nature, they affect many engineering applications such as skyscrapers, power transmission lines, underwater pipelines and bridge decks. The VIV phenomenon is highly dependent on the physical properties (such as mass, shape and condition of the surface) of the bluff body. The behavior of flow past a bluff body is entirely changed with the introduction of surface roughness, thus affecting the flow characteristics and VIV phenomenon. In real life, every surface is rough to some extent. In addition to surface roughness, the mass ratio has a significant impact on the flow characteristics and vortex formation.

Various studies have been conducted, both experimentally and numerically, to investigate the impact of surface texture on vortex-induced vibrations. Ramzi et al. [1]

experimentally investigated the VIV phenomenon on a short rigid cylinder and analyzed the dynamic responses at different surface roughness. The experimental study was performed in the range of 1 m/s to 8 m/s of wind speed in which recording and processing of amplitude response were performed using an accelerometer and LMS Test Xpress software, respectively. The study found that the higher surface roughness resulted in higher amplitude response reductions. Ghazali et al. [2] performed an experimental study in the subcritical range with roughness coefficients varying from Ks = 0.00019 D to 0.0057 D. They observed an increase in frequency vibration with an increase in surface roughness. Gao et al. [3] investigated the impact of the plan wall on the VIV phenomenon at Re = 5000. The results showed that the VIV trajectory of the cylinder is not significantly affected by variation in surface roughness. A cylinder with an initial small gap displayed a coalescing effect with the formation of the weak 2S vortex mode, while there was no coalescing effect for a cylinder with a large initial gap.

The analysis of Okajima et al. [4] covered the subcritical Reynolds number region, and the authors found that a coarse cylinder responds with a shorter amplitude than a polished one. Allen and Henning [5] carried out an experimental study at critical and subcritical regimes to investigate the impact of roughness. A low drag with no oscillation was found in the study. Bernitasas et al. [6] experimentally analyzed the VIV in the range of $8 \times 10^3 < \text{Re} < 2.0 \times 10^5$ with rough bands. The study showed that a cylinder's proximity to a fluid's flow behavior is highly sensitive to the roughness grit size and width and where the coarse ridges are located. The study showed that abrasion dispersion and scope can be utilized to control or maneuver the VIV response and the span of the lock-in region. Kiu et al. [7], in an experimental study in the range of 1.7×10^4 to 8.3×10^4, observed a decrease in oscillation and drag coefficient with an increase in roughness on the cylinder surface. It was also observed that a rough cylinder has greater Strouhal value compared to a smooth cylinder. Gao et al. [8] designed a physical experiment method in which the effect of face unevenness on a riser's VIV response was studied utilizing a physical study approach; the dislocation responses, friction, aerodynamic forces, tension, vortex-induced frequency and vibration frequency of the risers with various face roughnesses were compared. The VIV amplitude response is higher for the streamwise direction compared to the transverse direction at the low decreased velocities because the VIV lock-in phenomena happen earlier in the streamwise orientation than the transverse orientation. In comparison to the smooth riser, the coarse riser had a lesser VIV amplitude reaction, a greater vortex shedding frequency and a narrower "lock-in" zone.

Armin et al. [9] performed numerical studies to investigate the VIV phenomenon in multiple cylinders by developing a mathematical model. The study was validated with experimental results. The study also addressed the impact of the wake of the upstream cylinder on the downstream cylinder. The mathematical model developed in the study was capable of forecasting the lock-in regime and maximum oscillation in both leading and trailing cylinders. Liqun et al. [10] studied the mechanism by which harbor seals detect fish. The authors utilized the DNS for the VIV phenomenon on harbor seals. The study also compared the results with different shapes of the same characteristic length. The study evidenced low drag and reduced oscillation using the whisker model. Wang et al. [11] experimentally investigated the VIV phenomenon under influence of stretching in vertical risers. Badhurshah et al. [12] used the immersed boundary method to analyze the VIV phenomenon of a circular cylinder. Simulations were carried out with linear and bistable springs. The study evidenced a higher lock-in regime using bistable springs. Lin et al. [13] investigated the flow characteristics of tandem arranged cylinders (apart by 5 cm). The upstream cylinder was kept fixed whereas the downstream cylinder was made flexible. The study evidenced the strong impact of added mass correlation with the mode of vortex shedding in case of a single cylinder.

Tofa et al. [14] performed an experimental study on two identical cylinders in the subcritical range. They discussed the correlation of phase difference with cylinders. Smaller oscillations were observed at lower phase differences. Ming Zhao and Liang Cheng [15] numerically analyzed the impact of the free end on the vortex-induced vibration (VIV) of a rigid circular cylinder at Re = 300. It was observed that if a fixed cylinder's length is less than two cylinder diameters, vortex shedding is suppressed in the wake region. In another study, Zhao et al. [16] numerically analyzed the VIV behavior with multiple cylinders at Re = 150. The study showed the maximum and minimum amplitude at space ratios of 1.5 and 2.0, respectively. It is also found that minimum impact on the lock-in regime is observed when the space ratio is equal to or greater than 2.5, in the case of connected cylinders. Soti et al. [17] studied the VIV phenomenon at Re = 100 and 150 with a high mass ratio with variation in channel height. Oscillation amplitude is decreased with a decrease in the channel height. The study also showed that the initial branch almost disappears when channel height increased up to 2 D, where D is the diameter of the cylinder. In addition, with an increase in damping, the extracted energy is decreased, irrespective of the channel height. Han et al. [18] modified the law of wall to study the impact of surface roughness on the VIV phenomenon. Chen et al. [19] numerically analyzed the VIV phenomenon using DNS near a stationary wall with Re = 500. They observed significant variation in cylinder displacement due to vortex interaction with the proximity wall.

The study of Liu et al. [20] is related to the impact of mass ratio on the VIV phenomenon at a smaller Reynold number. The numerical results demonstrate that the mass ratio significantly affects the fluid force and vibration amplitude, particularly in the lock-in region with a mass ratio of less than 1.0. It is observed from the literature that most of the available analyses are carried out at smaller Reynold numbers (less than 5000) to investigate the impact of surface roughness, and limited research is available on high Reynold numbers.

The main objective of this article is to analyze the impact of surface roughness on the VIV phenomenon with mass ratio = 2.4 and 11 at Re = 10^4. For each mass ratio, six different reduced velocities in the range of 2 to 14 were used for this numerical study, and the results are compared with a smooth cylinder.

2. Domain Specifications

Since the flow domain size significantly affects the flow behavior, it is important to choose the domain size such that disturbance due to the boundary wall is avoided. As reviewed in the literature [21], various researchers used the domain size of up to 45 D (inflow direction) × 16 D (crossflow size): however, it has been concluded by Zdravkovich [22] and Zhao ([15,16,23]) that to avoid disturbance due to wall boundary, the blockage ratio of 5% should be ensured in numerical analysis.

In this study, the domain size of 45 D × 20 D is used, fulfilling the criteria mentioned in the literature (Figure 1). As depicted in Figure 1, the inlet is kept at a distance of 15 D and the top and bottom walls are kept at a distance of 10 D from the center of the cylinder. To achieve the Reynolds number = 10^4, the uniform velocity of 0.3149 m/s at inlet, a cylinder with a diameter equal to unity and a fluid with $\rho = 10^3$ kg/m^3 and $\upsilon = 0.03149$ kg/m^{-s} are used. The results obtained in the current study are compared with the available results of a smooth cylinder having the same parameters except for the roughness. The smooth cylinder results are taken from the experimental study by Hover [21] and numerical studies by Nguyen [24]) and Usman et al. [25].

For numerical stability, the working domain is split into two parts. The region immediately surrounding the cylinder is meshed using structured quadrilateral elements, while the remainder is meshed using triangular elements. The mesh is constructed such that the area next to the cylinder's wall has a very fine mesh, while the area farther out from the cylinder has a coarse mesh. Figure 2a,b represent the mesh and mesh close-up view near the cylinder. The dimensionless distance first layer cell height, known as the y+ value, should be less than or equal to unity in accurately solving the flow.

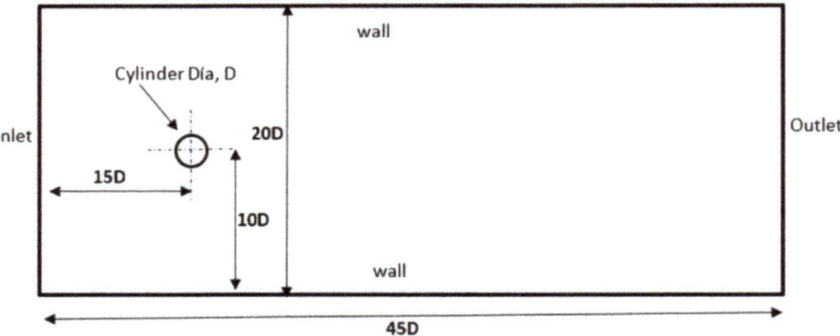

Figure 1. Computational domain.

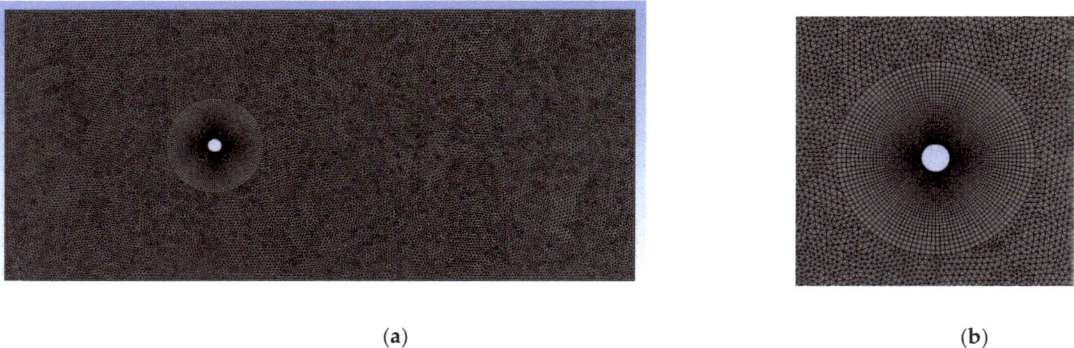

(a)   (b)

Figure 2. (a) Grid; (b) elements view (close-up).

To make sure the y+ is at unity (for the smooth case), the first cell layer height is placed at 1.4×10^{-3} D from the cylinder surface, where D is the diameter of the cylinder. A homogeneous pressure of 0 Pa is provided at the discharge boundary. The top and bottom walls of the flow field are both subject to a symmetrical boundary requirement. A no-slip condition is imposed on the cylinder surface which will ensure the capturing of the characteristic of boundary layer separation and vortex generating behavior. In order to capture the motion of the cylinder, a dynamic mesh is used with boundary motion. A UDF is utilized to extract the forces associated with the cylinder due to vortex shedding in each iteration. A diffusion-based dynamic smoothing method is used. Better mesh is produced through diffusion, which also enables high-amplitude cylinder oscillations without any restrictions on motion direction.

The VIV phenomenon for a rough cylinder is studied at two different mass ratios (m* = 2.4 and 11). For this study, the roughness height (Ks) 0.02 D is used. Figure 3 [26] shows the schematic diagram of the equivalent sand model. To incorporate this roughness height, the size of the roughness element should be less than the centroid of the first mesh node as shown in Figure 4. To incorporate this roughness height, the first layer for mesh is changed from 0.0014 to 0.006.

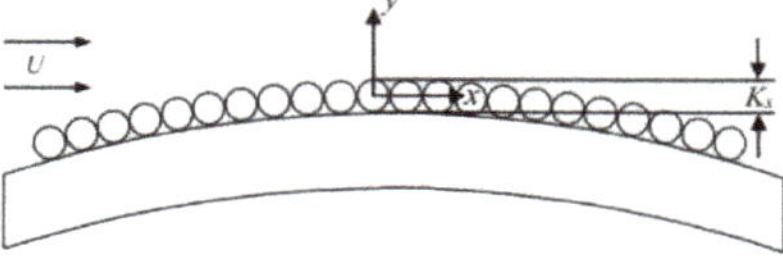

Figure 3. Schematic diagram of equivalent sand model [26].

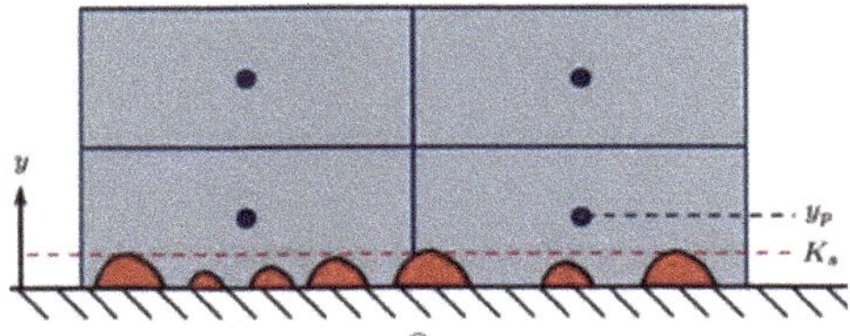

Figure 4. Schematic diagram for roughness height.

3. Results and Discussion

The case studies are performed for rough cylinders with a roughness height (Ks) of 0.02 D, where D is the diameter of the cylinder, having mass ratios of 2.4 and 11. In all the cases, Re = 10,000 has been maintained with an inlet velocity of 0.3149 m/s. The simulations are performed within the span of reduced velocity of $2 \leq Ur \leq 14$. Figure 5 shows amplitude response in the crossflow direction at various reduced velocities for mass ratios 2.4 and 2.11 with and without surface roughness.

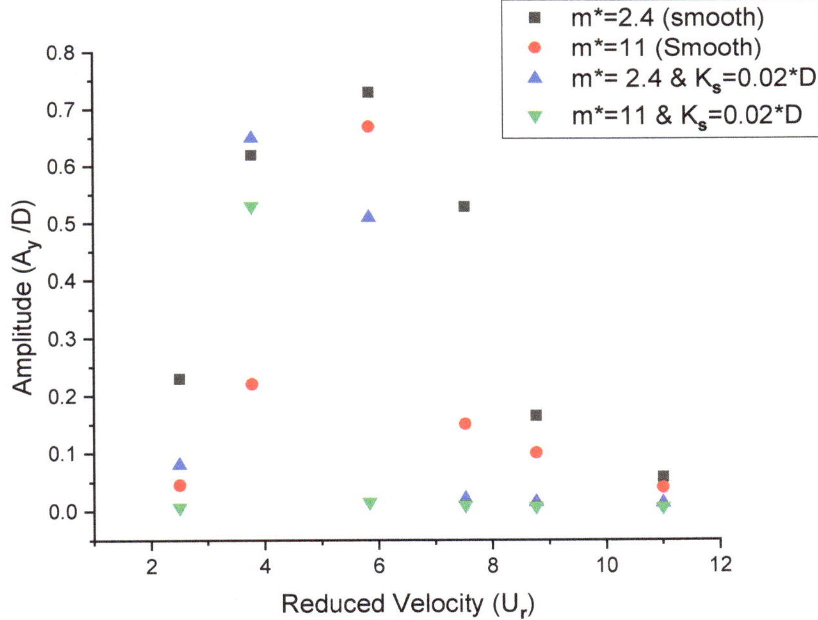

Figure 5. Comparison of amplitude responses for a smooth cylinder and a rough cylinder (i.e., Ks/D = 2×10^{-2}) with m* = 2.4 and 11.

With a surface roughness height of 0.02 D and a reduced velocity of 2.5, the amplitude response (Ay/D) of the cylinder was found to be Ay/D = 0.08 and 0.0065 for mass ratios of 2.4 and 2.11, respectively (Figure 5) which is relatively small in comparison to that of the smooth cylinder (where amplitude captured was 0.23 and 0.045 for mass ratios 2.4 and 2.11, respectively [25]). Figure 6 represents vortex mode at higher and lower mass ratio at different reduced velocities. Drag forces and lift forces acting on the cylinder were also observed to be small at the higher mass ratio (Figure 7).

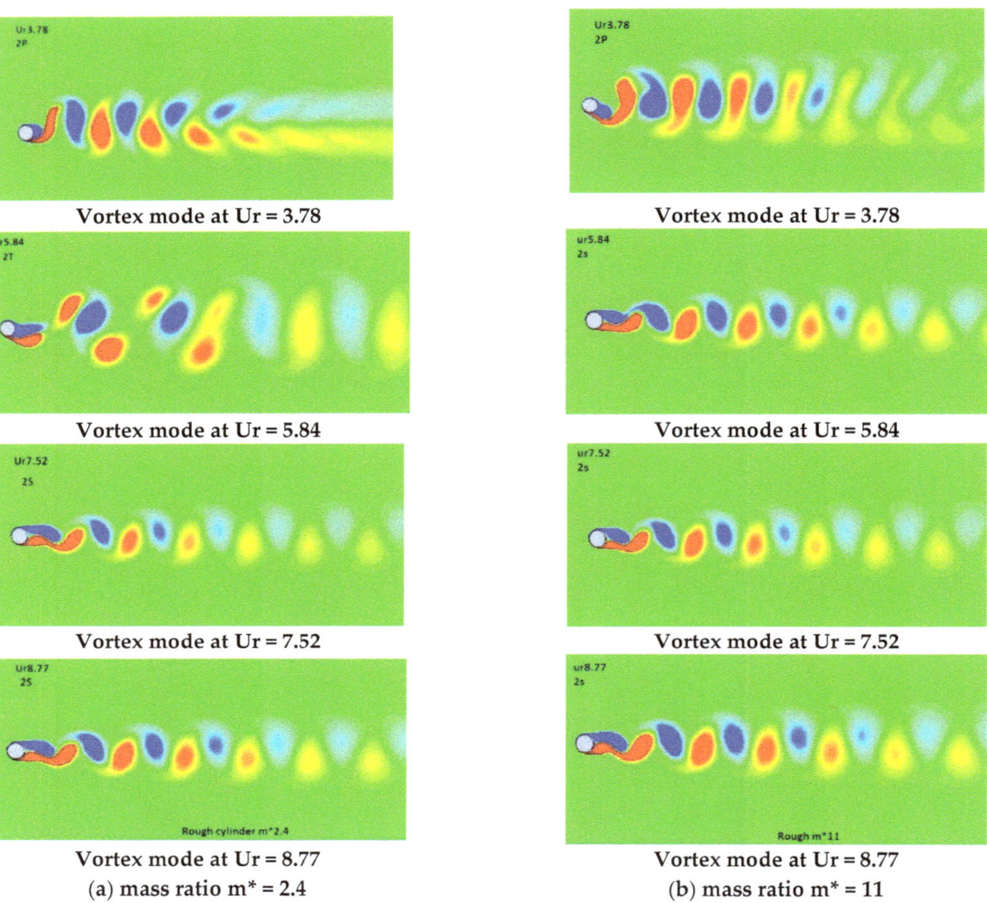

Figure 6. Vortex mode at different reduced velocities with roughness Ks = 0.02 D: (**a**) m* = 2.4 and (**b**) m* = 11.

As the reduced velocity was increased to 3.78, a significant increase in amplitude response (Ay/D = 0.65 and 0.53 for mass ratios 2.4 and 11, respectively) was measured, which is almost the same as that in the case of the smooth cylinder at m* = 2.4; however, roughness at the higher mass ratio result in a significant reduction in amplitude. The 2P vortex mode was observed at both low and high mass ratios (cylinder with roughness) in the wake region (Figure 6) which is completely different from the vortex mode observed in smooth cylinder case, i.e., P + S and 2S modes at lower and higher mass ratios [25]. A lower cd and a higher cl response were observed in comparison to a smooth cylinder at the lower mass ratio. At Ur = 5.84, amplitude responses Ay/D = 0.51 and 0.015 are recorded for lower and higher mass ratios, respectively; the amplitude response is relatively lower in the case of the lower mass ratio when compared with the smooth cylinder, whereas at the higher mass ratio, the roughness almost suppressed the oscillation. The 2T and 2S vortex modes were recorded in the wake region for lower and higher mass ratios, respectively, whereas the 2P vortex mode was observed in the case of the smooth cylinder [25]. The 2P mode is observed for the smooth case at lower and higher mass ratios.

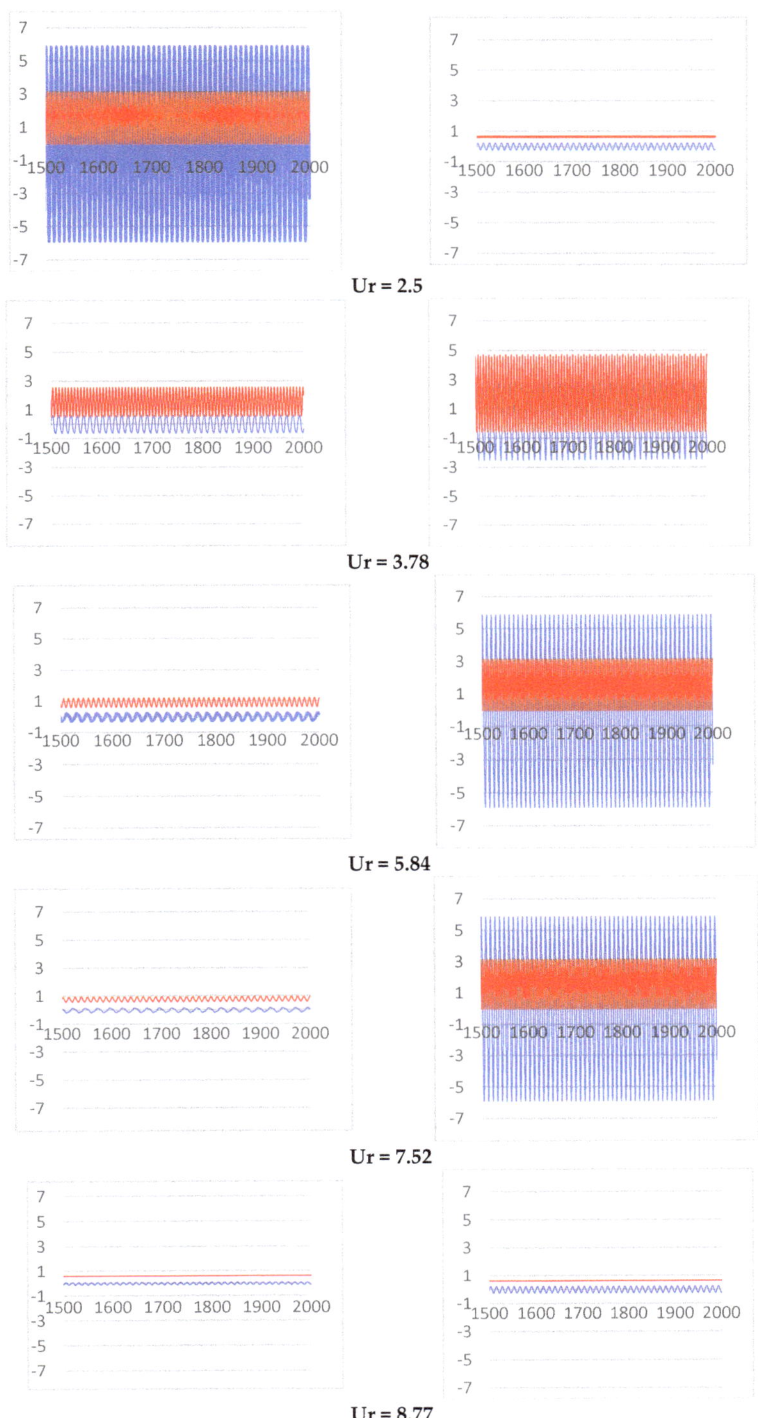

Figure 7. *Cont.*

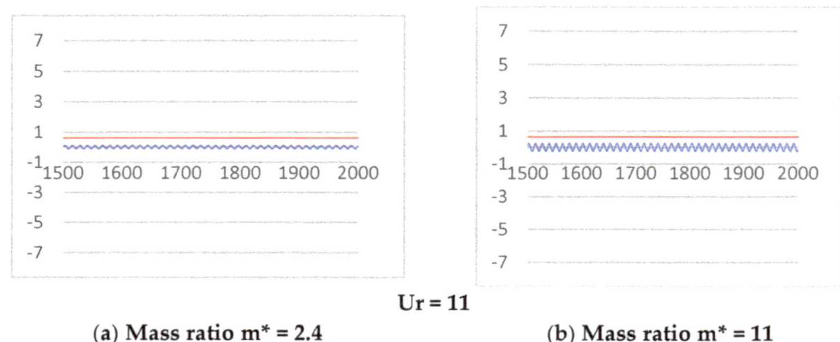

Ur = 11

(**a**) Mass ratio m* = 2.4   (**b**) Mass ratio m* = 11

Figure 7. Force coefficients Cd (Red) and Cl (Blue) at difference reduced velocities with surface roughness Ks = 0.02 D: (**a**) mass ratio m* = 2.4 and (**b**) mass ratio m* = 11.

At Ur = 7.52, the amplitude response computed is Ay/D = 0.021 and 0.0092 for m* = 2.4 and 11, respectively, which is very small in comparison to the smooth case [25]. At reduced velocity Ur = 8.77, the amplitude response reduced significantly to Ay/D = 0.015 and 0.0074 for lower and higher mass ratios, respectively. The 2S vortex mode was observed at Ur = 7.52 and Ur = 8.77 for both lower and higher mass ratios with roughness, whereas in the smooth cylinder case, 2P and 2S modes at Ur = 7.5 and 2T and 2S modes at Ur 8.77 are observed at lower and higher mass ratios, respectively. At Ur = 11, the amplitude response of Ay/D = 0.013 and 0.0064 for m* = 2.4 and 11, respectively, is computed, which is very small compared to the smooth cylinder case.

4. Conclusions

The study was performed to investigate the impact of surface roughness (i.e., Ks/D = 2 × 10^{-2}) on vortex shape, crossflow amplitude and lock-in region with lower (m* = 2.4) and higher (m* = 11) mass ratios. The results obtained from this study were compared with those for a smooth cylinder available in the literature. All the parameters in both cases were kept the same. It is concluded that roughness on a cylinder surface has a significant impact on the high mass ratio problem compared to the low mass ratio. The crossflow oscillation is significantly reduced due to surface roughness at both lower and higher mass ratios. However, higher mass ratio and lower mass ratio oscillations are almost completely suppressed (more than 90% with roughness) beyond reduced velocities of 5.84 and 3, respectively. At the higher mass ratio, drag forces were found to be high in the lock-in region and low outside of the lock-in region. It was found that surface roughness narrowed the lock-in region more in case of the higher mass ratio compared to the lower mass ratio. In addition, an overall decreasing trend in the cylinder amplitude response is recorded. Furthermore, the lock-in region is shifted toward the lower reduced velocities due to roughness in the higher mass ratio problem. Regarding force coefficients, it is observed that roughness reduced the drag coefficient significantly in case of the low mass ratio. In addition, the vortex mode is changed to 2S at all reduced velocities beyond Ur = 8.73 in higher mass ratio problems, whereas the same behavior is found in lower mass ratio problems beyond the reduced velocity of 7.5. Overall, it is concluded that higher mass ratio cases are significantly affected due to surface roughness.

Author Contributions: Conceptualization, M.U.A., N.B.K., M.A., A.M., M.N.B., M.J., M.F.R. and S.M.E.; methodology, M.U.A., N.B.K., M.A., A.M., M.N.B., M.J., M.F.R. and S.M.E.; software, M.U.A., N.B.K., M.A., A.M., M.N.B., M.J., M.F.R. and S.M.E.; validation, M.U.A., N.B.K., M.A., A.M., M.N.B., M.J., M.F.R. and S.M.E.; formal analysis, M.U.A., N.B.K., M.A., A.M., M.N.B., M.J., M.F.R. and S.M.E.; investigation, M.U.A., N.B.K., M.A., A.M., M.N.B., M.J., M.F.R. and S.M.E.; resources, M.U.A., N.B.K., M.A., A.M., M.N.B., M.J., M.F.R. and S.M.E.; data curation, M.U.A., N.B.K., M.A., A.M., M.N.B., M.J., M.F.R. and S.M.E.; writing—original draft preparation, M.U.A., N.B.K., M.A., A.M., M.N.B., M.J., M.F.R. and S.M.E.; writing—review and editing, M.U.A., N.B.K., M.A., A.M., M.N.B., M.J., M.F.R. and S.M.E.; visualization, M.U.A., N.B.K., M.A., A.M., M.N.B., M.J., M.F.R. and S.M.E.; supervision, M.U.A., N.B.K., M.A., A.M., M.N.B., M.J., M.F.R. and S.M.E.; project administration, M.U.A., N.B.K., M.A., A.M., M.N.B., M.J., M.F.R. and S.M.E.; funding acquisition, M.U.A., N.B.K., M.A., A.M., M.N.B., M.J., M.F.R. and S.M.E. All authors have read and agreed to the published version of the manuscript.

Funding: The authors extend their appreciation to the Deanship of Scientific Research at King Khalid University for funding this work through a large group project under grant number RGP. 2/83/43.

Institutional Review Board Statement: Not applicable.

Informed Consent Statement: Not applicable.

Data Availability Statement: The data that supports the findings of this study are available within the article.

Conflicts of Interest: The authors declare no conflict of interest.

References

1. Ramzi, N.A.; Quen, L.K.; Abu, A.; Siang, K.H.; Othman, N.A.; Ken, T.L.; Loon, S.C. Experimental analysis on vortex-induced vibration of a rigid cylinder with different surface roughness. *IOP Conf. Ser. Mater. Sci. Eng.* **2019**, *469*, 012003. [CrossRef]
2. Ghazali, M.K.; Shaharuddin, N.M.; Ali, A.; Siang, K.H.; Nasir, M.N.; Talib, M.H. Surface roughness effect on vortex-induced vibration phenomenon in cross-flow direction of a bluff body. *J. Adv. Res. Fluid Mech. Therm. Sci.* **2019**, *64*, 253–263.
3. Gao, Y.; Zhang, Z.; Zou, L.; Liu, L.; Yang, B. Effect of surface roughness and initial gap on the vortex-induced vibrations of a freely vibrating cylinder in the vicinity of a plane wall. *Mar. Struct.* **2020**, *69*, 102663. [CrossRef]
4. Okajima, A.; Nagamori, T.; Matsunaga, F.E.; Kiwata, T. Some experiments on flow-induced vibration of a circular cylinder with surface roughness. *J. Fluids Struct.* **1999**, *13*, 853–864. [CrossRef]
5. Allen, D.; Henning, D. Surface roughness effects on vortex-induced vibration of cylindrical structures at critical and supercritical Reynolds numbers. In Proceedings of the Offshore Technology Conference, Houston, TX, USA, 30 April–3 May 2001.
6. Bernitsas, M.M.; Raghavan, K.; Duchene, G. Induced separation and vorticity using roughness in VIV of circular cylinders at $8 \times 10^3 < \mathrm{Re} < 2.0 \times 10^5$. In Proceedings of the International Conference on Offshore Mechanics and Arctic Engineering, Estoril, Portugal, 15–20 June 2008.
7. Kiu, K.Y.; Stappenbelt, B.; Thiagarajan, K.P. Effects of uniform surface roughness on vortex-induced vibration of towed vertical cylinders. *J. Sound Vib.* **2011**, *330*, 4753–4763. [CrossRef]
8. Gao, Y.; Fu, S.; Wang, J.; Song, L.; Chen, Y. Experimental study of the effects of surface roughness on the vortex-induced vibration response of a flexible cylinder. *Ocean. Eng.* **2015**, *103*, 40–54. [CrossRef]
9. Armin, M.; Day, S.; Karimirad, M.; Khorasanchi, M. On the development of a nonlinear time-domain numerical method for describing vortex-induced vibration and wake interference of two cylinders using experimental results. *Nonlinear Dyn.* **2021**, *104*, 3517–3531. [CrossRef]
10. Song, L.; Ji, C.; Zhang, X. Vortex-induced vibration and wake tracing mechanism of harbor seal whisker: A direct numerical simulation. *Chin. J. Theor. Appl. Mech.* **2021**, *53*, 395–412.
11. Wang, C.; Wang, Y.; Liu, Y.; Li, P.; Zhang, X.; Wang, F. Experimental and numerical simulation investigation on vortex-induced vibration test system based on bare fiber Bragg grating sensor technology for vertical riser. *Int. J. Nav. Archit. Ocean Eng.* **2021**, *13*, 223–235. [CrossRef]
12. Badhurshah, R.; Bhardwaj, R.; Bhattacharyab, A. Numerical simulation of Vortex-Induced Vibration with bistable springs: Consistency with the Equilibrium Constraint. *J. Fluids Struct.* **2021**, *103*, 103280. [CrossRef]
13. Lin, K.; Jiasong, W.; Dixia, F.; Triantafyllou, M.S. Flow-induced cross-flow vibrations of long flexible cylinder with an upstream wake interference. *Phys. Fluids* **2021**, *33*, 065104. [CrossRef]
14. Tofa, M.M.; Maimun, A.; Ahmed, Y.M.; Jamei, S.; Kader, S.; Abyn, H. Vortex Induced Vibration on Two Equal Diameter Cylinders with low mass ratio in tandem. *J. Appl. Sci.* **2014**, *14*, 3578–3584.
15. Zhao, M.; Cheng, L. Vortex-induced vibration of a circular cylinder of finite length. *Phys. Fluids* **2014**, *26*, 015111. [CrossRef]
16. Zhao, M.; Kaja, K.; Xiang, Y.; Cheng, L. Vortex-induced vibration of four cylinders in an in-line square configuration. *Phys. Fluids* **2016**, *28*, 023602. [CrossRef]

17. Soti, A.K.; De, A. Vortex-induced vibrations of a confined circular cylinder for efficient flow power extraction. *Phys. Fluids* **2020**, *32*, 033603. [CrossRef]
18. Xiangxi, H.; Tang, Y.; Meng, Z.; Fu, F.; Qiu, A.; Gu, J.; Jiaming, W. Surface roughness effect on cylinder vortex-induced vibration at moderate Re regimes. *Ocean Eng.* **2021**, *224*, 108690.
19. Chen, W.; Chunning, J.; Dong, X.; Zhimeng, Z. Three-dimensional direct numerical simulations of vortex-induced vibrations of a circular cylinder in proximity to a stationary wall. *Phys. Rev. Fluids* **2022**, *7*, 044607. [CrossRef]
20. Liu, M.; Jin, R.; Wang, H. Numerical investigation of vortex induced vibration of a circular cylinder for mass ratio less than 1.0. *Ocean Eng.* **2022**, *251*, 111130. [CrossRef]
21. Hover, F.S.; Miller, S.N.; Triantafyllou, M.S. Vortex-induced vibration of marine cables: Experiments using force feedback. *J. Fluids Struct.* **1997**, *11*, 307–326. [CrossRef]
22. Zdravkovich, M.M. Conceptual overview of laminar and turbulent flows past smooth and rough circular cylinders. *J. Wind Eng. Ind. Aerodyn.* **1990**, *33*, 53–62. [CrossRef]
23. Zhao, M.; Cheng, L.; An, H.; Lu, L. Three-dimensional numerical simulation of vortex-induced vibration of an elastically mounted rigid circular cylinder in steady current. *J. Fluids Struct.* **2014**, *50*, 292–311. [CrossRef]
24. Nguyen, V.-T.; Nguyen, H.H. Detached eddy simulations of flow induced vibrations of circular cylinders at high Reynolds numbers. *J. Fluids Struct.* **2016**, *63*, 103–119. [CrossRef]
25. Anwar, M.U.; Lashin, M.M.; Khan, N.B.; Munir, A.; Jameel, M.; Muhammad, R.; Guedri, K.; Galal, A.M. Effect of Variation in the Mass Ratio on Vortex-Induced Vibration of a Circular Cylinder in Crossflow Direction at Reynold Number = 10^4: A Numerical Study Using RANS Model. *J. Mar. Sci. Eng.* **2022**, *10*, 1126. [CrossRef]
26. Gao, Y.; Zong, Z.; Zou, L.; Takagi, S.; Jiang, Z. Numerical simulation of vortex-induced vibration of a circular cylinder with different surface roughnesses. *Mar. Struct.* **2018**, *57*, 165–179. [CrossRef]

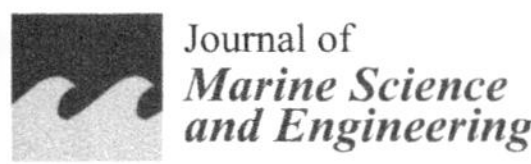

Article

Dynamic Response of DP Offshore Platform-Riser Multi-Body System Based on UKF-PID Control

Dapeng Zhang [1], Bowen Zhao [2], Yong Bai [3,*] and Keqiang Zhu [4]

[1]   Ship and Maritime College, Guangdong Ocean University, Zhanjiang 524088, China
[2]   Ocean College, Zhejiang University, Zhoushan 316021, China
[3]   College of Civil Engineering, Zhejiang University, Zhoushan 316000, China
[4]   Faculty of Maritime and Transportation, Ningbo University, Ningbo 315211, China
*   Correspondence: baiyong@zju.edu.cn

Abstract: The coupling effect between the offshore platform and the riser in the offshore platform-riser multi-body system might be greatly increased under heavy external maritime stresses. The system will become significantly more nonlinear. The partial secondary development of OrcaFlex is carried out considering the strong non-linearity of the dynamic positioning (DP) offshore platform-riser multi-body system, combined with the actual offshore construction engineering background and the lumped mass method, based on Python-language embedded programming with the basis of the operation principle of the application program interface (API) and the composition of its modules. To regulate the dynamic positioning of the offshore platform-riser multi-body coupling system, a UKF-PID control approach based on an unscented Kalman filter is presented. Based on the procedures described above, a classical calculation model is created, and the model's calculation results are compared to those of relevant references, confirming the method's validity and viability. Finally, the model of the PID-controlled dynamic positioning offshore platform-riser rigid–flexible multi-body system is developed, and a dynamic simulation is performed under specified sea conditions. The findings have implications for engineering practice.

Keywords: dynamic response; dynamic positioning; offshore platform-riser; Python; unscented Kalman filter; OrcaFlex; lumped mass method

Citation: Zhang, D.; Zhao, B.; Bai, Y.; Zhu, K. Dynamic Response of DP Offshore Platform-Riser Multi-Body System Based on UKF-PID Control. *J. Mar. Sci. Eng.* **2022**, *10*, 1596. https://doi.org/10.3390/jmse10111596

Academic Editor: Raúl Guanche García

Received: 26 September 2022
Accepted: 25 October 2022
Published: 28 October 2022

Publisher's Note: MDPI stays neutral with regard to jurisdictional claims in published maps and institutional affiliations.

1. Introduction

Large maritime constructions like floating offshore platforms and special engineering vessels that float in the water might typically have their location and spatial attitude determined by empirical calculations based on heading angles, velocity, and ocean currents [1–4]. In fact, the changes in a ship's position that can be roughly calculated by empirical formulas are not timely and not accurate enough, and for some construction projects that require high accuracy and fast feedback speed, they are far from meeting their technical requirements [5–7]. This is due to the instantaneity of wind, wave, and current load changes in the time domain, the randomness and mutability of changes in the height direction, and the changes in a ship's position that can be roughly calculated by empirical formulas [8–10]. The dynamic response of offshore platforms is nonlinear, time-varying, and uncertain, so it is a very difficult task to control the motion of ships and offshore platforms in actual engineering operations [11]. The dynamic positioning system is very important for the safety of ships and offshore platforms, especially large vessels and offshore platforms [12–15].

To put it simply, the dynamic positioning of an offshore platform is when the ship or offshore platform is at a low speed and the controller calculates three assumed values of longitudinal force, lateral force, and rotational torque [16]. The assumed forces and torques obtained from the controller are processed and transferred to the required propeller parameters, such as the propeller speed, blade setting angle, rudder angle, and propeller azimuth angle. Then, in combination with specific construction requirements, the position, angle, and space attitude of the ship or offshore platform are effectively adjusted. The dynamic positioning system has been developed to some extent. For, example, Fossen and Pettersen [17] suggested an underwater vehicle theory, asserting that the underwater vehicle's six-degrees-of-freedom dynamic positioning control was accomplished. They discovered that, as compared to surface dynamic positioning, additional degrees of freedom of underwater vehicles should be considered in dynamic positioning control due to the unpredictability of the underwater current and the wave and current stresses on the vehicles. In a subsequent study, Fossen [18] explicitly stated that in the motion operation of certain ships, due to the presence of many thrusters, the needed control actions could be executed in a variety of ways; other combinations of actuators can also provide the same control actions. To prevent linearization of the ship motion equation, it can be attempted to apply nonlinear control to the dynamic positioning system. Grovelen and Fossen [19] offered vector inversion as a viable solution to difficulties in dynamic positioning system control. This system is based on the effective filtering of measured ship position and heading data, with only white noise added as interference. The signals will, however, also be impacted in practical engineering applications by the waves hitting the hull. Thus, filtering is carried out using a so-called filtering technique before the calculated speed estimate is imported into the feedback loop. This involves extracting the wave frequency from the measured motion so that the controller can only obtain low-frequency signals and filter out high-frequency motion components and measurement noise. A motion control experiment on the motion of a physical model of a ship, Cybership II, was carried out in a shallow water tank at a laboratory in Trondheim to check the accuracy of Skjetne and Fossen's [20] adaptive control theory. The experimental results suggest that their hypothesis is quite useful in shallow water. Gierusz (2007) effectively built and evaluated a ship control system consisting of two separate controllers. It was discovered that the ship control system composed of these two controllers may increase the efficiency in a certain operation area.

A type of controller created by Morawski and Nguyen Cong [21] employs fuzzy logic to regulate the motion of the ship in the port. The adjustment and control of the ship's motion attitude in a confined space both benefit from this controller. Lee et al. examined the control effectiveness and control quality of a PID controller and a fuzzy controller throughout the process of a ship entering a port. They also outlined the benefits and drawbacks of the two control strategies [22]. Bui et al. suggested a technique to regulate the motions of four tugs throughout the maneuvering operation at the port [23]. By considering the filtering and reconstruction of low-frequency motion components, Fossen and Strand created an observer for dynamic positioning systems [24]. Asgeir J. Srensen et al. [25] developed a novel approach based on the design of the typical mooring auxiliary dynamic positioning FPSO, and the preliminary concepts put forward have been used to counteract the negative impacts of hazardous sea conditions on the placement of FPSO. Johann Wichers et al. [26] examined the use of mooring-assisted dynamic positioning for an FPSO in deep water and proposed a concept of an optimized design based on their study. Aalbers et al. [27] performed a hydrodynamic model test on a dynamically positioned ship under closed-loop control, which effectively tested the performance of the closed-loop system and determined the hydrodynamic parameters of the model ship under the action of the dynamic positioning system.

The method for studying the dynamic response of a DP offshore platform-riser multi-body system mainly focuses on the combination of numerical simulation and experiment. Sorheim [28] carried out computer modeling and simulations for a DP oil tanker under single-point mooring and preliminarily realized the computer simulations of the coupling response between the DP offshore platform and risers. The research showed that the thrust distribution of the DP offshore platform-riser multi-body system has a significant impact on the tension of the mooring system and the distribution of tension on each mooring line. By reasonably distributing the thrust of the DP platform's thrusters, the mooring tension borne by the mooring lines can be greatly reduced, thereby avoiding the conflict between the thrust system and the mooring system and improving the flexibility of the whole system. Wichers et al. [29] also conducted numerical simulations and experimental verification on the DP oil tanker under single-point mooring and found that the low-frequency partial viscous damping played an important role in the oscillating motion of the DP oil tanker. Lopez et al. [30] provides a description of the FPSO hull and station keeping system and the disconnectable turret-riser system and presented the preliminary results from a design study. Tannuri et al. [31] investigated how a large shuttle tanker responded dynamically to a dynamic positioning system. The above research mainly focuses on the interaction between the DP offshore platform and mooring lines. There are some differences between mooring lines and risers [32–35]. The bending stiffness and torsional stiffness of mooring lines are very small. For the risers with a certain bending stiffness and torsional stiffness, there is a certain difference between the risers and mooring lines [36]. Regarding the interaction between the riser and the platform or the underwater vehicles, Mai-The Vu [37–41] and Hyeung-Sik Choi [42–46] conducted a lot of research. A study on the hovering motion of an underwater vehicle with an umbilical cable was carried out by Mai-The Vu and Hyeung-Sik Choi [47,48]. A new full dynamics equation on the combined motions of the underwater vehicle and the umbilical cable was presented to analyze the dynamic performance of the underwater vehicle motion. The simulation results showed that the umbilical cable significantly affected the motion of the underwater vehicle during forward motion, sideward motion, and turning motion. However, the effect of the umbilical cable on the underwater vehicle motion during ascending motion was less significant, since the buoyancy force of the umbilical cable was assumed to be zero.

According to Wang et al. [49], the presence of a DP system can minimize mooring line stress and increase the positioning accuracy of dynamic positioning ships. However, when the external load and the rigid–flexible coupling degree grow in the rigid–flexible multi-body system of an offshore platform-riser, the nonlinearity of the entire system will increase. As a result, a novel control strategy that is more applicable under severe nonlinear situations is required. In this paper, a new PID control approach based on the unscented Kalman filter for the dynamic positioning offshore platform-riser multi-body system is created by merging the unscented Kalman filter with the traditional PID control, and the DP control of a rigid–flexible fluid coupling system composed of an offshore platform and risers is realized. Considering the influence of different control modes on the whole rigid–flexible multi-body coupling system, based on the offshore platform-riser multi-body system without a dynamic positioning system, the dynamic analysis model of the offshore platform-riser multi-body system under single PID control mode and unscented Kalman filter-PID (UKF-PID) control mode is established and compared. The rest of this paper is organized as follows. Section 2 introduces the PID control method and dynamic modeling of the DP offshore platform based on the unscented Kalman filter (UKF) in Python language. Section 3 verifies the correctness and reliability of the dynamic analysis model. Section 4 presents the results and discussion. Finally, the conclusions drawn from this paper are presented in Section 5.

2. Methodology

2.1. PID Control Method

Although there are many different process control methods, PID control is still the best option because of its smaller structural loop and higher resilience. However, the limited stability of the PID control is still a drawback. It is necessary to develop a PID control system based on the Kalman filter in order to enhance the system's compatibility and stability and make the response more reliable and accurate. The final form of the PID algorithm can be expressed as follows:

$$u(t) = K_p e(t) + K_i \int_0^t e(\tau)d\tau + K_d \frac{de(t)}{dt} \tag{1}$$

where K_p is the proportional gain, K_i is the integral gain, and K_d is the differential gain.

2.2. Unscented Kalman Filter Algorithm and Its Principle

The ideal state of a dynamic system may be determined via Kalman filtering. Even if the observed system state parameters contain noise and the observed values are not precise enough, Kalman filtering can complete the relative optimum estimation of the real-state value. It consists of the standard Kalman filter (KF), the extended Kalman filter (EKF), and the unscented Kalman filter (UKF). However, for non-linear situations, whether KF or EKF, there are issues with massive quantities of computations, and linear mistakes can easily influence the model's accuracy. As a result, the unscented transformation (an approximation approach for finding the moments of each order of nonlinear random variables, UT) may be introduced to tackle this problem based on the classic Kalman filter. It obtains the average value and variance mostly by frequent sampling and weighting, which employs the unscented Kalman filter (UKF). The impact of UKF can potentially produce effects that only second-order EKF is capable of producing since this approach has a greater approximation accuracy for statistical moments.

The guiding ideology of UT ensures that the mean and covariance of system sampling are $\bar{x}$ and P_{xx}, respectively. Then, a set of sigma points is selected, and the nonlinear transformation is applied to the sigma points of each sampling data to obtain the point set $\bar{y}$ and P_{yy} for the nonlinear transformation.

In the traditional sense, the algorithm of UT transformation is as follows.

According to the statistics of input variables $\bar{x}$ and P_{xx}, a sigma point set $\{X_i\}$ under a sampling strategy ($i = 0, 1, 2, \ldots, 2n$), and the corresponding weights W_i^m and W_i^c are selected, where i is the number of sigma points, W_i^m is the weight used for mean weighting, and W_i^c is the weight used for covariance weighting. Without proportional correction, $W_i^m = W_i^c = W_i$.

The corresponding weight W_i of $2n + 1$ sampling points is calculated:

$$X_0 = \bar{x}, \qquad W_0 = \frac{\sigma}{n + \sigma} \quad i = 0 \tag{2}$$

$$X_i = \bar{x} + \left(\sqrt{(n + \sigma)P_{xx}}\right)_i, \qquad W_i = \frac{1}{2(n + \sigma)} \quad i = 1, \cdots, n \tag{3}$$

$$X_i = \bar{x} - \left(\sqrt{(n + \sigma)P_{xx}}\right)_i, \qquad W_i = \frac{1}{2(n + \sigma)} \quad i = n + 1, \cdots, 2n \tag{4}$$

where σ is a fine-tuning parameter, which can only affect the deviation caused by the higher order moment after the second order; $\left(\sqrt{(n + \sigma)P_{xx}}\right)_i$ is the ith row or the ith column vector of the square root of the matrix $(n + \sigma)P_{xx}$, which can be calculated by Cholesky decomposition; W_i is the weight of the ith sigma point.

$$\sum_{i=0}^{2n} W_i = 1 \tag{5}$$

The square root of matrix P can be decomposed by Cholesky, which makes the calculation more stable and efficient. The sigma points $\{y_i\}$ of the sampled data are transformed nonlinearly under the following equation:

$$y_i = f(x_i) \quad i = 0, \cdots 2n \tag{6}$$

After weighting the point set $\{y_i\}$ obtained by the new transformation, the statistics $\bar{y}$ and P_{yy} of the output variable y are obtained.

$$\bar{y} = \sum_{i=0}^{2n} W_i^m y_i \quad P_{yy} = \sum_{i=0}^{2n} W_i^c (y_i - \bar{y}_i)(y_i - \bar{y}_i)^T \tag{7}$$

Combined with the UT transform described above to deal with the non-linear transfer of mean and covariance, the UKF algorithm can be realized. Due to the noise factor, the state equation for the system must be enlarged in the UKF algorithm.

$x^a = \left[x^T, v^T, n^T\right]^T$ is taken as the state estimation at k-time, and the specific algorithm flow is as follows:

$$\hat{x} = E(x_0) \tag{8}$$

$$P_0 = E((x_0 - \hat{x}_0)(x_0 - \hat{x}_0)^T) \tag{9}$$

Then, extend the initial condition of the state:

$$\hat{x}_0^a = E(x_0^a) = [\hat{x}_0^a; 0, 0] \tag{10}$$

$$P_0^a = E((x_0^a - \hat{x}_0^a)(x_0^a - \hat{x}_0^a)^T) = \begin{bmatrix} P_0 & 0 & 0 \\ 0 & Q & 0 \\ 0 & 0 & R \end{bmatrix} \tag{11}$$

Next, through an appropriate sampling strategy, the state estimation sigma point set at k-time is obtained ($i = 0, 1, 2, \dots, 2n$), and i is the number of sigma points of the sampled data. It should be noted that the state dimension is $n + q + m$ and X_i^x is the column vector composed of the n dimensions of X_{ia}; is the column vector from $n + 1$ to $n + q$ dimensions of X_i^a; and X_i^w is the column vector from $n + q + 1$ to $n + q + m$ dimensions of X_i^a. The sampling points will then be transferred using the system's state equation:

$$X_i^x(k + 1|k) = f\left[X_i^k(k|k), u(k), X_i^v\right] \tag{12}$$

As the transfer sampling is completed, the prediction sampling point $X_i^x(k + 1|k)$, weighting W_i^m and W_i^c, is used to calculate the predicted mean $\hat{x}(k + 1|k)$ and covariance $P(k + 1|k)$.

$$\hat{x}(k + 1|k) = \sum_{i=0}^{2n} W_i^m X_i^x(k + 1|k) \tag{13}$$

$$P(k + 1|k) = \sum_{i=0}^{2n} W_i^c (X_i^x(k + 1|k) - \hat{x}(k + 1|k))(\cdot)^T \tag{14}$$

The predicted forecast measurement sampling points can be expressed as:

$$z_i(k + 1|k) = h[X_i^x(k + 1|k), u(k), X_i^w(k + 1)] \tag{15}$$

Then, predict the measured value and covariance:

$$\hat{z}(k + 1|k) = \sum_{i=0}^{2n} W_i^m z_i(k + 1|k) \tag{16}$$

$$P_{zz}(k + 1|k) = \sum_{i=0}^{2n} W_i^c (z_i(k + 1|k) - \hat{z}(k + 1|k))(\cdot)^T \tag{17}$$

$$P_{zz}(k + 1|k) = \sum_{i=0}^{2n} W_i^c (X_i^x(k + 1|k) - \hat{x}(k + 1|k))(z_i(k + 1|k) - \hat{z}(k + 1|k))^T \tag{18}$$

The state vector and variance are updated when the aforementioned procedures have been finished, and the UKF gain is determined:

$$K(k + 1) = P_{xz}(k + 1|k)P_{zz}^{-1}(k + 1|k) \tag{19}$$

$$\hat{x}(k + 1|k + 1) = \hat{x}(k + 1|k) + K(k + 1)(z(k + 1) - \hat{z}(k + 1|k)) \tag{20}$$

$$P(k + 1|k + 1) = P(k + 1|k) - K(k + 1)P_{zz}(k + 1|k)K^T(k + 1) \tag{21}$$

In other words, any state in the UKF may be represented by numerous sigma points. When creating a new nonlinear function, just insert the sigma points into the function value, and then compute a new state from the new function value.

2.3. State Estimation and Filtering Model of the Dynamic Positioning Offshore Platform under UKF

Typically, just the surge, sway, and yaw movements of the offshore platform must be addressed for dynamic placement. Two coordinate systems must be specified in order to characterize the movements of the dynamic positioning offshore platform, as follows: (i) the earth inertial coordinate system, also known as the global coordinate system O-xyz, which is mainly used to describe the position and heading vector of the offshore platform $\eta = [x,y,\Psi]^T$; and (ii) the local coordinate system O_0-$x_0y_0z_0$ of the offshore platform, which is fixed on the offshore platform and moves with it and its coordinate origin is usually set at the center of gravity of the offshore platform to describe its velocity vector $v = [u,v,r]^T$. Two coordinate systems are shown in Figure 1.

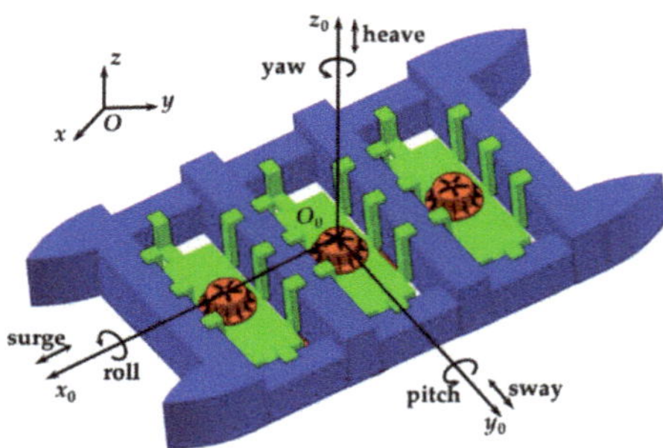

Figure 1. Two coordinate systems.

The low-frequency motion model of the dynamic positioning offshore platform can be expressed as follows:

$$\dot{\eta} = R(\psi)v \tag{22}$$

$$M\dot{v} = -Dv + u + R^T(\psi)b + w_1 \tag{23}$$

$$\dot{b} = -T_b^{-1}b + w_2 \tag{24}$$

where M is the inertial force matrix of the offshore platform with additional mass and D is the linear damping matrix of the offshore platform, u is the control force matrix of the platform, b is a deviation term describing low-frequency environmental disturbances and un-modeled dynamics, and w_1 and w_2 are random processes of zero mean white noise. $R(\Psi)$ is the rotation matrix of an offshore platform and can be defined as follows:

$$R(\psi) = \begin{bmatrix} \cos\psi & -\sin\psi & 0 \\ \sin\psi & \cos\psi & 0 \\ 0 & 0 & 1 \end{bmatrix} \tag{25}$$

The motion model of an offshore platform is non-linear due to the presence of a rotation matrix; nonetheless, in order to simplify the issue, it has been linearized. The offshore platform's parallel coordinate system is introduced. The axes of the coordinate system are parallel to the local coordinate system of the offshore platform, and its origin is the same as that of the earth's coordinate system. The motion vector of the offshore platform in the parallel coordinate system is represented by η_p, and the motion transformational relation between the parallel coordinate system and the earth coordinate system can be expressed as:

$$\eta_p = R^T(\psi)\eta \tag{26}$$

$$\dot{\eta}_p = v \tag{27}$$

The linear low-frequency motion state space model of the offshore platform in the parallel coordinate system is obtained by:

$$\begin{cases} \dot{x}_L = A_L x_L + B_L u + E_L w_L \\ y_L = H_L x_L + n_L \end{cases} \tag{28}$$

where $x_L = [\eta_p{}^T, v]^T$ is the state variable, y_L is the controlled output variable, w_L is the disturbance vector including the deviation term b, and n_L is the measurement of Gauss white noise. The matrices in Equation (28) are defined as follows:

$$A_L = \begin{bmatrix} 0_{3\times3} & I_{3\times3} \\ 0_{3\times3} & -M^{-1}D \end{bmatrix}, B_L = \begin{bmatrix} 0_{3\times3} \\ M^{-1} \end{bmatrix}$$
$$E_L = \begin{bmatrix} 0_{3\times1} \\ M^{-1} \end{bmatrix}, C_L = \begin{bmatrix} I_{3\times3} & 0_{3\times3} \end{bmatrix} \tag{29}$$

Based on the low-frequency motion model, the high-frequency motion of the offshore platform caused by the first-order wave force should also be considered. The spatial model of the high-frequency motion state of the dynamic positioning offshore platform can be expressed as follows:

$$\dot{\xi} = A_\omega \xi + E_\omega w_3 \tag{30}$$

$$\eta_\omega = C_\omega \xi \tag{31}$$

The vector of wave motion ξ can be written as:

$$\xi = [\xi_x, \xi_y, \xi_z, x_w, y_w, \psi_w]^T \tag{32}$$

w_3 zero mean white noise process, $\eta_w = [x_w, y_w, \Psi_w]^T$, is the high-frequency motion vector of waves. The matrices in Equations (30) and (31) are defined as follows:

$$A_\omega = \begin{bmatrix} 0_{3\times3} & I_{3\times3} \\ -\Omega_{3\times3} & -\Lambda_{3\times3} \end{bmatrix}, E_\omega = \begin{bmatrix} 0_{3\times1} \\ I_{3\times1} \end{bmatrix}, C_\omega = \begin{bmatrix} 0_{3\times1} & I_{3\times1} \end{bmatrix} \tag{33}$$

$$\Omega = diag\left\{\omega_{01}^2, \omega_{02}^2, \omega_{06}^2\right\} \tag{34}$$

$$\Lambda = diag\{2\zeta_1\omega_{01}, 2\zeta_2\omega_{02}, 2\zeta_6\omega_{06}\} \tag{35}$$

where ω_{0i} is the peak frequency of the wave and ζ_i is the relative damping coefficient, which is usually taken from 0.05 to 0.2.

The state estimation and filtering model of the dynamic positioning offshore platform might be obtained by considering the offshore platform low- and high-frequency motion model in Equations (22)–(24) and Equations (30) and (31), respectively:

$$\begin{cases} \dot{x} = Ax + Bu + Ew \\ y = Hx + n \end{cases} \tag{36}$$

The state vector x and the noise vector w can be written as the following, respectively:

$$x = [\xi^T, \eta^T, b^T, v^T]^T \tag{37}$$

$$w = [w_1{}^T, w_2{}^T, w_3{}^T]^T \tag{38}$$

$y = \eta + \eta_w + n$ is the measurement model, including the high- and low-frequency motion of the offshore platform, and n is the measurement noise of the system. Based on the state estimation and filtering model of the dynamic positioning offshore platform, Equation (36) can be rewritten as a time-discrete form:

$$\begin{cases} x_{k+1}=\Phi x_k + \Delta u_k + \Gamma w_k \\ \quad = f(x_k, u_k) + \Gamma w_k \\ y_k=H x_k + n_k = h(x_k) + n_k \end{cases} \tag{39}$$

$$\Phi = \exp(A \cdot \Delta t), \Delta = A^{-1}(\Phi - I)B, \Gamma = A^{-1}(\Phi - I)E \tag{40}$$

where Δt is the sampling time; x_k is the state vector of the system at K time step; and y_k is the state measurement of the system at K time step. The system noise w_k is Gaussian white noise with a nonzero mean and its variance is Q_k; the measurement noise n_k is a white Gauss noise independent of system noise and its variance is R_k.

Based on the dynamic positioning offshore platform nonlinear discrete system expressed in Equation (39), the detailed process of state estimation and filters using the unscented Kalman filter is as follows:

(1) Define the state and covariance initial value:

$$\hat{x}_0 = E(x_0), P_0 = E[(x_0 - \hat{x}_0)(x_0 - \hat{x}_0)^T] \tag{41}$$

(2) Calculate sigma points at K time step and select appropriate weights:

$$\chi_{k-1} = \begin{bmatrix} \hat{x}_{k-1} & \hat{x}_{k-1} + \sqrt{n+\lambda}\sqrt{P_{k-1}} \\ & \hat{x}_{k-1} - \sqrt{n+\lambda}\sqrt{P_{k-1}} \end{bmatrix} \tag{42}$$

$$\begin{cases} W_0^m = \lambda/(n+\lambda), \\ W_0^c = \lambda/(n+\lambda) + (1 - \alpha^2 + \beta) \\ W_i^m = W_i^c = \lambda/[2(n+\lambda)], i = 1,2,\ldots,2n \end{cases} \tag{43}$$

where W_i^m and W_i^c (i = 0,1, ..., 2n) represent the weights of sigma point mean and variance, respectively; n is the dimensionality of the system; $\lambda = [\alpha^2(n+k) - n]$ is the scale factor; α determines the distribution of sigma points around the mean point, usually between 10^{-4} and 1; k is usually taken as 0; and β determines the distribution state of prior state estimation. For a Gaussian distribution, $\beta = 2$ is the best.

(3) Time update: the state and measurement Equation (39) is used to carry out UT on the sigma point to obtain the prior state and the predicted value of the measurement output, and the predicted value of its covariance.

$$\chi_{k|k-1}^i = f(\chi_{k-1}, u_{k-1}), i = 0, 1, \ldots, 2n \tag{44}$$

$$\hat{x}_{k|k-1} = \sum_{i=0}^{2n} W_i^m \chi_{k|k-1}^i \tag{45}$$

$$P_{k|k-1} = \sum_{i=0}^{2n} W_i^c (\chi_{k|k-1}^i - \hat{x}_{k|k-1})(\chi_{k|k-1}^i - \hat{x}_{k|k-1})^T \\ + \Gamma_{k-1}Q_{k-1}\Gamma_{k-1}^T \tag{46}$$

$$y_{k|k-1}^i = h(\chi_{k|k-1}^i), i = 0, 1, \ldots, 2n \tag{47}$$

$$\hat{y}_{k|k-1} = \sum_{i=0}^{2n} W_i^m y_{k|k-1}^i \tag{48}$$

$$P_k^{yy} = \sum_{i=0}^{2n} W_i^c (y_{k|k-1}^i - \hat{y}_{k|k-1})(y_{k|k-1}^i - \hat{y}_{k|k-1})^T + R_k \tag{49}$$

$$P_k^{xy} = \sum_{i=0}^{2n} W_i^c (\chi_{k|k-1}^i - \hat{x}_{k|k-1})(y_{k|k-1}^i - \hat{y}_{k|k-1})^T \tag{50}$$

(4) Measurement update: calculate UKF gain matrix K_k, state estimation value, and state error covariance matrix P_k.

$$K_k = P_k^{xy}(P_k^{yy})^{-1} \tag{51}$$

State estimation value:

$$\hat{x}_k = \hat{x}_{k|k-1} + K_k(y_k - \hat{y}_{k|k-1}) \tag{52}$$

$$P_k = P_{k|k-1} - K_k P_k^{yy} K_k^T \tag{53}$$

The above procedures are performed continuously at each sample period, and the UKF state estimate filter value at each time is used as the starting point for forecasting future offshore platform dynamics. By modifying and updating Q and R simultaneously, it is simple to produce the divergence of filtering results when both the process noise Q and the measurement noise R are unknown. The direct distinction between the state transition and the actual process is represented by Q, which is the covariance of the process excitation noise.

The selection of process noise Q is frequently problematic since the process cannot be observed directly. In engineering issues, the choice of process noise is sometimes a trade-off based on experience between convergence rate and steady-state accuracy. When the noise value in the process is high, the filtering convergence speed is rapid but the stability is weak, and vice versa. In the target tracking and positioning problem, on the one hand, we want to track the response fast when the target is navigating, so we should pick a bigger process noise, but when the target is moving smoothly, the filtering error will be greater. On the other hand, if we choose a lower process noise to enhance the steady-state estimate accuracy, the tracking error will increase abruptly owing to the underestimation of the maneuverability of the target. The offshore platform in this research belongs to the position-holding model rather than the target's large-scale maneuvering model because it is positioned in a specific spatial coordinate location based on specific operating characteristics. Therefore, a relatively small process noise covariance Q could be set to reduce the influence of model error under the premise of fast convergence speed after repeated debugging. For R, both too-large values and too-small values will make the filtering effect worse. The smaller the value of R, the faster the convergence will be. Therefore, considering the characteristics of Q and R, in order to improve the accuracy and ensure that the filtering results do not diverge, Q and R are generally adjusted and updated in inverse proportion. In this work, Q is considered as 0.01, and R is taken as 10 after repeated debugging and weighing.

In practical engineering, the true values of x_0 and P_0 are often uncertain or unknown, so they can only be assumed. The problem of filter stability is to investigate the effect of filter initial value selection on filter stability; that is, as filter time increases, the state estimation value x_k and the state error covariance matrix P_k become increasingly independent of the initial estimation values x_0 and P_0, respectively. According to the researcher's hypothesis, if P_0 cannot be determined precisely, we can utilize its higher, more conservative value by selecting a range of probable values. The actual filtering error matrix can fulfill the requirements as long as the estimate error covariance matrix derived in the filtering computation meets the requirements. Therefore, if the system model is assumed to be accurate, only the effects of the initial state value x_0, initial covariance value P_0, process noise Q, and measurement noise R are considered. Since the UKF algorithm has strong robustness, it is insensitive to filtering parameters, especially for more complex nonlinear systems. Therefore, the setting of P_0 is relatively loose. In general, as long as it is greater than 0, it can converge and have no effect on the filtering effect. As a result, when there is little confidence in the starting value of the parameters, picking the bigger one can increase the convergence speed and accuracy of parameter identification. Under normal circumstances $P_0 = k \times I_{3\times3}$, k can be taken as a large number. In this paper, the k of P_0 is taken as 100 and x_0 is the positioning target position of the offshore platform (Target X = +30 m, Target Y = −20 m, Target Yaw = 90°).

2.4. Calling up the Unscented Kalman Filter in Python

Python programming might be used to accomplish the selection of sigma points and the computation of weights in accordance with the specific calculation procedures and formulae provided above. At the same time, we may utilize the pre-made tool FilterPy in Python because Python includes several built-in function libraries and tool libraries. After installing the PIP tool, one can directly enter the command "PIP install filterpy, PIP install numpy" to complete the installation of filterpy and numpy. Subsequently, downloading and installing the numpy, scipy, numpydoc, and nose modules together, it is very simple and convenient to use Python to call it, since it has good convergence. By using "from pykalman import unscentedkalman filter", the unscented Kalman filter module can be called out quickly.

2.5. Modeling of DP Offshore Platform-Riser Multi-Body System Based on UKF-PID Control with Python Language Embedded in OrcaFlex

As an object-oriented dynamic scripting language, Python does not require any compiler or linking steps when accessing the application program interface (API). Due to the flexibility of Python in processing data types, the Python interface is designed as a wrapper to access the internal functions of the dynamic link library (Dll). This may enhance program performance more effectively and be used for storage management and casting in the Python interface without the need for initial variable declaration. The OrcaFlex API module uses Python to compress many interfaces, classes, and functions for the C API. At the same time, the object data names in OrcaFlex can also be completely copied to the objects of the Python interface, and then its functions can be transferred to the C API of OrcaFlex. The definition and assignment of some coefficients can also be carried out at the same time by changing the names of specific attribute values. The motion system of the subframe in relation to the parent frame may be used to describe ship motion. For a moving ship, the subframe is its own local coordinate system, while the parent frame is the overall coordinate system. Python external functions can be used to confine and control the mobility of a floating offshore platform or ship. OrcaFlex will automatically find four methods defined by the interface in Python external functions: initialize(), calculate(), storestate(), and finalize(). These methods will be called directly if they exist.

Since the function algorithm in PID control incorporates state information, the state information must be stored using the TExtFnInfo structure. Its state information is stored in the data variable of the TExtFnInfo structure to guarantee that it may be promptly and effectively loaded when reloading a partially finished model. When the model starts to run, OrcaFlex will set up a Python environment for Python coding and import the modules in OrcFxAPI and External Function into the environment, then it will create each function unit in the External Function interface in the OrcaFlex model and call related objective functions. It should be pointed out that when Python and OrcaFlex are combined in OrcaFlexAPI, for each new time step, the parameter attribute of info.NewTimeStep is set to True at the beginning of OrcaFlex. Only the implicit integration approach may be utilized in this situation to embed and compile the Python language in OrcaFlex, as the explicit integration method is inapplicable. The control system compares the sway, surge, and yaw of the DP vessel with the target value through the external function. Through the calculation of the controlled equation, the required reaction force and reaction moment of the DP vessel are obtained, and the thrust distribution is carried out according to the relevant principles of thrust distribution.

$$F_{x,y} = f(e_{x,y}) \tag{4}$$

$$M_z = f(e_\theta) \tag{55}$$

$$f(e_{x,y,\theta}) = K_p e + K_i \int e\, dt + K_d de/dt + F_w(a_w, v_w) \tag{56}$$

where $F_{x,y}$ is the restoring resultant force that contains the longitudinal restoring force F_x of the vessel's surge motion and the lateral restoring force F_y of the vessel's sway motion; M_z is the restoring moment of the ship's yaw motion; e_x is the difference between the surge and the target value, e_y is the difference between the sway and the target value, and e_θ is the difference between the vessel's yaw angle and the target value; K_p is the proportional gain, K_i is the integral gain, and K_d is the differential gain; and a_w is the wind angle, v_w is the wind speed, and F_w is the force or moment of the wind acting on the vessel. In this paper, the required restoring force of the platform is 1000 kN for every 1 m of displacement, and the required restoring moment of the platform is 1000 kN. m for every $1°$ of deflection in the yaw direction. The proportional coefficient of force K_{pf} is 10, the proportional coefficient of moment K_{pm} is 8, K_i is 0.02, and K_d is 5. Through the cut-and-trial method, the three coefficients of the PID control are adjusted and calculated. The specific process is: when the initial offset of the platform is large, the proportional gain coefficient K_p is mainly adjusted, and then the integral gain coefficient K_i is adjusted after the offset of the platform reaches a certain degree. After the two coefficients are adjusted to a certain extent, the two coefficients are kept unchanged within a certain range, and then the differential gain coefficient K_d is slightly adjusted by using the control variable method until the positioning effect reaches a satisfactory level.

The riser is discretized into a model with lumped mass parameters. The published literature by Bai et al. [50] provides a full introduction to the lumped mass approach and the coupling modeling method of the riser and offshore platform, which will not be explored here. The modeling of the dynamic positioning offshore platform-riser multi-body coupling system with UKF-PID control based on Python-language embedded programming in OrcaFlex is mostly complete after the procedures outlined above.

3. Validation

The corresponding model is established, the theory put forth in this paper is combined with the specific parameters in the dynamic positioning offshore platform-riser model system created by Sorensen et al. [51] and Leira et al. [52], and the results of the model are compared with the model by Asgeir J. Sorensen to assess the method's rationality and accuracy. Most of the offshore platform characteristics are the same; the main difference is that the dynamic positioning approach described in this paper is used. The total length of the riser used for simulation is 1000 m, the radius of the riser is 0.25 m, the wall thickness of the riser is 0.025 m, the modulus of elasticity E = 2.12×10^8 Pa, the top tension is 2500 kN, and the bottom tension is 1200 kN (in this paper, the same effect can be achieved directly by applying the corresponding pretension at the top and bottom in the initial stage of simulation). The riser is divided into 10 segment units.

The specific ocean environment load parameters are as follows: the average surface current velocity after regression is V_C = 1 m/s, and it is assumed that the current velocity at the middle layer is 75% of the surface current velocity, the current velocity on the seabed is 15% of the surface current velocity, and the current direction is $30°$. The profile of the current in the water depth direction is shown in Figure 2. The significant wave height is 7 m, the peak period is 14 s, and the direction is $20°$. The average wind speed is 15 m/s and the wind direction is $20°$.

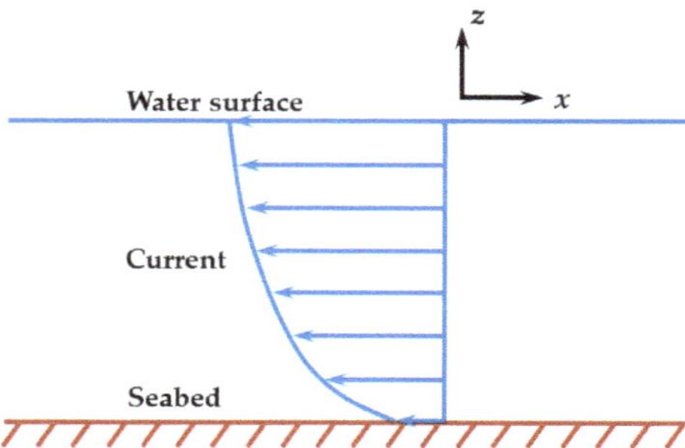

Figure 2. View of the current profile.

Just as shown in Figure 3, The angle of the riser changes as the offset of the offshore platform changes. The appropriate riser angle and platform offset curves produced from the theory in this study and those obtained from the theory in Leira et al. [52] are compared.

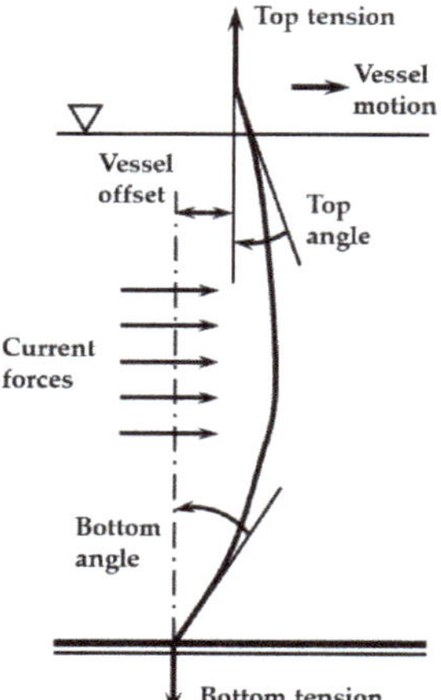

Figure 3. Validation model of the riser.

The curves in Figure 4a,b demonstrate that, within the acceptable range of errors, the general trend of curve changes and the overall shape distribution are substantially coincident with the results of the work by Leira et al. [52], with minimal variation in values. This demonstrates the feasibility of the theory and approach suggested in this work. The trajectory curve of the Tai'an Kou semi-submersible vessel in the X direction is compared with that of the Tai'an Kou semi-submersible vessel in the X direction with the traditional Kalman filter mode in Dr. Liu's paper to compare the effect of the unscented Kalman filter and traditional Kalman filter, with reference to the specific parameters of the Tai'an Kou semi-submersible vessel in the Ph.D. thesis of Liu [53]. The dimensionless hydrodynamic coefficient, the inertial mass matrix, and the damping matrix of the semi-submersible vessel are all introduced in detail in Dr. Liu's thesis, along with the displacement and other pertinent characteristics of the Tai'an Kou semi-submersible vessel. In the simulation process, the force and torque input by the ship's dynamic positioning system change with time, just as shown in Equation (57):

$$\tau = \begin{bmatrix} 9000|\sin(0.03\,t)| \\ 7500|\sin(0.03\,t)| \\ 4500|\sin(0.03\,t)| \end{bmatrix} \tag{57}$$

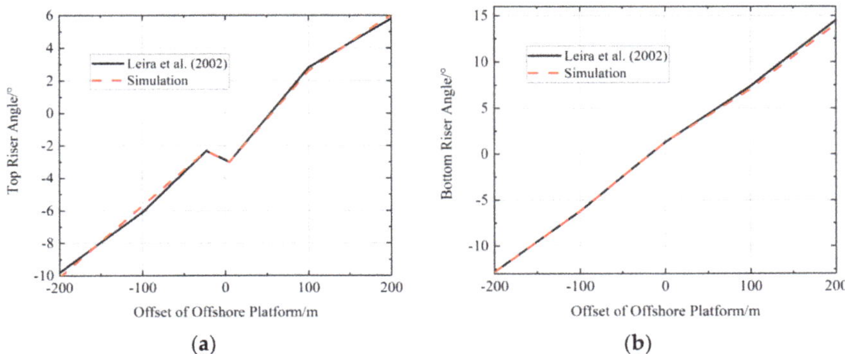

Figure 4. (**a**) Top riser angle versus offset of offshore platform. (**b**) Bottom riser angle versus offset of offshore platform[52].

The motion trajectory curve of the Tai'an Kou semi-submersible vessel in the X direction in the unscented Kalman filter mode and the motion trajectory curve in the X direction in the traditional Kalman filter mode in Dr. Liu's thesis are compared.

The trajectories of the vessel in the X direction under the two filtering modes (conventional Kalman filter and unscented Kalman filter) are practically identical at the beginning stage of the simulation, as shown in Figure 5. With the extension of time, the ship trajectory in UKF mode is closer to the ideal trajectory, and the trajectory curves in the two filtering modes begin to have obvious differences at 800 s. This difference shows a trend of further expansion with the increase in time. After more than 800 s, the traditional Kalman filter has a divergent tendency as time goes on. Therefore, the trajectory of the vessel with unscented Kalman filter mode in the X direction is closer to the ideal trajectory. The ship trajectory in UKF mode approaches the optimum trajectory as time passes, and the trajectory curves in the two filtering modes begin to diverge at 800 s. This disparity demonstrates a pattern of increasing enlargement as time passes. The standard Kalman filter has a diverging tendency after more than 800 s. As a result, the overall trajectory of the ship using the unscented Kalman filter mode in the X direction is more similar to the ideal trajectory. The dynamic response speed of this method is relatively fast based on the time-domain fluctuation of the vessel's translation curve in the X direction, and the fast-positioning dynamic response speed means that the time required for the control process will be relatively shortened, which has some reference significance and practical value for practical engineering, and it can also improve the safety margin in practical engineering to some extent. At the same time, it has the potential to increase the economic advantages of offshore positioning operations.

To demonstrate the advantages of the unscented Kalman filter, it must be compared to the extended Kalman filter mode. Unfortunately, most of the parameters presented in the literature on the use of EKF in ship dynamic positioning systems that can currently be discovered are inadequate; therefore, the model reduction comparison analysis cannot be performed in the same way as the standard Kalman filter can. However, with further research and effort, a more idealistic approach of contrasting the extended Kalman filtering mode with the unscented Kalman filtering mode was discovered in Bao's master's dissertation [54]. In this case, the approach described by Bao was immediately used, and the trajectory data of the two filtering modes generated by Bao were collected again, analyzed, and compared. The impacts of the two filtering mechanisms were analyzed and compared in this indirect manner. To acquire the filtering effect, Bao simulated the state space equation of the generic nonlinear system, which is represented by the following formula:

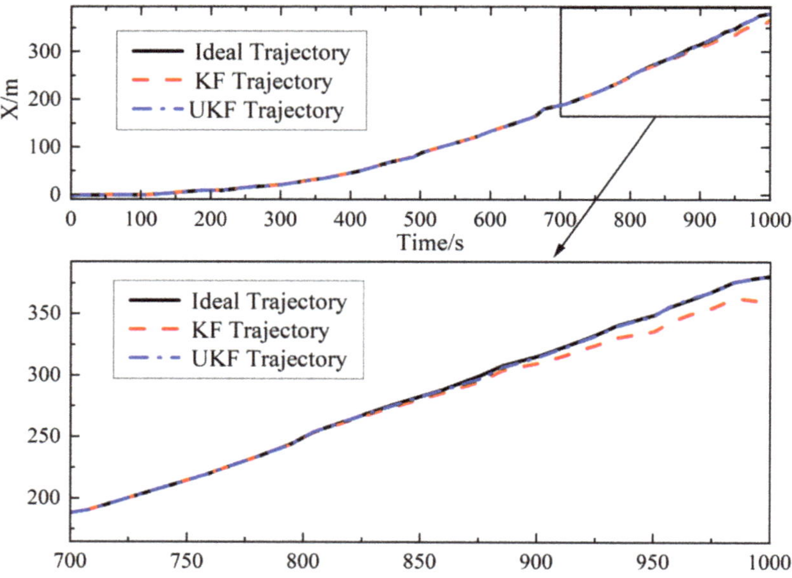

Figure 5. Trajectories of "Tai'an Kou" in the time domain with UKF and KF.

$$\begin{cases} X(k+1) = AX(k) + BU(k) \\ \qquad Z(k) = HX(k) \end{cases} \tag{58}$$

To evaluate the effect, Bao [54] assumes that the variance between the observation noise and the process noise of the preceding nonlinear process is 5 and that the noise is created at random. In Figure 6, it is found that the error correction effect in unscented Kalman filter mode is superior to that in extended Kalman filter mode. The trajectory in the unscented Kalman filter mode is closer to the ideal trajectory in the whole-time domain. This is demonstrated by the fact that the position estimate deviation with unscented Kalman filter mode is less than that with extended Kalman filter mode at any moment in the whole-time domain. Although the deviation of the trajectory in both filtering modes increases at the start of the simulation, the divergence of the trajectory in the unscented Kalman filter mode decreases fast and becomes stable, and the filtering becomes stable. However, the extended Kalman filter mode's trajectory deviation will develop swiftly and continually, and the extended Kalman filter has a diverging tendency.

When the trajectory deviation is steady in the time domain, the unscented Kalman filter may minimize the trajectory deviation by 80% when compared to the extended Kalman filter mode, and the correction impact will steadily rise with time. Even at the start of the simulation, when the divergence in both filtering modes is large, the unscented Kalman filter may reduce the trajectory deviation by 25%. To summarize, because the dynamic positioning process of ships and offshore platforms is non-linear, the classic Kalman filter and extended Kalman filter ignore high-order elements while linearizing the non-linear process, resulting in varying degrees of model mismatch. The unscented Kalman filter is more insensitive to noise items than the traditional Kalman filter and extended Kalman filter, and UT transformation skips the phase of system linearization, reducing the possibility of model mismatch problems during linearization, so its advantage is more obvious when dealing with non-linear problems.

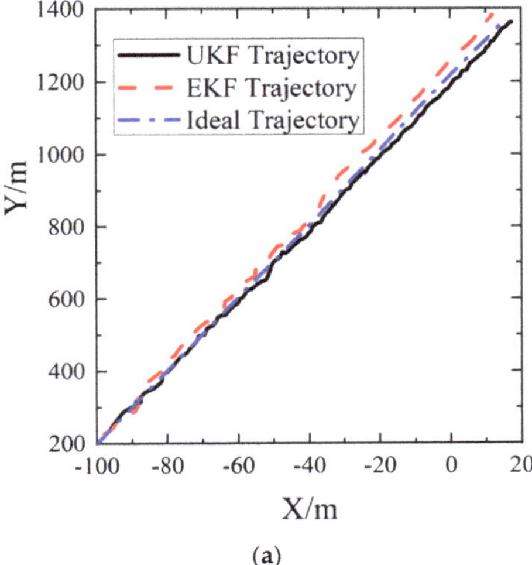
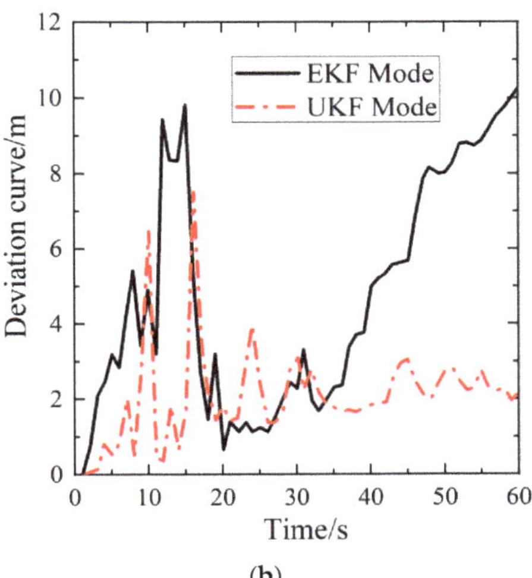

Figure 6. (**a**) Trajectories with EKF and UKF. (**b**) Deviation curves with EKF and UKF.

The above DP platform-riser system considered neither the coupling effects of riser and platform motion nor the effects of waves. Therefore, the above verification has some limitations, and it is impossible to determine the applicability of the coupling theory proposed in this paper in the presence of waves. For this reason, the establishment of the platform-riser multi-body coupling theory based on the lumped mass method proposed in this paper is compared with the same coupling model of the offshore platform riser multi-body system based on the SESAM software proposed by Gu et al. [55]. Because they have compared the calculation results with the model test results in the wind-wave tank, the correctness of the calculation results of the same coupling model of the offshore platform riser multi-body system proposed by them based on the SESAM software has been verified. Therefore, if the calculation results of the verification model are compared with the results of the same coupling model of the platform-riser multi-body system proposed by them based on the SESAM software, the correctness of the theory proposed in this paper can be further verified under wave load.

The quadrilateral tension leg platform is shown in Figure 7. The platform consists of four square buoys and four square pontoons. The specific parameters of the offshore platform-riser system are as follows. The design draft is 25.25 m, the diameter of the buoy is 17.4 m, and the center distance of the pontoons is 51.4 m. The length, width, and height of the pontoon are 34 m, 11.6 m, and 8.7 m, respectively. The displacement is 45536 t. In order to be completely consistent with the parameter settings in [55], when the heave, sway, and roll are verified, the wave height is set as 8 m and the wave period is set as 10 s. When the tension is verified, the wave height is set as 12 m and the wave period is 10 s. The water depth is 550 m. Since the RAO of a 270° wave direction is clearly given in [55], the wave direction is set as 270°. However, it should be noted that although the RAO in [55] is given when the wave direction is 270°, the authors did not give the all calculation results when the wave direction is 270°. After further careful review, we found that the calculation results in the original literature were relatively complete when the wave direction was 90°. Since the platform is a quadrilateral platform that is completely symmetrical on the left, right, front, and back, and the layout of risers and tension legs is also completely symmetrical on the left, right, and front, the mechanical parameters of the four risers are identical, and the mechanical parameters of eight tension legs are also identical. The difference between a 90° wave direction and a 270° wave direction is 180°, which is on the same straight line but in opposite directions. Therefore, the cases of 90° wave direction and 270° wave direction are completely symmetrical for the hydrodynamic characteristics of the system. Therefore, the calculation results in these two cases have symmetry.

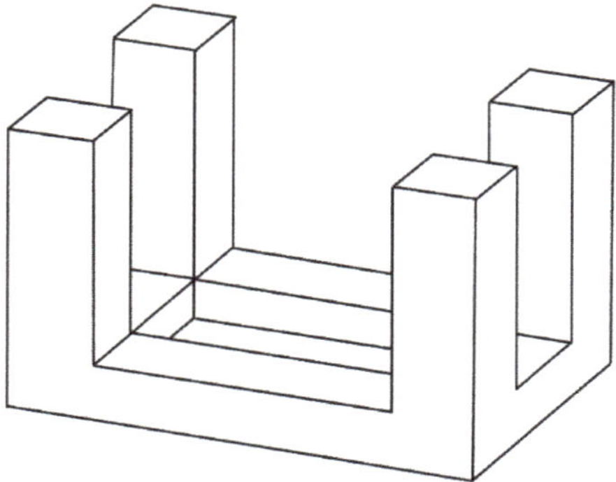

Figure 7. Validation model of a platform.

The validation results are shown in Figure 8. The time-domain curve of heave under the 270° wave direction of the simulation comparison model coincides with that of the 90° wave direction of [55], and its heave amplitude increases by 4% compared with reference [55]. The time-domain curve of sway under the 270° wave direction in the simulation is basically the same as the value on the time-domain curve of sway in the 90° wave direction in [55], and the direction is opposite (the absolute value difference of the down sway value of the two kinds of waves is less than 3%, and the absolute value of the simulated sway is larger). The time-domain curve of roll under the 270° wave is basically the same as the value on the roll time-domain curve of the platform in the 90° wave direction in [55], and the direction is opposite (the difference of the roll value in the two kinds of waves is less than 5%, and the value of the roll is larger). The time-domain curve of tension under the 270° wave direction in the simulation is basically the same as that under the 90° wave direction in [55], and the tension value simulated at each time step is about 7–9% larger than that in the reference [55].

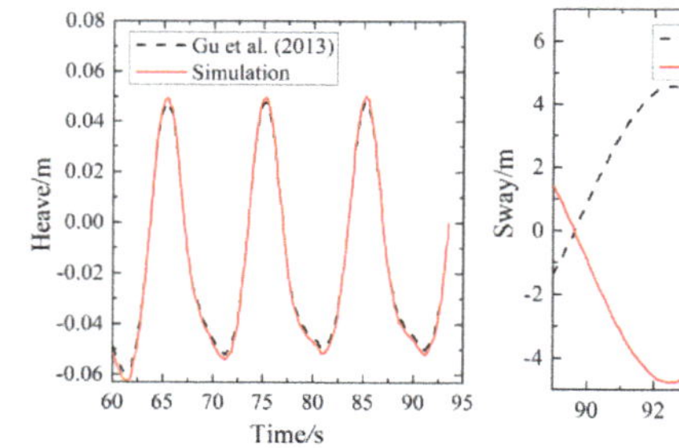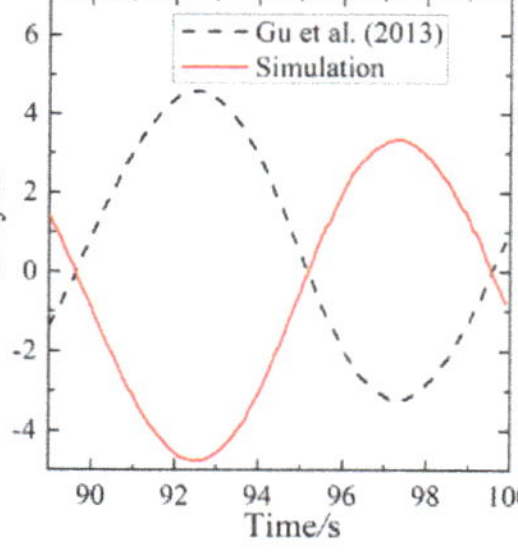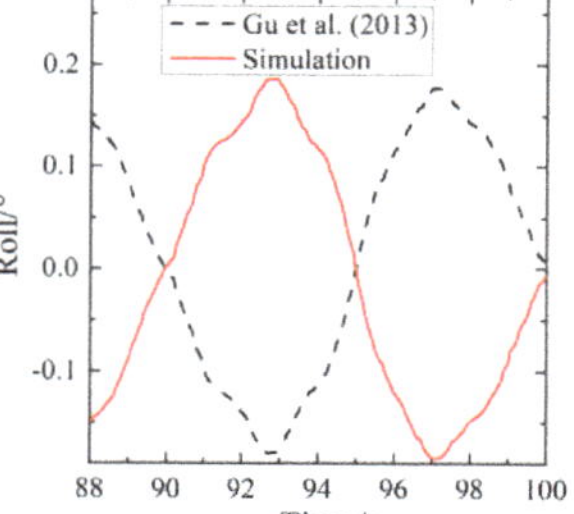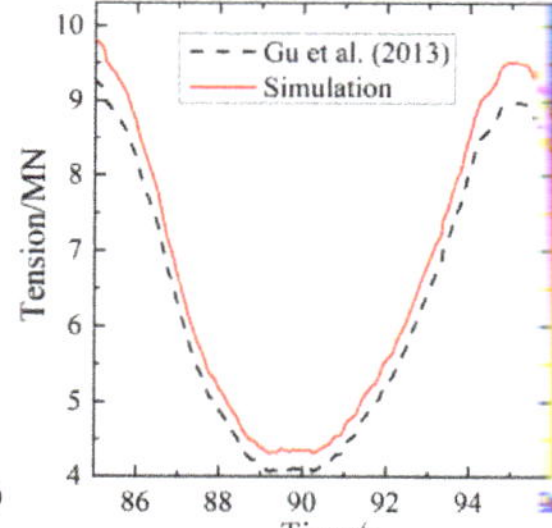

Figure 8. Bending moment and curvature of the riser under the action of UKF-PID and single PID [55].

The reason why the tension of the riser in the simulation model is slightly larger at this time is that because the amplitude of the heave, sway, and roll of the platform in the simulation comparison model is slightly increased, the increase in these three motion amplitudes of the platform will lead to the enhancement of the pulling effect on the riser, so the amplitude error of the riser tension compared with the tension value in the literature is slightly larger than the amplitude error of the platform motion. However, the small increase in the platform motion amplitude should be caused by the error that may occur when the RAO data are obtained by point sampling and regression according to the RAO curve, which makes the RAO values increase by a very small margin. In addition, because the method for risers in this paper is discretized into a lumped mass model, compared with the method for risers by Gu et al. [55], this discrete method for risers can better reflect the flexibility of offshore pipelines and make its dynamic response more obvious when the top of the riser moves.

4. Results and Discussion

4.1. Model Establishment

The total length of the riser is 531 m and is discretized by the lumped mass method. The stress-sensitive zone near the upper end of the riser is subdivided. The 0–51 m area is discretized into several special segments: the top 11 m of the riser connected to the offshore platform is lumped into a slip joint; the middle 40 m of the riser is divided into 20 segment units, 2 m for each segment unit; and the bottom 480 m of the riser is discretized into 96 segments with 5 m segmentation units. Except for the top 11 m, the rest of the 520 m is the main part of the riser.

One of the primary characteristics of the riser is that the top tension is reasonably steady, and the effective tension at the bottom end must be positive. As a result, four link elements in the standard cruciform configuration are utilized to adjust and buffer the tension at 11 m from the riser where the top end is positioned, allowing the fluctuation amplitude of the top tension to be within a defined range. The parameters of the main part of the riser are as follows: the outer diameter is 0.65 m, the inner diameter is 0.6 m, Poisson's ratio is 0.3, Young's modulus is 21,200 MPa, and the density is 7.85 t/m^3. The parameters of the slip joint are as follows: the outer diameter is 0.65 m, the inner diameter is 0.6 m, the bending stiffness is 508,944 kN·m^2, the axial stiffness is 1 kN, Poisson's ratio is 0.3, the torsional stiffness is 391,500 kN·m^2, and the linear density is 0.385 t/m. The environmental parameters are as follows: the seawater density is 1025 kg/m^3; the water depth is 500 m; the seabed is flat; the JONSWAP spectrum is chosen for the modeling; the peak enhancement factor γ is 4.7934; a is 0.0086, which is calculated by the specified H_s and T_z; σ_1 is 0.07 and σ_2 is 0.09; the significant wave height H_s is 7 m; the zero crossing period T_z is 9 s; the wave direction is 270°; the current is 0.5 m/s; and the direction of the current is the same as the wave direction. The offshore platform has eight pontoons, the draught of the offshore platform is 24.38 m, the displacement is 10,000 t, and the moments of inertia are $I_x = 500,000$ t.m^2, $I_y = 7,000,000$ t.m^2, $I_z = 7,000,000$ t.m^2, respectively. The positioning coordinates of the offshore platform are (target X = +30 m, target Y = −20 m), target heading = 90°. The schematic diagram of the model is shown in Figure 9.

The UKF-PID control of the offshore platform motion in the system is realized by combining the OrcaFlex API with a PID dynamic positioning control statement in UKF mode based on Python programming.

4.2. Calculation Analysis

4.2.1. Effective Tension of Riser under the UKF-PID and Single PID

Figure 10 depicts the distributions of effective tension in the length direction of the riser under the action of the UKF-PID (PID control with unscented Kalman filter) dynamic positioning system and single PID dynamic positioning system. It is found that the curve configurations and values of the effective tension of the riser under PID control in unscented Kalman filter mode in the length direction have little difference, which means that the effective tension of the riser does not change much in the length direction in this mode. The presence of an unscented Kalman filter has no effect on the distribution of the effective tension of the riser in the length direction, and its influence on the change of tension at a specific position on the riser is quite weak. The curves of the effective tension spectral density at the three ends of the riser and the standard deviation distribution of the effective tension in the length direction are also very similar to the corresponding riser curves in the dynamic positioning offshore platform-riser multi-body system under a single PID control. This demonstrates that the inclusion of an unscented Kalman filter has no effect on the magnitude of riser tension, time-domain fluctuation, distribution of riser tension throughout the length of the riser, coordination, and synchronism of riser tension variation. In other words, the inclusion of the unscented Kalman filter has less of an impact on the riser's tension variation than the multi-body offshore platform riser controlled by a single PID, which is completely represented in the riser's fluctuation, coordination, and synchronization of tension variation.

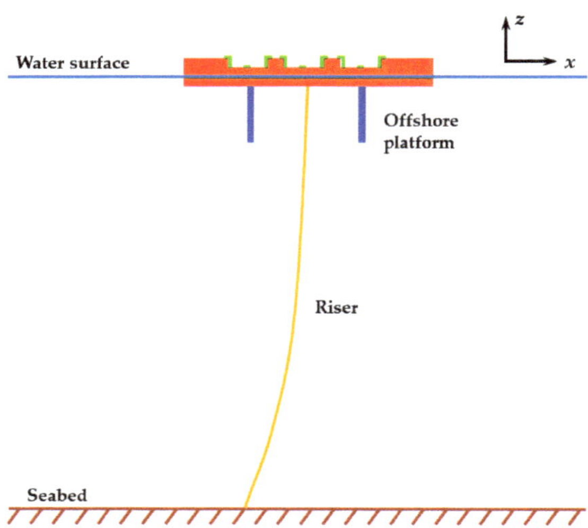

Figure 9. Model of the rigid–flexible fluid multi-body system with a dynamic positioning system of UKF-PID.

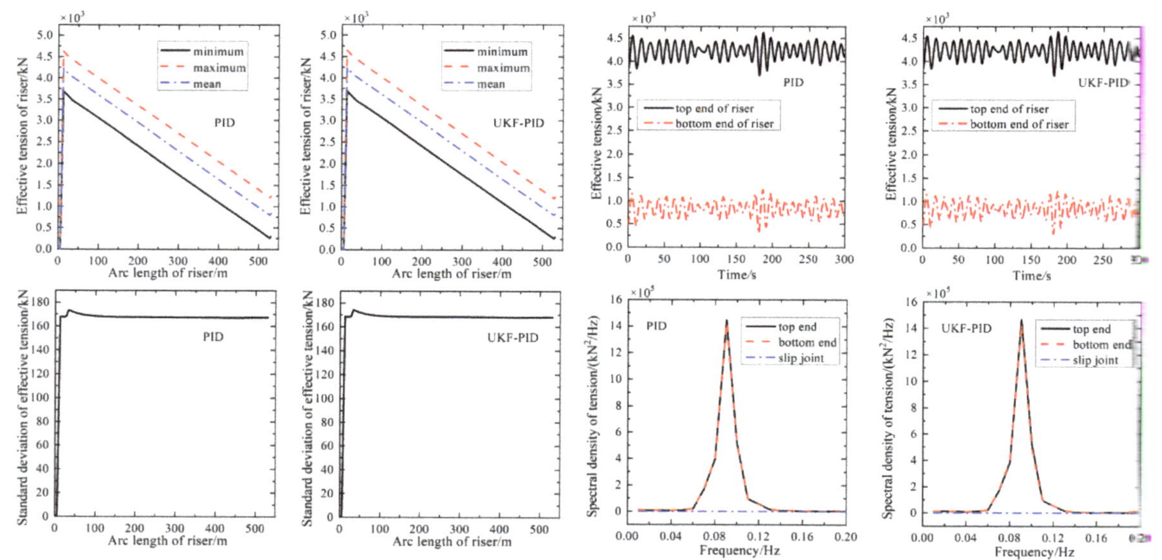

Figure 10. Dynamic response of the effective tension of the riser under the action of UKF-PID and single PID.

4.2.2. Bending Moment and Curvature of Riser under UKF-PID and Single PID

Figure 11 depicts the curves of the distributions of the bending moment and curvature in the length direction of the riser under the action of the UKF-PID dynamic positioning system and single PID dynamic positioning system. It is discovered that the bending moment of the riser under the UKF-PID dynamic positioning system is greater than the position under a single PID control at a certain position in the length direction, but the position where the riser is subjected to greater bending does not change in the whole length direction, implying that the bending moment of the riser increases greatly in the whole-length direction with the addition of the unscented Kalman filter but does not change the position. It can be seen from the variation of the effective tension of the riser under the action of the UKF-PID dynamic positioning system that the existence of the UKF-PID dynamic positioning system does not change the distribution and variation of the axial tension of the riser, but changes the distribution and variation of the lateral bending load of the riser. The standard deviations of the riser curvature under the influence of the UKF-PID dynamic positioning system not only increase throughout the entire-length direction but also the distributions of the curvature standard deviation curves in the length direction of the riser have changed significantly. The curvature standard deviation curves exhibit stronger stepped distributive behavior compared with the corresponding calculation results of the multi-body system under a single PID control. This demonstrates how the addition of the unscented Kalman filter increases the non-coordination of the riser's curvature fluctuation over the full-length direction and its non-linearity.

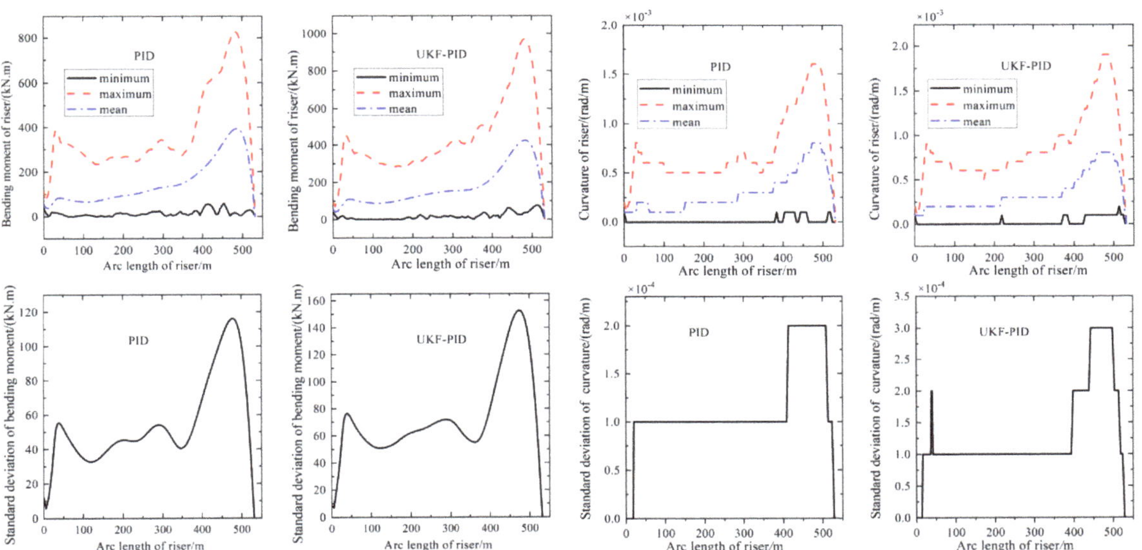

Figure 11. Bending moment and curvature of the riser under the action of UKF-PID and single PID.

The non-synchronization and non-coordination of the variation of the curvature of the riser, as well as the stepped distribution in the length direction, are increasingly apparent as the standard deviation of the curvature of the riser's standard deviation increases. This indicates that when the unscented Kalman is included, the riser's synchronization of the curvature variation of the riser diminishes in the whole-length direction and tends to have a stronger and more discrete stepped distribution. As a result, the strong nonlinearity of the riser's bending is better reflected with the addition of the unscented Kalman filter, which is beneficial for the visual capture of the nonlinear characteristics in the engineering practice, the investigation of potential safety hazards, and the adoption of pertinent measures to improve security.

4.2.3. Rotation Angle of Riser under the Action of UKF-PID and Single PID

The distribution of the Rx, Ry, and Rz angles in the length direction of the riser under the action of the UKF-ID dynamic positioning system is found to be larger than those at the same position of the riser of the offshore platform-riser system under the control mode of the single PID dynamic positioning system, but the distribution of Rx angles in the length direction of the riser has no significant change. This implies that, with the addition of the unscented Kalman filter, the filtering effect enhances the motion component of the offshore platform in a specific direction, resulting in the nonlinear bending of the riser in a specific direction, which makes the riser endure bending and strengthens the bending around its own local coordinate system in the *y*-axis direction, and the degree of the bending around the *x*-axis has increased.

Figure 12 depicts the rotation angle in the length direction of the riser under the action of the UKF-PID dynamic positioning system and single PID dynamic positioning system. The torsional impact of the riser along its own axial axis z is nevertheless magnified when compared to the calculation results under the single PID control condition, but the Rz angles induced by this action are still small when compared to Rx and Ry. This behavior is caused by the insertion of an unscented Kalman filter, which nonlinearly enhances the whole motion of the offshore platform-riser rigid flexible system. The nonlinear amplification of the entire system motion increases the motion response to change more quickly when the system overcomes a change in the external environment load, which also boosts the riser's bending and twisting. The variation of the Ry angle dramatically increases with the addition of the unscented Kalman filter, but the synchronization and coordination of the variation of the Ry angle in the length direction of the riser basically halve when the distributions of the standard deviations of the Ry angle of the riser under the action of UKF-PID dynamic positioning system and those under the action of a single PID dynamic positioning system are compared.

The synchronization and coordination of the variation of the riser's Rz angle under the action of the single PID dynamic positioning system have a stepped distribution and abrupt change in the length direction when comparing the distributions of the standard deviations of the Ry angle of the riser under the action of UKF-PID dynamic positioning system to those under the action of the single PID dynamic positioning system. However, the coordination and synchronization of the variation of the Rz angle of the riser under the action of the UKF-PID dynamic positioning system does not change abruptly in the length direction of the riser, but rather slowly. This indicates that the addition of an unscented Kalman filter increases the flexibility of the entire system by synchronizing and coordinating the variation of the Rz angle in a manner that is orderly and gradually decreases and grows in the length direction. The Rx, Ry, and Rz angles and their related standard deviations under the control of the UKF-PID dynamic positioning system do not vary appreciably as the velocity of the content flow increases.

4.2.4. Six Degrees of Freedom of the Offshore Platform under the Action of UKF-PID and Single PID

The three-degrees-of-freedom translational motion of the offshore platform under the action of the UKF-PID dynamic positioning system has been observed using time-domain curves, and it is discovered that this motion is altered significantly when compared to that under the action of the single PID dynamic positioning system. The maximum sway amplitude is significantly lowered before the sway motion achieves a steady state, and the time needed to reach the steady state of its sway motion is further shortened, but the fluctuation in the surge direction somewhat rises. When compared to the single PID dynamic positioning system, there is not much of a difference in the time-domain response in the sway direction. When compared to a single PID dynamic positioning system, the main change in the time-domain curves of the three-degrees-of-freedom rotational motion of the offshore platform under the action of the UKF-PID dynamic positioning system is reflected in the change of the yaw angle, in which case, the yaw angle of the offshore platform has a small sustained fluctuation in the subsequent time step with a short stagnation at the initial time step in the time domain. The cause of this phenomenon is likewise connected to the system's overall nonlinear increase with the inclusion of the unscented Kalman filter.

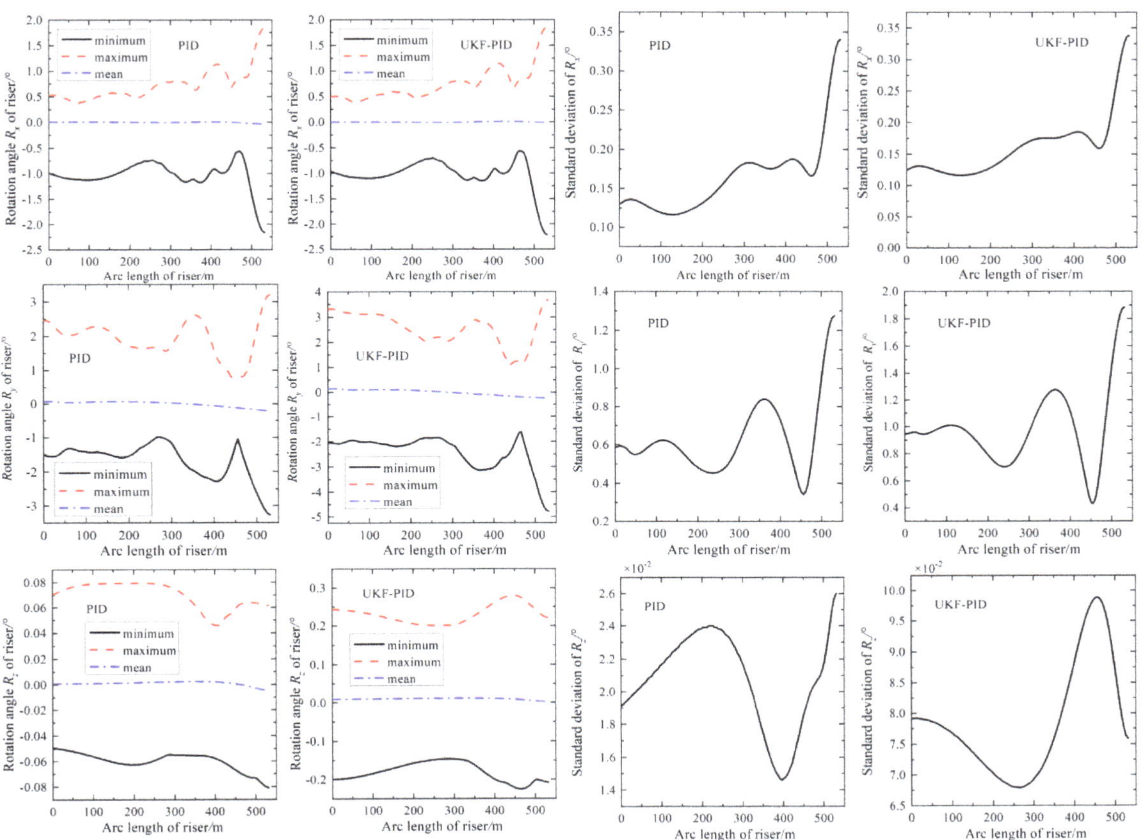

Figure 12. Dynamic response of the rotation angle of the riser under the action of UKF-PID and single PID.

Figure 13 depicts the six degrees of freedom of the offshore platform under the action of the UKF-PID dynamic positioning system and single PID dynamic positioning system. The direction of the surge motion is where the offshore platform's translational motion spectral density changes the most. The offshore platform surge motion spectral density in the single PID dynamic positioning system contains two low peaks, one of which is lower than the peak frequency of the heave motion spectral density, and the other of which is higher than the peak frequency of the heave motion spectral density. However, the surge motion spectral density of the offshore platform has just one high peak value under the action of the UKF-PID dynamic positioning system, and its corresponding frequency is larger than that of the heave motion spectral density. In other words, the superharmonic resonance of the offshore platform in the direction of the surge motion vanishes under the influence of the UKF-PID dynamic positioning system, which helps to boost the safety of the entire offshore platform-riser rigid–flexible system and reduces the risk of fatigue damage. The three-degrees-of-freedom rotational motion spectral density curves of the offshore platform under the action of UKF-PID dynamic positioning are essentially the same as those of the offshore platform under the action of single PID dynamic positioning.

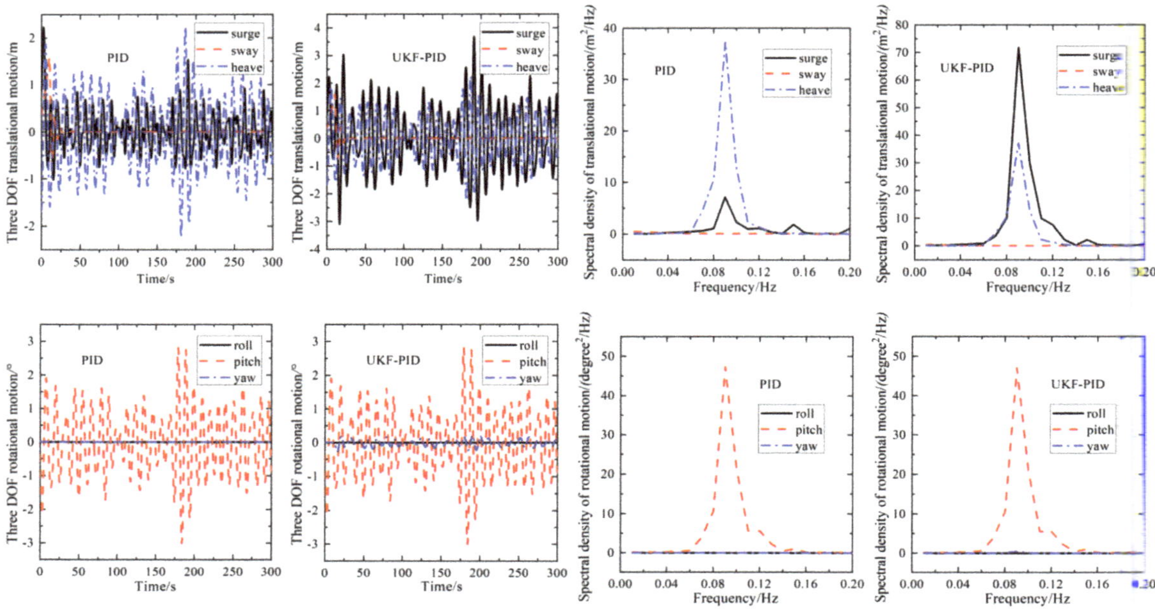

Figure 13. Six degrees of freedom of the offshore platform under UKF-PID and single PID.

5. Conclusions

In this paper, a new PID control approach based on the unscented Kalman filter for the dynamic positioning offshore platform-riser multi-body system is created by merging the unscented Kalman filter with the traditional PID control. The DP control of a rigid–flexible fluid coupling system composed of an offshore platform and risers is realized. Compared with the traditional PID control, the calculation results of UKF-PID control demonstrate the following:

(1) The effect of the UKF-PID dynamic positioning system on the variation of the effective tension of the riser is not significant, which is fully reflected in the fluctuation, coordination, and synchronism of the variation. Compared with the multi-body system controlled by a single PID static positioning system, the bending moment of the platform riser becomes larger when the positioning system is changed. With the addition of the unscented Kalman filter, the strong nonlinearity of riser bending change in the whole system is better reflected. This is helpful for the visual capture of this nonlinearity in engineering practice and the investigation of potential safety hazards and measures to improve safety.

(2) Compared with the dynamic positioning system under the control of a single PID, the bending moment of the riser in the UKF-PID dynamic positioning system and the transmission of bending along the length of the riser will change. This change is mainly reflected in a relatively larger bending moment. The bending moment of the riser at a certain position will be more severe, but the relative hysteresis of the bending moment and curvature still exists. In addition, the hysteresis of load and strain transfer will be further enhanced at some locations. In this case, the overall synchronization and coordination of riser curvature changes along the length direction will be reduced.

(3) The proposed control approach of this paper improves the nonlinearity of the three-degrees-of-freedom translational movements of the offshore platform and modifies the

energy distribution in three translation motions of the offshore platform. Under UKF-PID control, the overall motion nonlinearity of the offshore platform-riser multi-body system has been significantly improved. The enhancement of nonlinearity makes the system more sensitive to the change of its own motion response when overcoming the change of external environmental load, which also leads to the increase in the Ry and Rz angles of the riser.

Due to the complexity limit of the rigid–flexible multi-body system, the forces and moments of the DP offshore platform under external environmental loads are exerted on the center of mass of the offshore platform to achieve its positioning control. This method is feasible when considering the interaction between the riser and DP offshore platform in the overall analysis and can greatly improve the calculation efficiency. However, in practical engineering, the dynamic positioning of offshore platforms is achieved by arranging different thrusters at different coordinate positions and then providing different thrusts and torques or changing the direction of thrusts and moments by adjusting the speed and spatial azimuth of the thruster. Therefore, the research in this paper is still simplified. Future work will focus on how to add the component force and spatial azimuth provided by each thruster to the dynamic model to better conform to the real engineering situation.

Author Contributions: Conceptualization, D.Z. and B.Z.; methodology, D.Z. and Y.B.; software, K.Z.; validation, D.Z. and K.Z.; formal analysis, D.Z. and B.Z.; investigation, K.Z. and B.Z.; resources, D.Z.; data curation, K.Z.; writing—original draft preparation, D.Z.; writing—review and editing, D.Z.; visualization, K.Z.; supervision, D.Z. and Y.B.; funding acquisition, D.Z. All authors have read and agreed to the published version of the manuscript.

Funding: This research was funded by the Program for Scientific Research Start-up Funds of Guangdong Ocean University, grant number 060302072101, and Zhanjiang Marine Youth Talent Project-Comparative Study and Optimization of Horizontal Lifting of Subsea Pipeline, grant number 2021E5011.

Institutional Review Board Statement: Not applicable.

Informed Consent Statement: Not applicable.

Data Availability Statement: Not applicable.

Conflicts of Interest: The authors declare no conflict of interest.

References

1. Altuzarra, J.; Herrera, A.; Matías, O.; Urbano, J.; Romero, C.; Wang, S.; Guedes Soares, C. Mooring System Transport and Installation Logistics for a Floating Offshore Wind Farm in Lannion, France. *J. Mar. Sci. Eng.* **2022**, *10*, 1354. [CrossRef]
2. Amaechi, C.V.; Reda, A.; Butler, H.O.; Ja'E, I.A.; An, C. Review on fixed and floating offshore structures. Part I: Types of platforms with some applications. *J. Mar. Sci. Eng.* **2022**, *10*, 1074. [CrossRef]
3. Amaechi, C.V.; Reda, A.; Butler, H.O.; Ja'E, I.A.; An, C. Review on fixed and floating offshore structures. Part II: Sustainable design approaches and project management. *J. Mar. Sci. Eng.* **2022**, *10*, 973. [CrossRef]
4. Li, C.B.; Chen, M.; Choung, J. The Quasi-Static Response of Moored Floating Structures Based on Minimization of Mechanical Energy. *J. Mar. Sci. Eng.* **2021**, *9*, 960. [CrossRef]
5. Chen, C.; Shiotani, S.; Sasa, K. Numerical ship navigation based on weather and ocean simulation. *Ocean Eng.* **2013**, *69*, 44–53. [CrossRef]
6. Nguyen, V.S. A novel approach to determine the ship position with an azimuth of celestial body and factors of ship route. *Int. J. Civil. Eng. Technol.* **2019**, *10*, 1162–1167.
7. Lee, D.; Lee, S.J. Motion predictive control for DPS using predicted drifted ship position based on deep learning and replay buffer. *Int. J. Nav. Arch. Ocean* **2020**, *12*, 768–783. [CrossRef]
8. Cheng, Z.; Gao, Z.; Moan, T. Numerical modeling and dynamic analysis of a floating bridge subjected to wind, wave, and current loads. *J. Offshore Mech. Arct. Eng.* **2019**, *141*, 011601. [CrossRef]
9. Nagavinothini, R.; Chandrasekaran, S. Dynamic analyses of offshore triceratops in ultra-deep waters under wind, wave, and current. *Structures* **2019**, *20*, 279–289. [CrossRef]
10. Zhang, D.; Bai, Y.; Soares, C.G. Dynamic analysis of an array of semi-rigid "sea station" fish cages subjected to waves. *Aquac. Eng.* **2021**, *94*, 102172. [CrossRef]

11. Loria, A.; Fossen, T.I.; Panteley, E. A separation principle for dynamic positioning of ships: Theoretical and experimental results. *IEEE Trans. Control Syst. Technol.* **2000**, *8*, 332–343. [CrossRef]

12. Dutopo, H.; Marwanto, A.; Alifah, S.; Hidayah, M. Improved ship position stability on offshore-based dynamic position maintenance with the PID method. *Tech. Rom. J. Appl. Sci. Technol.* **2022**, *4*, 99–112. [CrossRef]

13. Hao, L.; Zhang, H.; Li, T.; Lin, B.; Chen, C.P. Fault tolerant control for dynamic positioning of unmanned marine vehicles based on TS fuzzy model with unknown membership functions. *IEEE Trans. Veh. Technol.* **2021**, *70*, 146–157. [CrossRef]

14. Deng, F.; Yang, H.; Wang, L. Adaptive unscented Kalman filter based estimation and filtering for dynamic positioning with model uncertainties. *Int. J. Control. Autom. Syst.* **2019**, *17*, 667–678. [CrossRef]

15. Liu, L.; Zhang, W.; Wang, D.; Peng, Z. Event-triggered extended state observers design for dynamic positioning vessels subject to unknown sea loads. *Ocean Eng.* **2020**, *209*, 107242. [CrossRef]

16. Witkowska, A.; Śmierzchalski, R. Adaptive dynamic control allocation for dynamic positioning of marine vessel based on backstepping method and sequential quadratic programming. *Ocean Eng.* **2018**, *163*, 570–582. [CrossRef]

17. Pettersen, K.Y.; Fossen, T.I. Under actuated dynamic positioning of a ship-experimental results. *IEEE Trans. Control Syst. Technol.* **2000**, *8*, 856–863. [CrossRef]

18. Fossen, T.I. Guidance and Control of Ocean Vehicles. Doctors Thesis, University of Trondheim, Trondheim, Norway, 1994.

19. Fossen, T.I.; Grovlen, A. Nonlinear output feedback control of dynamically positioned ships using vectorial observer backstepping. *IEEE Trans. Control Syst. Technol.* **1998**, *6*, 121–128. [CrossRef]

20. Skjetne, R.; Fossen, T.I.; Petar, V. Kokotović. Adaptive maneuvering, with experiments, for a model ship in a marine control laboratory. *Automatica* **2005**, *41*, 289–298. [CrossRef]

21. Morawski, L.; Nguyen, C.V. Ship control in maneuvering situation with fuzzy logic controllers. *TransNav Int. J. Mar. Navig. Saf. Od Sea Transp.* **2008**, *2*, 77–84.

22. Lee, G.; Surendran, S.; Kim, S.H. Algorithms to control the moving ship during harbour entry. *Appl. Math. Model.* **2009**, *33*, 2474–2490. [CrossRef]

23. Bui, V.P.; Kawai, H.; Kim, Y.B.; Lee, K.S. A ship berthing system design with four tug boats. *J. Mech. Sci. Technol.* **2011**, *25*, 1257–1264. [CrossRef]

24. Fossen, T.I.; Strand, J.P. Passive nonlinear observer design for ships using lyapunov methods: Experimental Results with a Supply Vessel. *Automatica* **1999**, *35*, 3–16. [CrossRef]

25. Sorensen, A.J.; Strand, J.P.; Fossen, T.I. Thruster assisted position mooring system for turret-anchored FPSOs. In Proceedings of the IEEE International Conference on Control Applications, Kohala Coast, HI, USA, 22–27 August 1999.

26. Wicher, J.; Van, D.R. Benefits of using assisted DP for deepwater mooring systems. In Proceedings of the Offshore Technology Conference, Houston, TX, USA, 3–6 May 1999.

27. Aalbers, A.B.; Merchant, A.A. The hydrodynamic model testing for closed loop DP assisted mooring. In Proceedings of the Offshore Technology Conference, Houston, TX, USA, 3–6 May 1996.

28. Sørheim, H.R. Analysis of motion in single-point mooring systems. *Modeling Ident Control* **1980**, *1*, 165–186. [CrossRef]

29. Wichers, J.E. On the slow motions of tankers moored to single point mooring systems. *J. Petro. Technol.* **1978**, *30*, 13–33. [CrossRef]

30. Lopez-Cortijo, J.; Duggal, A.S.; Van Dijk, R.R.; Matos, S. DP FPSO-A Fully Dynamically Positioned FPSO For Ultra Deep Waters. In Proceedings of the Thirteenth International Offshore and Polar Engineering Conference, Honolulu, HI, USA, 25 May 2003.

31. Tannuri, E.A.; Saad, A.C.; Morishita, H.M. Offloading operation with a DP shuttle tanker: Comparison between full scale measurements and numerical simulation results. *IFAC Proc. Vol.* **2009**, *42*, 249–254. [CrossRef]

32. Ja'E, I.A.; Ali, M.O.A.; Yenduri, A.; Nizamani, Z.; Nakayama, A. Optimisation of mooring line parameters for offshore floating structures: A review paper. *Ocean Eng.* **2022**, *247*, 110644. [CrossRef]

33. Yan, J.; Qiao, D.; Li, B.; Wang, B.; Liang, H.; Ning, D.; Ou, J. An improved method of mooring damping estimation considering mooring line segments contribution. *Ocean Eng.* **2021**, *239*, 109887. [CrossRef]

34. He, X.; Zhao, Z.; Su, J.; Yang, Q.; Zhu, D. Adaptive inverse control of a vibrating coupled vessel-riser system with input backlash. *IEEE Trans Syst. Man Cybern. Syst.* **2019**, *51*, 4706–4715. [CrossRef]

35. Liu, X.; Liu, Z.; Wang, X.; Zhang, N.; Qiu, N.; Chang, Y.; Chen, G. Recoil control of deepwater drilling riser system based on optimal control theory. *Ocean Eng.* **2021**, *220*, 108473. [CrossRef]

36. Yu, J.; Wang, F.; Yu, Y.; Liu, X.; Liu, P.; Su, Y. Test System Development and Experimental Study on the Fatigue of a Full-Scale Steel Catenary Riser. *J. Mar. Sci. Eng.* **2022**, *10*, 1325. [CrossRef]

37. Aly, A.A.; The Vu, M.; El-Sousy, F.F.; Alotaibi, A.; Mousa, G.; Le, D.; Mobayen, S. Fuzzy-Based Fixed-Time Nonsingular Tracker of Exoskeleton Robots for Disabilities Using Sliding Mode State Observer. *Mathematics* **2022**, *10*, 3147. [CrossRef]

38. Vu, M.T.; Choi, H.; Nhat, T.Q.M.; Nguyen, N.D.; Lee, S.; Le, T.; Sur, J. Docking assessment algorithm for autonomous underwater vehicles. *Appl. Ocean Res.* **2020**, *100*, 102180. [CrossRef]

39. Vu, M.T.; Choi, H.; Nguyen, N.D.; Kim, S. Analytical design of an underwater construction robot on the slope with an up-cutting mode operation of a cutter bar. *Appl. Ocean Res.* **2019**, *86*, 289–309. [CrossRef]

40. Vu, M.T.; Jeong, S.; Choi, H.; Oh, J.; Ji, D. Study on down-cutting ladder trencher of an underwater construction robot for seabed application. *Appl. Ocean Res.* **2018**, *71*, 90–104. [CrossRef]

41. Mobayen, S.; Bayat, F.; Din, S.U.; Vu, M.T. Barrier function-based adaptive nonsingular terminal sliding mode control technique for a class of disturbed nonlinear systems. *ISA Trans.* **2022**, *8*, 006. [CrossRef]

42. Huang, J.; Choi, H.; Vu, M.T.; Jung, D.; Choo, K.; Cho, H.; Nam Anh, P.H.; Zhang, R.; Park, J.; Kim, J. Study on Position and Shape Effect of the Wings on Motion of Underwater Gliders. *J. Mar. Sci. Eng.* **2022**, *10*, 891. [CrossRef]
43. Jin, H.; Cho, H.; Jiafeng, H.; Lee, J.; Kim, M.; Jeong, S.; Ji, D.; Joo, K.; Jung, D.; Choi, H. Hovering control of UUV through underwater object detection based on deep learning. *Ocean Eng.* **2022**, *253*, 111321. [CrossRef]
44. Cho, H.; Jeong, S.; Ji, D.; Tran, N.; Vu, M.T.; Choi, H. Study on control system of integrated unmanned surface vehicle and underwater vehicle. *Sensors* **2020**, *20*, 2633. [CrossRef]
45. Nguyen, N.; Choi, H.; Lee, S. Robust adaptive heading control for a ray-type hybrid underwater glider with propellers. *J. Mar. Sci. Eng.* **2019**, *7*, 363. [CrossRef]
46. Vu, M.T.; Choi, H.; Kim, J.; Tran, N.H. A study on an underwater tracked vehicle with a ladder trencher. *Ocean Eng.* **2016**, *127*, 90–102. [CrossRef]
47. Vu, M.T.; Choi, H.; Kang, J.; Ji, D.; Jeong, S. A study on hovering motion of the underwater vehicle with umbilical cable. *Ocean Eng.* **2017**, *135*, 137–157.
48. Vu, M.T.; Van, M.; Bui, D.H.P.; Do, Q.T.; Huynh, T.; Lee, S.; Choi, H. Study on dynamic behavior of unmanned surface vehicle-linked unmanned underwater vehicle system for underwater exploration. *Sensors* **2020**, *20*, 1329. [CrossRef] [PubMed]
49. Wang, S. On the assessment of thruster assisted mooring. In Proceedings of the Offshore Technology Conference, Houston, TX, USA, 3–6 May 2010.
50. Bai, Y.; Zhang, D.; Zhu, K.; Zhang, T. Dynamic analysis of umbilical cable under interference with riser. *Ships Offshore Struct.* **2018**, *13*, 809–821. [CrossRef]
51. Sørensen, A.J.; Leira, B.; Peter, S.J.; Larsen, C.M. Optimal setpoint chasing in dynamic positioning of deep-water drilling and intervention vessels. *Rob. Nonlinear Control* **2001**, *11*, 1187–1205. [CrossRef]
52. Leira, B.; Chen, Q.; Sørensen, A.J.; Larsen, C.M. Modeling of riser response for DP control. *J. Offshore Mech. Arct. Eng.* **2002**, *124*, 219–225. [CrossRef]
53. Liu, F. *Research on Modeling and Simulating of Dynamic Positioning Control System of the Semi-Submersible Vessel*; Wuhan University of Technology: Wuhan, China, 2011. (In Chinese)
54. Bao, W. *Research on Time Domain Analysis of Dynamic Positioning System for Drilling Platform*; Jiangsu University of Science and Technology: Zhenjiang, China, 2018. (In Chinese)
55. Gu, J.; Lu, H.; Yang, J. Studies on coupling dynamic response of TLP in stochastic waves. *J. Ship. Mech.* **2013**, *17*, 888–900. (In Chinese)

MDPI

St. Alban-Anlage 66

4052 Basel

Switzerland

www.mdpi.com

Journal of Marine Science and Engineering Editorial Office

E-mail: jmse@mdpi.com

www.mdpi.com/journal/jmse